Lecture Notes in Computational Vision and Biomechanics

Volume 38

Research related to the analysis of living structures (Biomechanics) has been carried out extensively in several distinct areas of science, such as, for example, mathematics, mechanical, physics, informatics, medicine and sports. However, for its successful achievement, numerous research topics should be considered, such as image processing and analysis, geometric and numerical modelling, biomechanics, experimental analysis, mechanobiology and enhanced visualization, and their application on real cases must be developed and more investigation is needed. Additionally, enhanced hardware solutions and less invasive devices are demanded. On the other hand, Image Analysis (Computational Vision) aims to extract a high level of information from static images or dynamical image sequences. An example of applications involving Image Analysis can be found in the study of the motion of structures from image sequences, shape reconstruction from images and medical diagnosis. As a multidisciplinary area, Computational Vision considers techniques and methods from other disciplines, like from Artificial Intelligence, Signal Processing, mathematics, physics and informatics. Despite the work that has been done in this area, more robust and efficient methods of Computational Imaging are still demanded in many application domains, such as in medicine, and their validation in real scenarios needs to be examined urgently. Recently, these two branches of science have been increasingly seen as being strongly connected and related, but no book series or journal has contemplated this increasingly strong association. Hence, the main goal of this book series in Computational Vision and Biomechanics (LNCV&B) consists in the provision of a comprehensive forum for discussion on the current state-of-the-art in these fields by emphasizing their connection. The book series covers (but is not limited to):

• Applications of Computational Vision and Biomechanics
• Biometrics and Biomedical Pattern Analysis
• Cellular Imaging and Cellular Mechanics
• Clinical Biomechanics
• Computational Bioimaging and Visualization
• Computational Biology in Biomedical Imaging
• Development of Biomechanical Devices
• Device and Technique Development for Biomedical Imaging
• Experimental Biomechanics
• Gait & Posture Mechanics
• Grid and High Performance Computing on Computational Vision and Biomechanics
• Image Processing and Analysis
• Image processing and visualization in Biofluids
• Image Understanding
• Material Models
• Mechanobiology
• Medical Image Analysis
• Molecular Mechanics
• Multi-modal Image Systems
• Multiscale Biosensors in Biomedical Imaging
• Multiscale Devices and BioMEMS for Biomedical Imaging
• Musculoskeletal Biomechanics
• Multiscale Analysis in Biomechanics
• Neuromuscular Biomechanics
• Numerical Methods for Living Tissues
• Numerical Simulation
• Software Development on Computational Vision and Biomechanics
• Sport Biomechanics
• Virtual Reality in Biomechanics
• Vision Systems
• Image-based Geometric Modeling and Mesh Generation
• Digital Geometry Algorithms for Computational Vision and Visualization

In order to match the scope of the Book Series, each book has to include contents relating, or combining both Image Analysis and mechanics.

Indexed by Scopus.

More information about this series at https://link.springer.com/bookseries/8910

João Manuel R. S. Tavares ·
Christoph Bourauel · Liesbet Geris ·
Jos Vander Slote
Editors

Computer Methods, Imaging and Visualization in Biomechanics and Biomedical Engineering II

Selected Papers from the 17th International Symposium CMBBE and 5th Conference on Imaging and Visualization, September 7–9, 2021

Springer

Editors
João Manuel R. S. Tavares
Faculdade de Engenharia
Universidade do Porto
Porto, Portugal

Liesbet Geris
Biomechanics Research Unit
University of Liège
Liège, Belgium

Christoph Bourauel
University Hospital Bonn
Bonn, Germany

Jos Vander Slote
Biomechanics
Katholieke Universiteit Leuven
Leuven, Belgium

ISSN 2212-9391 ISSN 2212-9413 (electronic)
Lecture Notes in Computational Vision and Biomechanics
ISBN 978-3-031-10017-8 ISBN 978-3-031-10015-4 (eBook)
https://doi.org/10.1007/978-3-031-10015-4

This Springer imprint is published by the registered company Springer Nature Switzerland AG
The registered company address is: Gewerbestrasse 11, 6330 Cham, Switzerland

Preface

The 17th International Symposium on Computer Methods in Biomechanics and Biomedical Engineering and the 5th Conference on Imaging and Visualization (CMBBE 2021) were organized and planned together in Bonn, Germany, from September 7 to 9, 2021, and took place online as a result of the ongoing pandemic. Despite being held virtually, CMBBE 2021 managed to keep its traditionally strong impact on the development of computational biomechanics and biomedical imaging and visualization; particularly by identifying emerging areas of research and promoting the collaboration between participants. Actually, CMBBE 2021 program included more than 250 podium sessions running in 5 parallel streams, 9 workshops and 75 poster presentations. Additionally, four eminent researchers delivered important plenary lectures, addressing current challenges in computational biomechanics and biomedical imaging. This book includes the extended versions of selected works presented during the program.

Briefly, the included 24 contributions address topics in the following areas: *Bioengineering in the fight against COVID-19, Biomechanics of movement and rehabilitation, Brain biomechanics, Cardiovascular fluid dynamics, Cellular and molecular biomechanics, Computational analysis of the bone-implant interface, Fluid biomechanics, Growth and remodeling, inverse problems and parameter identification, Imaging & Visualization, Implants/orthotics/prosthetics/devices/biologics, Intelligent rehabilitation technologies, Machine learning, Musculoskeletal dynamics and neuromuscular control, Sports biomechanics, Injury, Optimal control of human movement, Patient-specific modeling, Skin biomechanics and Soft tissue mechanics, damage, remodeling.*

Different applications are addressed and described throughout the book, including advanced reality, computational simulation, flow analysis, medical diagnosis and rehabilitation, medical imaging, numerical analysis, and stress and strain analysis. Therefore, this book is of high interest for researchers, students, end users, and manufacturers from several multidisciplinary fields, as the ones related with 3D printing, bioengineering, biology, biomechanics, computational mechanics, computer graphics, computer sciences, mathematics, mechanobiology, medical imaging, medicine, physics.

The editors would like to seize this opportunity to thank the authors of the published contributions for sharing their works, experiences, and knowledge, and making possible their dissemination through this book.

Christoph Bourauel
João Manuel R. S. Tavares
Co-editors and Chairs of CMBBE 2021

Organization

CMBBE 2021 Chairs

Christoph Bourauel	Oral Technology Center of Dento-Maxillo-Facial Medicine, Faculty of Medicine, University of Bonn, Germany
Liesbet Geris	Skeletal Biology and Engineering Research Center, KU Leuven and University of Liège, Executive Director of VPH Institute, Belgium
Jos Vander Sloten	Department of Mechanical Engineering, Division BMe, KU Leuven, Belgium
João Manuel R. S. Tavares	Departamento de Engenharia Mecânica, Faculdade de Engenharia, Universidade do Porto, Portugal

CMBBE 2021 Technical Advisory Panel

Gerard Ateshian	Columbia University, USA
Lynne Bilston	University of New South Wales, Australia
Christoph Bourauel	University of Bonn, Germany
Valentin Brimkov	State University of New York, USA
Silvia Budday	Friedrich-Alexander-Universität, Germany
Deva D. Chan	Purdue University, USA
Estevam de las Casas	Universidade Federal de Minas Gerais, Brazil
Scott L. Delp	Stanford University, USA
Paolo Di Giamberardino	Sapienza University of Rome, Italy
Manuel Doblare	University of Zaragoza, Spain
Sonia Duprey	University Lyon, France
Stephen Ferguson	Institute for Biomechanics, Switzerland
Justin Fernandez	University of Auckland, New Zealand
Paulo R. Fernandes	Technical University of Lisbon, Portugal

José Manuel García Aznar	University of Zaragoza, Spain
Amit Gefen	Tel Aviv University, Israel
Liesbet Geris	Katholieke Universiteit Leuven, Belgium
Manuel González Hidalgo	Balearic Islands University, Spain
Marek Gzik	Silesian University of Technology, Poland
Ridha Hambli	Orleans University, France
Marie-Christine Ho Ba Tho	Université de Technologie de Compiègne, France
Andrew Hopkins	Zimmer Biomet, Switzerland
Thomas Hughes	Aerospace Engineering and Engineering Mechanics, USA
Daniel Hurtado	Pontificia Universidad Católica de Chile, Chile
Daniela Iacoviello	Università degli Studi di Roma 'La Sapienza,' Italy
Khan M. Iftekharuddin	Old Dominion University, USA
Leo Joskowicz	The Hebrew University of Jerusalem, Israel
Rodger D. Kamm	Massachusetts Institute of Technology, USA
Seungbum Koo	Korea Advanced Institute of Science and Technology, South Korea
Ellen Kuhl	Stanford University, USA
Paola Lecca	University of Trento, Italy
Harry van Lenthe	University of Leuven, Belgium
Shuo Li	Western University, ON, Canada
Cristian A. Linte	Rochester Institute of Technology, USA
Paige Little	Queensland University of Technology, Australia
Sandra Loerakker	Eindhoven University of Technology, the Netherlands
Richard Lopata	Eindhoven University of Technology, the Netherlands
June Madete	Kenyatta University, Kenya
Eduardo Mazza	ETH Zürich, Switzerland
Kristin M. Myers	Columbia University, USA
Martyn Nash	University of Auckland, New Zealand
Glen Niebur	University of Notre Dame, USA
Jerome Noailly	Pompeu Fabra University, Spain
Fatima Nunes	Information Technology Center of São Paulo, Brazil
Han van Oosterwyck	University of Leuven, Belgium
Michelle L. Oyen	East Carolina University, USA
João Paulo Papa	Universidade Estadual Paulista, Brazil
Areti Papastavrou	Technische Hochschule Nürnberg, Germany
Anne M. Robertson	University of Pittsburgh, USA
Michael S. Sacks	The University of Texas at Austin, USA
Francisco Sahli	Pontificia Universidad Católica de Chile, Chile
Fiorella Sagallari	University of Bologna, Italia
Saeed Shirazi-Adl	École Polytechnique de Montréal, Canada

Wafa Skalli	Laboratoire de Biomécanique, France
Eduardo Soudah	International Center for Numerical Methods in Engineering, Spain
Fong-Chin Su	National Cheng Kung University, Taiwan
David A. Steinman	University of Toronto, Canada
Demetri Terzopoulos	University of California, Los Angeles, USA
Jos Vander Sloten	Katholieke Universiteit Leuven, Belgium
Daniela Valdez-Jasso	University of California, USA
Maria del C. Valdes Hernandez	University of Edinburgh, UK
Pascal Verdonck	University of Ghent, Belgium
Shigeo Wada	Osaka University, Japan
Lizhen Wang	Beihang University, China
Jeffrey A. Weiss	University of Utah, USA
Hiroshi Yamada	Kyushu Institute of Technology, Japan
Jessica Zhang	Carnegie Mellon University, USA
Ming Zhang	The Hong Kong Polytechnic University, Hong Kong
Jun Zhao	Shanghai Jiao Tong University, China
Philippe K. Zysset	University of Bern, Switzerland

Contents

A Spatial Markov Chain Cellular Automata Model for the Spread of Viruses . . . 1
Jenny Lu and Fred Vermolen

A Novel Review of Temporomandibular Joint Replacement Options . . . 24
Christine Walck, Yeram Lim, and Seth Rosenstein

Could an Exoskeleton-Driven Rehabilitation Treatment Improve Muscle Forces Generation in PD? - a Pilot Study . . . 36
Marco Romanato, Fulvia Fichera, Fabiola Spolaor, Daniele Volpe, and Zimi Sawacha

EMG Signals as a Way to Control Soft Actuators . . . 50
António Diogo André, Ana Margarida Teixeira, and Pedro Martins

Movement Optimization Through Musculoskeletal Modeling and Multidimensional Surface Interpolation . . . 59
Christine Walck, Christopher Lamb, and Pablo Vilches

Simulating the Dynamics of a Human-Exoskeleton System Using Kinematic Data with Misalignment Between the Human and Exoskeleton Joints . . . 65
Divyaksh Subhash Chander, Max Böhme, Michael Skipper Andersen, John Rasmussen, and Maria Pia Cavatorta

A 2D-FEM Model of Nonlinear Ultrasound Propagation in Trans-cranial MRgFUS Technique . . . 74
Fabiano Bini, Andrada Pica, Maurizio Marrale, Cesare Gagliardo, and Franco Marinozzi

Predicting Neurological Effects Associated with Traumatic Brain Injuries Using Video Analysis and Finite Element Modeling . . . 90
Bianca Acot, Branko Glisic, Annegret Dettwiler, and Michael D. Gilchrist

Dense-Discrete Phase Simulations of Blood Flow in a Stenotic Coronary ... 101
Violeta Carvalho, Nelson Rodrigues, José C. Teixeira, Rui Lima, and Senhorinha Teixeira

Predicting the Efficacy of Stalk Cells Following Leading Cells Through a Micro-Channel Using Morphoelasticity and a Cell Shape Evolution Model ... 112
Q. Peng, F. J. Vermolen, and D. Weihs

Simulation of Cell Proliferation Using a Meshless Tool ... 123
M. I. A. Barbosa, J. Belinha, R. M. Natal Jorge, and A. Carvalho

Explicit Non-linear Finite Element Analysis for Prediction of Primary Stability in Uncemented Total Hip Arthroplasty ... 128
Marzieh Ovesy and Philippe K. Zysset

Hemodynamic Effects of Entry Versus Exit Tear Size and Tissue Stiffness in Simulations of Aortic Dissection ... 143
Kathrin Bäumler, Judith Zimmermann, Daniel B. Ennis, Alison L. Marsden, and Dominik Fleischmann

Reproducibility of in Vivo Constitutive Parameter Identification Based on 4D Ultrasound Strain Imaging ... 153
Andreas Wittek, Claus-Peter Fritzen, Armin Huß, and Christopher Blase

A Systematic Review of the Uses and Benefits of 3-D Printing in Orthopaedic Surgery ... 170
Firas Nasr and Caroline Hing

Assessing Methodological Uncertainty of *In-Vitro* Digital Volume Correlation Bone Strain Measurements in Total Shoulder Arthroplasties ... 183
Jakub Targosinski, Jonathan Kusins, Nicole Martensson, Andrew Nelson, Nikolas Knowles, and Louis Ferreira

Evidence for the Applicability of Musculoskeletal Human Models to Improve Outcomes of Total Hip Arthroplasty ... 194
David Scherb, Christopher Fleischmann, Stefan Sesselmann, Jörg Miehling, and Sandro Wartzack

Development of a Silent Speech Interface for Augmented Reality Applications ... 208
Christine Walck, Tania Rivas, Riley Flanagan, and Michael Fornito

Hand Gesture Recognition for Sign Languages Using 3DCNN for Efficient Detection ... 215
Taranya Elangovan, R. Arockia Xavier Annie, Keerthana Sundaresan, and J. D. Pradhakshya

Simulation Study to Investigate the Accuracy of in Vivo Motor-Unit Twitch Force Measurements 234
Harnoor Saini and Oliver Röhrle

On the Use of Mesh-Based Joint Contact Models Within Simulations Using Automatic Differentiation 244
Gil Serrancolí, Jordi Torner, Simone Perelli, and Joan Carles Monllau

Lower Limb Joint Load Comparison from Subject Specific Musculoskeletal Model Simulation and Direct Measurements on Different Subject with Instrumented Implant During Normal and Abnormal Gait 250
Carlos Rodrigues, Miguel Correia, João Abrantes, Marco Rodrigues, and Jurandir Nadal

An Anisotropic Hyperelastic Model for Human Skin: Finite Element Modeling, Identification of Parameters, Mechanical Tests 271
Wael Alliliche, Christine Renaud, Jean-Michel Cros, and Zhi-Qiang Feng

Elastic Properties of Normal Breast Tissues Using an Indentation Protocol - Preliminary Study 281
Ana Margarida Teixeira, António André, Rossana Correia, Maria da Luz Barroso, Horácio Costa, and Pedro Martins

Author Index 291

A Spatial Markov Chain Cellular Automata Model for the Spread of Viruses

Jenny Lu[1] and Fred Vermolen[2,1](✉)

[1] Delft Institute of Applied Mathematics, Delft University of Technology, Mekelweg 4, 2628 CD Delft, The Netherlands
J.Lu@student.tudelft.nl

[2] Computational Mathematics Group, Department of Mathematics and Statistics, University of Hasselt, Agoralaan Building D, 3590 Diepenbeek, Belgium
Fred.Vermolen@uhasselt.be

Abstract. In this paper a Spatial Markov Chain Cellular Automata model for the spread of viruses is proposed. The model is based on a graph with connected nodes, where the nodes represent individuals and the connections between the nodes denote the relations between humans. In this way, a graph is connected where the probability of infectious spread from person to person is determined by the intensity of interpersonal contact. Infectious transfer is determined by chance. The model is extended to incorporate various lockdown scenarios. Simulations with different lockdowns are provided. In addition, under logistic regression, the probability of death as a function of age and gender is estimated, as well as the duration of the disease given that an individual dies from it. The estimations have been done based on actual data of RIVM (from the Netherlands).

1 Introduction

At the time of writing, the COVID-19 crisis has been affecting the global human population worldwide for almost two years. The impact of the virus has been enormous, almost five million people have died [1], and many countries have been in various degrees of lockdowns and economies have been hit hard. The lockdown policies have caused countries to close their borders, ban travel, people to live in isolation and companies to suffer great losses. Sadly, despite the vaccinations, the pandemic is far from over. There are new mutated versions of the virus, like the beta, gamma, delta and omikron versions. In addition, recovered people can lose their immunity against the virus in a couple of months, hence making reinfection a problem as well.

The disease, COVID-19, is renamed as the strain Severe Acute Respiratory Syndrome Coronavirus 2 (SARS-CoV-2) by the World Health Organisation [2]. It is often characterised by flu-like symptoms, which in some cases lead to excessive fever or even to lung inflammations. One of the serious problems regarding

J. M. R. S. Tavares et al. (Eds.): CMBBE 2021, LNCVB 38, pp. 1–23, 2023.
https://doi.org/10.1007/978-3-031-10015-4_1

this disease is the high infection rate from person to person. In addition, the disease affects every individual. Even in young people, COVID-19 can cause strokes, seizures and Guillain-Barre syndrome—a condition that causes temporary paralysis. COVID-19 may also increase the risk of developing Parkinson's disease and Alzheimer's disease according to the Mayo Clinic [3]. The three organs that are impacted most by the disease are the heart, lungs and brain. The long term effects of the disease are yet still unknown.

In order to predict the dynamics of the spread, death rate and recovery rate of the coronavirus, many different strategies are used. A very common model is the so-called SIR model, see [4] for the original paper. More modern elaborations on the SIR principle have been presented in [5,6] and [7]. In particular, the model in [7] bears some similarities with the approach that is presented in the current paper. This model simulates a homogeneous population that is exposed to a virus. It contains a susceptible, infected, recovered and dead fraction of the population. Many more advanced models are variations on this strategy. One attempts to include spatial spread by the incorporation of diffusion terms, which are justified by random (unpredictable) migration and interaction of individuals. Other extensions are based on the incorporation of networks, which allows so-called jump processes so that airborne communication can be taken into account. The models described in [8] distinguish several challenges for network modelling in epidemics. A review on mathematical modelling of epidemics has been written in [9]. This review considers the different practices and limitations of modelling global spread of diseases. Duan et al. [10] wrote a review about epidemic modelling where models of different nature are discussed. First the S(E)IR models based on ordinary differential equations are introduced, and this is followed by network models that are based on stochastic principles. It is reported that stochastic (network) models yield very realistic results [10]. Stochastic network models for epidemics have been presented in [11], where an exact final size distribution is constructed using recursive formulas. Further, the impact of vaccination is quantified in [11]. A Bayesian inference for stochastic epidemic models has been presented in [12]. Both [12] and [11] favour the use of stochastic network models because of the huge flexibility from temporal and random effects that the models are able to handle. The current model elaborates on the influence of the topology of the network on the evolution of the epidemic. The results that are presented in the current paper should be classified as preliminary in the sense that the results are based on simulation with hypothetical input parameters. However, later in the paper we do present an estimation for the recovery rate parameter based on actual data of the National Institute for Public Health and the Environment or short RIVM of the Netherlands. The infection parameter is hard to estimate based on the data, due to the many factors that influence the possibility of infection and because of the implementation of measures that change over time.

The model presented in this paper is based on cellular automata, in which the nodes of the grid represent individuals who are connected to each other by means of a graph. Each individual is assigned a state at every time instance, these states are: susceptible, infected, recovered or dead. The stochastic nature of the model can be seen by the probability of infection as well as death. Contact between individuals does not always lead to infection, and hence here a stochastic process is considered. The probability that an individual infects another is determined by the intensity of the contacts that the individuals have. Lockdown policies have been implemented in the model by adjusting a pre-specified parameter. Next to being infected, recovery is incorporated and once an individual has recovered, then it is assumed that the individual is immune to the disease. We are aware of possible reinfection, and we study this topic, but for the current manuscript, reinfection is omitted. Since COVID-19 can be a lethal disease in some cases, death has been incorporated as well. The model has been extended for modelling lockdown policies that certain governments have adopted. One of the advantages of the current approach is the small number of input parameters needed. A further innovation is the uncertainty quantification and the statistical assessment of the results.

2 The Mathematical Model

In this section the mathematical model will be derived and explained. First a basic model is given, which is later extended to a more realistic model.

To begin, consider a graph with nodes and vertices. The nodes represent individuals that can be in one of four different states: susceptible, infected, recovered, dead. If a person is susceptible, then this individual can only be infected. Once the individual is infected, then, the person can either recover or die. If (s)he recovers then this person is assumed to be resistant. If a person is susceptible, dead or resistant, then (s)he will not spread the virus to other people (although this assumption may be subject to discussion because a non-infected could spread the virus via the hands or other objects, however, this effect is neglected in the current modelling). The interpersonal relations are represented by connecting line segments in the graph. The connection is subject to an intensity, which represents the frequency that two individuals physically interact. This intensity and connection can be interpreted in a generic sense regarding relations and geographical distances. This connection determines the probability that, if one of the two individuals is infected, the disease is transferred from one another. Furthermore, infected individuals may recover or die.

Mathematically, this can be written as follows. The population consists of n individuals, which at every time instance is denoted by a vector of length n, where each entry in this vector contains the state of individual. This vector is denoted by $\mathbf{v}$, where the value of v_i contains the integer states: $v_i \in \{1, 2, 3, 4\}$, where $v_i = 1$, $v_i = 2$, $v_i = 3$ and $v_i = 4$, respectively, correspond to the susceptible, infected, recovered and dead states. All individuals are connected to other individuals by vertices between nodes (or individuals). The connection between person i and j is denoted by $a_{ij}(t)$, where $a_{ij}(t) = 0$ represents the case that individuals i and j have no physical contact. The entries $a_{ij}(t)$ are assembled into the contact intensity matrix $A(t)$. Note that the entries are dependent on time t as the intensity of the contact between two individuals changes over time. Large values of $a_{ij}(t)$ represent a high intensity of the contacts, while lower values mean that the people have less contact. The dynamics of the spread of the disease is discussed in the coming subsections.

2.1 The Transfer of the Virus from Individual to Individual

Suppose that individual i is infected and that individual j is susceptible. The time between going from the susceptible state to the infected state is assumed to follow an exponential distribution with infection parameter rate $\lambda_{ij}(t)$ in a given time interval denoted by τ. The time interval τ is assumed to be small. Hence, the probability that person j becomes infected in the small time interval τ, given that person i is infected, is given by:

$$P(v_j(t+\tau)=2|v_j(t)=1,v_i(t)=2)=\int_t^{t+\tau}\lambda_{ij}(s)e^{-\lambda_{ij}(s)(s-t)}ds. \qquad (1)$$

Since it is only possible to go from the susceptible state to the infected state, the probability that person j stays susceptible in a small time interval τ, given that he/she was susceptible at time t and person i is infected at time t, is given by:

$$P(v_j(t+\tau)=1|v_j(t)=1,v_i(t)=2)=1-\int_t^{t+\tau}\lambda_{ij}(s)e^{-\lambda_{ij}(s)(s-t)}ds. \qquad (2)$$

Therefore, the probability that this non-infected person j dies or recovers from the disease is zero. Hence:

$$P(v_j(t+\tau)\in\{3,4\}|v_j(t)=1,v_i(t)=2)=0. \qquad (3)$$

The infection rate parameter $\lambda_{ij}(t)$ is assumed to be of the following form:

$$\lambda_{ij}(t)=\lambda_g\ a_{ij}(t), \qquad (4)$$

where λ_g is a general infection rate parameter that is assumed to be the same for every individual.

Next we consider the set of people an individual is in contact with. Define the set N_j of individuals that is in contact with person j by:

$$N_j(t)=\{k\in\{1,...,n\}:a_{kj}(t)>0\}, \qquad (5)$$

where the set $N_j(t)$ represents the 'neighbours' of person j; these are the of individuals are in contact with person j at time t. This set can be reduced to a set where we only consider all the neighbours of individual j that are in the infected state. This subset is denoted by N_j^I:

$$N_j^I(t)=\{k\in N_j(t):v_k(t)=2\}, \qquad (6)$$

where the superscript I denotes the infected individuals.

Next, assume that all the states of the individuals in the 'neighbour' set N_j or N_j^I are independent of each other. Hence to obtain the probability of not being infected, the product of all the probabilities can be used. Therefore, the probability that node j stays susceptible is as follows:

$$P(v_j(t+\tau) = 1|v_j(t) = 1) = \prod_{k \in N_j^I(t)} (1 - \int_t^{t+\tau} \lambda_{kj}(s) e^{-\lambda_{kj}(s)(s-t)} ds). \quad (7)$$

As a direct consequence, the probability that node j becomes infected is given by:

$$P(v_j(t+\tau) = 2|v_j(t) = 1) = 1 - \prod_{k \in N_j^I(t)} (1 - \int_t^{t+\tau} \lambda_{kj}(s) e^{-\lambda_{kj}(s)(s-t)} ds). \quad (8)$$

During the time interval $[t, t+\tau]$ where $s \in [t, t+\tau]$ we take λ_{kj} constant, hence $\lambda_{kj}(s) = \lambda_{kj}(t)$. Then Eq. (8) can be rewritten as:

$$P(v_j(t+\tau) = 2|v_j(t) = 1) = 1 - \prod_{k \in N_j^I(t)} e^{-\lambda_{kj}(t)\tau} = 1 - e^{-\tau \sum_{k \in N_j^I(t)} \lambda_{kj}(t)}. \quad (9)$$

Substituting the definition of $\lambda_{kj}(t)$ gives:

$$P(v_j(t+\tau) = 2|v_j(t) = 1) = 1 - \prod_{k \in N_j^I(t)} e^{-\lambda_{kj}(t)\tau} = 1 - e^{-\tau \sum_{k \in N_j^I(t)} a_{kj}(t)\lambda_g}. \quad (10)$$

From equation (10), it can be seen that there is an effective transfer probability rate that can defined for each node v_j, namely:

$$\lambda_j^{eff} = \lambda_g \sum_{k \in N_j^I(t)} a_{kj}(t). \quad (11)$$

This is summarised in Theorem 1, of which a similar version was proved in [13]

Theorem 1: *Let node i possess neigbours $\mathcal{N}_i^I(t)$ that are infected. Then, assuming the contact intensity matrix not to change during the time interval $(t, t+\tau)$, the effective probability rate in the exponential distribution for node i to become infected is given by*

$$\lambda_i^{eff} = \lambda_g \sum_{j \in \mathcal{N}_i^I(t)} a_{ij}(t).$$

2.2 Transition to Recovery and Death

People that are in the infected state await two different scenarios: recovery with being resistant or death. Some people recover very quickly after having had (very) mild or even no symptoms, whereas other people need a long time to recover or pass away. In the current modelling, it is assumed that the recovery time follows an exponential distribution with probability rate parameter $\mu > 0$, that is

$$P(v_i(t+\tau) = 3|v_i(t) = 2) = \int_t^{t+\tau} \mu e^{-\mu(s-t)} ds. \tag{12}$$

It has been assumed that μ is constant, which is not realistic as μ is be subject to temporal changes due to improvements of medical therapies against the disease as well as the health conditions of each individual or even seasonal effects. Later this assumption is relaxed. The probability that a person stays infected is then given by:

$$P(v_i(t+\tau) = 2|v_i(t) = 2) = 1 - \int_t^{t+\tau} \mu e^{-\mu(s-t)} ds. \tag{13}$$

The expected recovery time from the moment that the patient was infected is determined by

$$T_r = \frac{1}{\mu}, \tag{14}$$

which follows from the properties of the exponential distribution. In this model it is assumed that if a person has been infected during a time-interval that exceeds a threshold, say $T_{death} = M \times \tau$, where $M > 0$ is some positive integer value, then the person dies with probability one. This is not always the case in reality, but it is a reasonable assumption. The rationale behind this assumption is that a long lasting exposure to the disease potentially damages the patient's vital organs so much that he/she dies.

To develop our intuition behind the relation between the recovery rate and time interval of death T_d, we assume that the probability that someone dies from the disease is given by α. Hence all patients that have been ill over a period that exceeds T_d are assumed to die. Then if person i was infected at time t, then the time interval of death and the probability to die are related by:

$$1 - \int_t^{t+T_d} \mu e^{-\mu(s-t)} ds = \alpha \tag{15}$$

We can rewrite it as:

$$\int_0^{T_d} \mu e^{-\mu(s-t)} ds = 1 - \alpha \Longrightarrow T_d = -\frac{1}{\mu}\log(\alpha) = -\log(\alpha) T_r,$$

where log is the natural logarithm and α is a very small probability that is at most 2–3 %.

If it is assumed that person i got infected at time $t_{inf} = \min_{t>0}\{v_i(t) = 2\}$ we can write the following mathematically:

$$v_i(t_{inf} + \theta) = \begin{cases} 3, & \text{if } \theta < T_d \\ 4, & \text{if } \theta \geq T_d \end{cases} \tag{16}$$

where θ is amount of time that person i has been infected.

3 Computational Implementation

In all the simulations, a constant population size is assumed. The current preliminary computations involve a simplified square topology, in which each node has at most four connections. It is easy to revise this topology. A uniform transmission probability rate λ_g to obtain the probability that the node changes from susceptible to infected during the time step τ. To model transmission, the effective transmission probability rate is computed by the use of the contact intensity matrix. Subsequently for each susceptible node a random number, ξ, from the standard uniform distribution (between zero and one) is sampled, that is $\xi \sim U(0,1)$. If the number is smaller than the probability of transmission from susceptible to infected then the state is changed from susceptible to infected, that is v_i is changed from 1 to 2, that is

$$v_i(t+\tau) = \begin{cases} 2, & \xi < P(v_i(t+\tau) = 2 | v_i(t) = 1), \\ \\ 1, & \xi > P(v_i(t+\tau) = 1 | v_i(t) = 1). \end{cases}$$

Otherwise, it stays in the susceptible state.

The same is done for the transition from the infected state to the resistant or dead state. However, we keep track of the time-interval that a node has remained in the infected state by adding the time step τ to the time-interval that a node is in the infected state. If the total length of the time interval that a nodal point stays in the infected state exceeds the length T_d, then the node is moved to the dead state. As long as this time-interval has not been exceeded, the node either stays in the infected state or it is transferred to the resistant state analogously to the treatment of susceptible nodes, but with a different number $\xi_2 \sim U(0,1)$ drawn from the uniform distribution.

Since interpersonal contacts are often fluctuating (like going to shops, meeting friends, working, etc.), randomised values for the contact intensity matrix $a_{ij}(t)$ are used, that is, considering person i:

$$\text{For } s \in (t, t+\tau) \ : \ a_{ij}(s) \sim U(0,1), \qquad \text{if } j \in \mathcal{N}_i.$$

In the case of lockdown, the contact intensity matrix is multiplied by a factor, $\beta(t)$, whose value ranges between zero and one. Small values of the factor $\beta(t)$ represent severe lockdown policies. Hence, the contact intensity matrix is redefined by

$$\hat{A}(t) = \beta(t)A(t),$$

and in all expressions given earlier, $A(t)$ and its entries are replaced with $\hat{A}(t)$ and its corresponding entries are given by $\hat{a}_{ij}(t) = \beta(t)a_{ij}(t)$. Note that the lockdown policy depends on t, and therefore $\beta = \beta(t)$, where $\beta : \mathbb{R}^+ \longrightarrow [0,1]$.

In order to compute the fractions of susceptible, dead, infected and resistant people, we introduce the standard Kronecker Delta Function:

$$\delta_{p,k} : \mathbb{N} \times \mathbb{N} \longrightarrow \{0,1\} \ : \qquad \delta_{p,k} = \begin{cases} 1, & \text{if } k = p, \\ 0, & \text{else.} \end{cases}$$

The fraction of individuals in state $p \in \{1,2,3,4\}$ is given by

$$f_p(t) = \frac{1}{n} \sum_{j=1}^{n} \delta_{p,v_j(t)}.$$

This definition reproduces that $\sum_{p \in \{1,2,3,4\}} f_p = 1$. The computations are terminated as soon as the number of infected people equals zero.

4 Simulation Results

In the previous sections the mathematical model has been derived and the numerical implementation has been explained. In this section the results of various simulations will be shown and examined.

In all of these simulations, a rectangular arrangement of 100×100 nodes is taken. Every internal node will only have four neighbours (left, right, up, down), every boundary node will have three neighbours and every corner node will only have two neighbours. Initially all the nodes are in susceptible state, except for a node in the lower left corner, that node is infected (this is a random choice, one could have made any node in the population infected). Without an infected person, the model will never predict the spread of the virus. The reason why it has been chosen in the lower left corner is that this grid can be seen as one of four 'quadrants', so the results can be reflected towards the other three quadrants. It is to be noted that the results of the current simulations are hypothetical as we have only used hypothetical values. Later on in this paper, parameter estimation will be done in order to find the input parameters of the model, given a simulated data set. This enables an estimate of the input parameters based on observed data and therewith the model can be made predictive (Table 1).

4.1 Simulation 1: No Lockdown

We begin by examining the scenario where there is no lockdown, hence in the model this means that $\beta(t) = 1$. It is expected that the virus will be able to spread rapidly in the population, causing many active cases in a short period of time. The virus rapidly spreads within the population until eventually everybody has either remained susceptible, recovered or died. We will refer to the different

Table 1. Table of parameters used in the simulations

Parameter	Value
Grid ($n_x \times n_y$)	100×100
τ	1
λ_g	0.5
μ	0.1
end time	100τ
Time lockdown start	15τ
Time lockdown end	60τ
Time to death T_d	8

groups of people in the population as the *Susceptible, Infected, Recovered* and *Dead* sub-populations. These sub-populations consider the number susceptible, infected, recovered and dead people respectively at a certain time interval in the simulation. Capital letters have been chosen as we refer to this specific sub-population group. To get a better understanding of how the virus spreads, the sub-populations have been plotted against the time in Fig. 1, the Recovered and Dead sub-populations have been plotted cumulative against the time. The Susceptible and Infected sub-populations have not been plotted cumulative, because the number of susceptible people only decreases over time and eventually all the infected people recover or die.

Figure 1 shows that the virus spreads exponentially in the beginning, causing a rapid growth in the number of active cases and a drop in the number of susceptible people. The Infected sub-population graph looks almost like a bell-curve shape. It starts with exponential growth up until time 15τ and then it makes a round turn and has a more parabola shape afterwards, until it flattens out as there are no more infected people left.

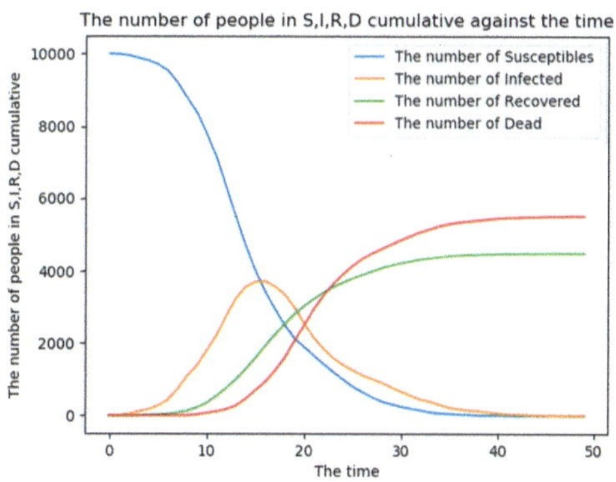

Fig. 1. Graph of the number of people in Susceptible, Infected, Recovered, Dead sub-populations cumulative against the time with no lockdown.

The reason why there is a turn in the number of infected people is because at time 15τ we do not have 'enough' susceptible people left to obtain a higher peak than before, therefore the number of infections decreases afterwards. The Susceptible sub-population graph has a very steep negative slope, suggesting that the number of susceptible people rapidly decreases. The number of infected people follow a similar trend, but then reversed (so a rapid increase). After the number of infected people have peaked, the decrease in the number of susceptible people also slows down, until there are no more susceptible people remaining. The cumulative graphs of the Recovered and Dead sub-populations in Fig. 1 look like logistic growth. The Recovered sub-population graph starts to rise earlier compared to the Dead sub-population graph, as it takes some time for infected people to recover. In this situation there are more dead people than recovered people, but that is due to the way the parameters in the simulation have been chosen. If a different set of parameters were chosen, the outcome would be different.

Uncertainties in the Simulations when there is no Lockdown

The mathematical model that is presented is based on random processes, therefore each simulation will have a different graph of all the four sub-populations. To see the possible 'bandwidth' that a sub-population might have, the simulation has been carried out a hundred times and the various sub-populations have been plotted against the time. These graphs are found in Fig. 2.

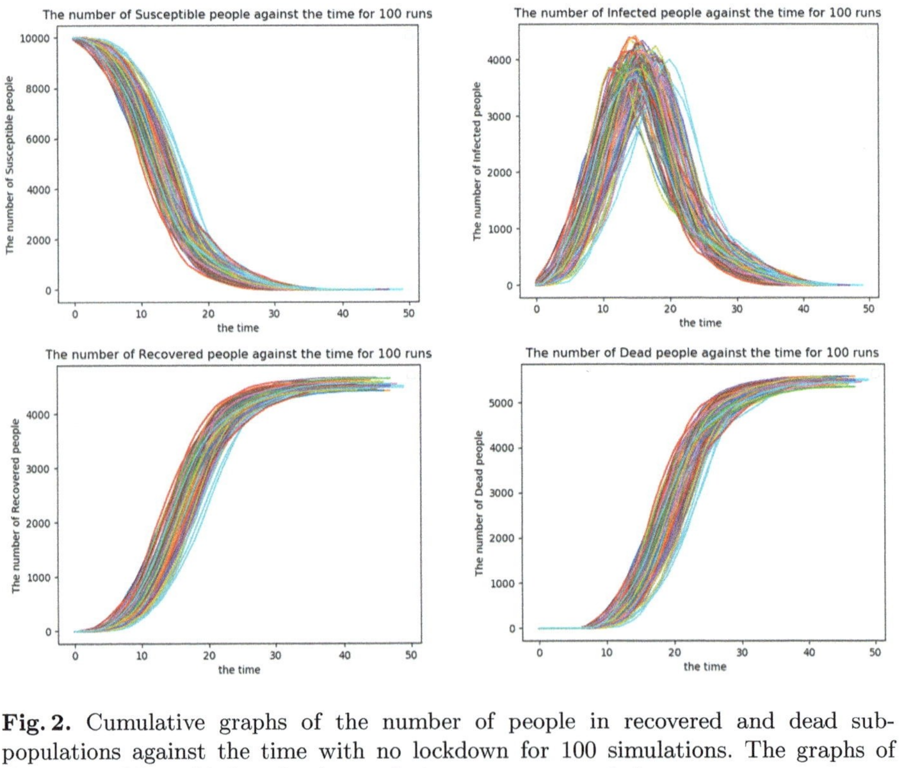

Fig. 2. Cumulative graphs of the number of people in recovered and dead sub-populations against the time with no lockdown for 100 simulations. The graphs of the susceptible and infected sub-populations are not cumulative.

If we consider the top two figures in Fig. 2 more closely, then it is noticeable that the peak of the Infected sub-population graphs is around time 15τ if there is no lockdown implemented. This also corresponds to the steepest slope in the Susceptible people sub-population graphs, suggesting it might be interesting to incorporate a lockdown scenario at time 15τ. This will be done later in this section.

From Fig. 2 it can be seen that the number of recovered people fluctuates around the 4000 to 4500 (when considering a total population size of 10000 people), which is roughly 40–45% of the total population. The number of dead people ranges between 5000 and 5500, which is roughly 50–55% of the total population. This is of course not the ideal outcome, but it is the result of the parameters chosen in the simulations. Figure 2 shows that all the simulation results are consistent, but possess some variation.

4.2 Applying a Temporary Lockdown

In this section some simulations will be shown where the lockdown is not kept constant over time or lifted immediately, but there is a step in between. First a case is considered, where a severe lockdown of $\beta(t) = 0.1$ is implemented, then loosened to $\beta(t) = 0.5$ and finally to $\beta(t) = 1$ (no lockdown). In addition, another case is considered where first a heavy lockdown of $\beta(t) = 0.3$ is implemented and then lifted to a lockdown of $\beta(t) = 0.6$ or $\beta(t) = 0.7$ to see what impact such lockdown has. The parameters that are used in the simulations are found in Table 2.

Table 2. Table of parameters used in the simulations

Parameter	Value
Grid ($n_x \times n_y$)	100×100
τ	1
λ_g	0.5
μ	0.1
End time	100τ
Time lockdown 1 start	15τ
Time lockdown 2 start	35τ
Time lockdown end	60τ
Time to death T_d	8τ

Case 1: Severe Lockdown of $\beta(t) = 0.1$ to Medium Lockdown of $\beta(t) = 0.5$ and then no Lockdown

We start by examining a simulation where first a severe lockdown of $\beta(t) = 0.1$

has been implemented. This lockdown is later lifted to a medium lockdown of $\beta(t) = 0.5$ at time 35τ and at time 60τ it is entirely lifted. Therefore:

$$\beta(t) = \begin{cases} \tilde{\beta}(t) = 0.1, & \text{for } t \in T_{ld_1} = (15\tau, 35\tau) \\ \tilde{\beta}(t) = 0.5, & \text{for } t \in T_{ld_2} = (35\tau, 60\tau) \\ 1, & \text{else.} \end{cases}$$

This could refer to the case where a country has first implemented a strict protocol where people are only allowed to go outside for one hour a day and all public restaurants/events/bars/shops are closed. Then the rules are lifted to a medium lockdown where shops are open again but only for limited amount of customers and everybody has to wear a face mask, and at time 60τ everything is back to normal again. At time 100τ, when the simulation ended, there is still a significant amount of susceptible people left as well as infected people. This indicates that the virus is still spreading amongst the population. In order to understand what consequences this lockdown implementation has brought to the population, consider the graphs of the various sub-populations in the first figure in Fig. 3. From this figure some interesting events as a result of the lift of the lockdown rules can be seen. Until time 15τ everything is like before as the virus is free to spread. At time 15τ, the consequence of the severe lockdown of $\beta(t) = 0.1$ can be seen (just like in Case 3 of the previous subsections). Due to the relaxation of the lockdown at time 35τ, a second outbreak occurs, which causes a rise in the number of infections. However, the number of susceptible people still decreases at a relatively constant rate, until time 60τ. At time 60τ the lockdown is lifted and we notice a third peak in the number of active cases. The second and third peak are nevertheless relatively small compared to the first peak. This is not only a result of the lockdown rules, but also a result of less susceptible people in the population. The most striking observation in the Recovered and Dead graphs is that the third peak is not really visible, suggesting that the relaxation of a medium lockdown to no lockdown does not affect the number of recovered or dead people significantly. However, this might be explained by the population size. If a larger population size would be taken (say 100 million) this would be more significant.

Case 2: Heavy Lockdown of $\beta(t) = 0.3$ to Mild Lockdown of $\beta(t) = 0.6$ and then no Lockdown

The next scenario is as follows: a country has first implemented a heavy lockdown of $\beta(t) = 0.3$, then relaxed the rules to a mild lockdown of $\beta(t) = 0.6$ and after that the lockdown is entirely lifted. The results can be seen in the second graph in Fig. 3.

$$\beta(t) = \begin{cases} \tilde{\beta}(t) = 0.3, & \text{for } t \in T_{ld_1} = (15\tau, 35\tau) \\ \tilde{\beta}(t) = 0.6, & \text{for } t \in T_{ld_2} = (35\tau, 60\tau) \\ 1, & \text{else.} \end{cases}$$

The effect of the first lockdown of $\beta(t) = 0.3$ is significantly present as can be seen in the number of active cases. The loosening of the rules at time 35τ causes another rise in the number of infections from time 35τ until time 42τ. It is quite interesting to see that the first peak looks almost triangular while the second peak looks like half an oval (shape wise). This is probably due to the fact that the first peak is caused by no lockdown rules, so we first have an exponential growth of infected people. Then suddenly a heavy lockdown is implemented which causes the amount of infected people to drop drastically. The second lockdown is mild, which causes a rise in the number of active cases, but since there are few susceptible people remaining in the population, it quickly turns and starts decreasing again. The peaks in the number of active cases correspond to the rapid increases in the Recovered and Dead sub-population graphs. In addition, another simulation has been run, but in this case, the second lockdown was lifted till 0.7 instead of 0.6 to see if this makes any difference. The results can be seen the third graph in Fig. 3. The difference is not significant since the same trends as in the second case are observed.

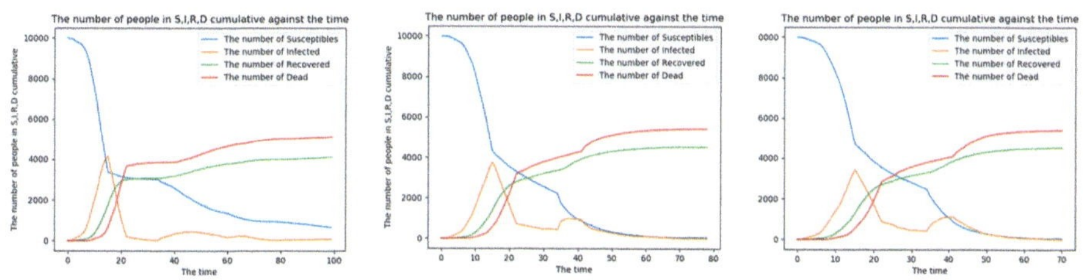

Fig. 3. Cumulative graph of the sub-populations against the time. The first graph refers to case 1, where $\beta(t) = 0.1$ to 0.5 to 1.0, the second to case 2, where $\beta(t) = 0.3$ changes to 0.6 to 1.0, the last graph refers to $\beta(t) = 0.3$ to 0.7 to 1.0.

5 Estimation of Recovery and Death

In this section an estimate for the recovery rate (hence also the death rate) and an estimate of the duration of COVID-19, given that the patient dies, are provided. The estimates are done by the use of statistical models, such as logistic regression for the probability of recovery as a function of age and gender, linear regression, log-linear regression and log-Poisson regression for the duration of the disease given that the patient dies as a function of age of the patient. Similar approaches have been carried out by Alleman et al. [6] and by Chen [14]. The estimations are based on the data provided by the Dutch Institute for Health and Environment (RIVM). First some data processing is done, followed by the underlying mathematical model and the parameter estimation. The estimation of the infection parameter λ_g is not done based on real data due to the fact that it is so highly dependent on the lockdown degree (which was changed almost every two weeks in the Netherlands), the testing rate, the number of contacts, age, vaccination etc.

5.1 Data Processing

The data provided by the RIVM consists of a couple of variables, defined as follows:

- the date that the person reported that he/she got infected by COVID at the GGD (Municipal Health Services of the Netherlands).
- the gender of the patient, which in this case is Male, Female or Unreported.
- the age of the patient in years.
- first date that the patient experienced symptoms of COVID.
- the date that the patient died because of COVID.

There are a few things to be noted. If an individual is infected with COVID-19, then he/she has to report it to the government. However, there is no obligation to report if a person actually died after the infection. Hence, the data that is obtained is under-reported. In addition, the deaths that are reported may not always be a direct consequence of COVID-19. The person might also have died of other health complications or simply because of an old age. This is seen back in the data set, as some people only die after hundreds of days, which is not likely to be a direct cause of the COVID-19 infection. In order to work with this data set, there have been a few columns added, namely:

- the duration of the illness of the patient, if he/she died.
- the boolean parameter 'death', which is 0 if the person recovered and 1 if he/she died. Note that in this case it is assumed that if death of a person has not been reported, then the person is assumed to have recovered.

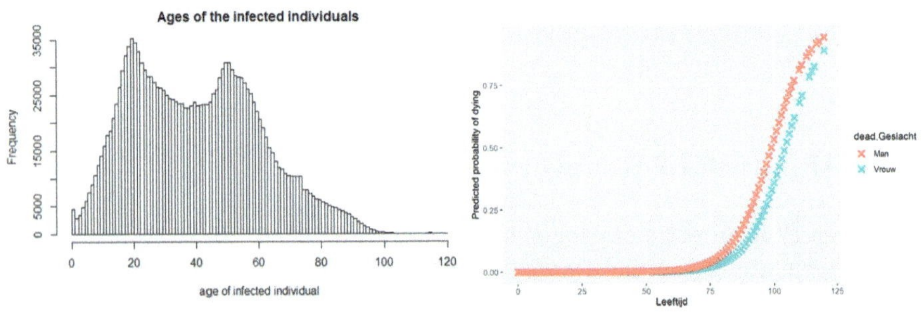

Fig. 4. Left: histogram of the age of all the infected individuals. Right: predicted probability of dying against the age of the patient for the genders male, female and undefined

The data ranges from 2020-06-01 until 2021-06-28 and consists of 1666371 observations. The ages range between 0 and 120 years old. There were 98 people who did not report the age. In order to deal with the missing values of the age variable, a random number between 0 and 120 was drawn as the age ranged from 0 till 120. As 98 out of 1666371 observations is a small portion, this missing information did not contaminate the data much. The histogram of the age of the infected individuals is shown in Fig. 4 (Left).

5.2 Mathematical Model of the Recovery Rate Estimation

The observations in the data are all assumed to be independent events. In addition, the following variables are defined:

- l_i: the age of patient i
- f_i: the gender of patient i, $f_i = 1$ represents female, $f_i = 0$ represents male
- δ_i: $\delta_i = 1$ if the person dies from this COVID-19 infection and $\delta_i = 0$ if the person recovers.
- x_i: the duration of the COVID-19 infection after the person got infected, $x_i = y_i$ the duration of the illness and $x_i = \infty$ if the person recovers. So observe that $x_i = \infty \iff \delta_i = 0$. The reason why the duration is set to infinity is because the person recovers from the illness.

The dichotomous variable δ_i follows a Bernoulli distribution with parameter p_{l_i,f_i}, where p_{l_i,f_i} is the probability of dying given the age l_i and gender f_i. A logistic regression model is used to estimate this parameter $p_{l_i} = p(l_i, f_i)$, which is actually a function of the age as well. Moreover, it is assumed that given that patient i dies from COVID-19, the duration until death follows an exponential distribution with parameter $\mu_{l_i} = \mu(l_i)$, function of the age. Generalised linear models will be used to estimate this parameter. To begin, we fit a logistic regression model to the data and we apply regression on both age and gender. From the summary we see that both variables, age and gender, are significant as they have very small p-values, suggesting a strong association of the gender as well as the age of a person with the probability of dying. For this case, the following model is obtained for the log odds of death from COVID :

$$\log(\frac{p_{l_i}}{1 - p_{l_i}}) = -13.19 + 0.13 l_i - 0.71 f_i,$$

where l_i is the age of person i and $f_i = 1$ if the individual is female and $f_i = 0$ if the individual is male. The probability of recovery from COVID-19 is $1 - p(l, f)$. Furthermore, using the fitted model of the log-model, predictions can be made. In Fig. 4 right, predictions of the probability of dying are made based on the age of a person as well as their gender. From these figures it can be seen that males tend to have a higher probability of dying compared to females. In addition, the older the individual, the higher the probability of dying.

Duration of COVID and Age

Next, we will condition on the fact that the individual has died from COVID, hence all the recovered people are removed from the data frame. In this way the time that it takes for an individual to die, given that the person will die, can be estimated. In the model it is assumed that given that the patient died as a result of this COVID infection, the duration of the illness is exponentially distributed with parameter μ_{l_i}, which depend on the age l_i. In this analysis we do not distinguish between gender. The duration of the illness in the data set has the following descriptors:

- Min: 1 day
- 1st Quartile: 7 days
- Median: 11 days
- Mean: 13.71 days
- 3rd Quartile: 16 days
- Max: 391 days.

From these descriptors it can be deduced that individuals who are ill for more than 30 d might not have died as a direct result of the COVID-19 infection. The virus may have caused other health related issues or that the person died because of other reasons than COVID-19. However, in this model it is still assumed that all the individuals who died of COVID-19 have this infection as a cause.

To begin, a linear regression model is fitted to the data. In this case, the assumption is that the observation x_i is drawn from a Normal (Gaussian) distribution with a mean μ_i, that depends on the age l_i and a constant variance σ^2 across all ages. So $x_i \sim N(\mu_i, \sigma^2)$ and $\mathbb{E}(x_i) = \mu_i = \alpha + \beta l_i$, $\forall i$. Hence the residuals $\epsilon_i = x_i - \mu_i \sim N(0, \sigma^2)$. So actually $x_i = \alpha + \beta l_i + \epsilon_i = 31.70 - 0.22 l_i$, where $\sigma = 12.45$.

Next, a log-transformed linear regression model is applied. This models the duration of the illness on a logarithmic scale, where the model is given by: $\log(x_i) \sim N(\mu_i, \sigma^2)$ and $\mathbb{E}(\log(x_i)) = \mu_i = \alpha + \beta l_i$. The model assumption is that the duration follows a log-normal distribution, so $x_i \sim \log(N(\mu_i, \sigma^2))$ with $\mathbb{E}(x_i) = exp(\mu_i + \sigma^2/2)$.

Finally, a Poisson regression is performed. In this case the response variable x_i is assumed to have a Poisson distribution and it assumes that the $\log(\mathbb{E}(x_i))$ is a linear combination of the unknown parameters. The reason why we use this Poisson regression model is that this model is very common for count data (that is data consisting of natural numbers $\{0, 1, 2, 3, \ldots\}$). In this case, we count the number of days that the individual is ill until the person dies. Although the Poisson model, like the log-transformed, is based on the assumption that the $\log(\mathbb{E}(x_i))$ is a linear combination of the unknown parameters, the main difference with the log-transformed model is that the response variable follows a Poisson distribution, whereas in the log-linear regression model above, the response variable is assumed to follow a Gaussian distribution.

The Poisson distribution only has one parameter μ_i, which is also the expected value. The model is given by: $x_i \sim Poisson(\mu_i)$, $\log(\mu_i) = \alpha + \beta l_i$ and $\mathbb{E}(x_i) = exp(\alpha + \beta l_i)$. The link function is the 'log' function in this case. Note that the mean and the variance are the same for the Poisson probability distribution. The reason why the Poisson distribution is used, is because it will generate integer numbers, which is in line with the actual duration of the data given (even though in reality we do not have whole days of course, but more days + hours + minutes + seconds). In Fig. 5 left, a plot with all the observations, as well as the three different fitted models, is shown. From this plot we see that actually all three models seem to be similar. Interestingly, the data is still quite scattered containing quite a few outliers, which are not detected by any of the models. In order to choose the best model, the Akaike Information

Criterion (AIC) and R^2 values of the three models are compared. The values can be found in Table 3, and judging from the AIC and R^2–values, it is clear that the log-linear model has the most favourable characteristics.

Table 3. Table of the estimated coefficients for the linear, log-linear and Poisson model as well as their null deviance, residual deviance and if present the residual standard deviation. The AIC and R^2 values are reported too.

Model	Linear	Log-linear	Poisson
Intercept	31.70	3.31	3.81
Age coefficient	–0.22	–0.01	–0.01
Null deviance	1325070.9	4665.2	59160.7
Residual deviance	1282943.4	4545.8	56262.1
Residual sd sigma	12.45	0.74	–
AIC	65294.64	18544.87	91199.97
R^2	0.0318	0.0256	0.04890

Statistical Simulations

Since all the models have a certain distribution associated to it, we subsequently simulate the duration of illness based on the respective three (normal, log-normal and Poisson) probability distributions with the estimated parameters on the actual ages and compare it to the real data. In Fig. 5 right, a plot of the simulated data as well as the real data is shown. From this plot it is clear that the log-transformed linear model performed best (which is consistent with our earlier findings regarding the R^2 and AIC statistics) as it is able to also predict higher illness duration (probably due to the variance in the model). Logically, the four outliers of the data are not predicted, but since these are only four points in the data, fitting them into the model as well, would lead to overfitting. Hence from the simulated data it can be seen that a log-linear model is the most suitable model, which is in line with the lowest AIC value.

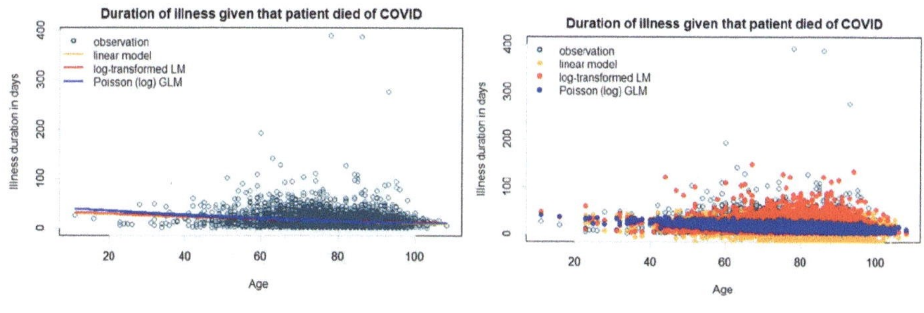

Fig. 5. Left: plot of the observations together with the three fitted models: linear, log-linear and Poisson, Right: Simulated duration of the illness duration compared to the actual duration from the data

Therefore, we obtain the following for the duration of illness before death (in days):

$$x_i = \begin{cases} e^{3.31-0.012l_i}, & \text{if the person dies} \\ \infty, & \text{if the person recovers.} \end{cases}$$

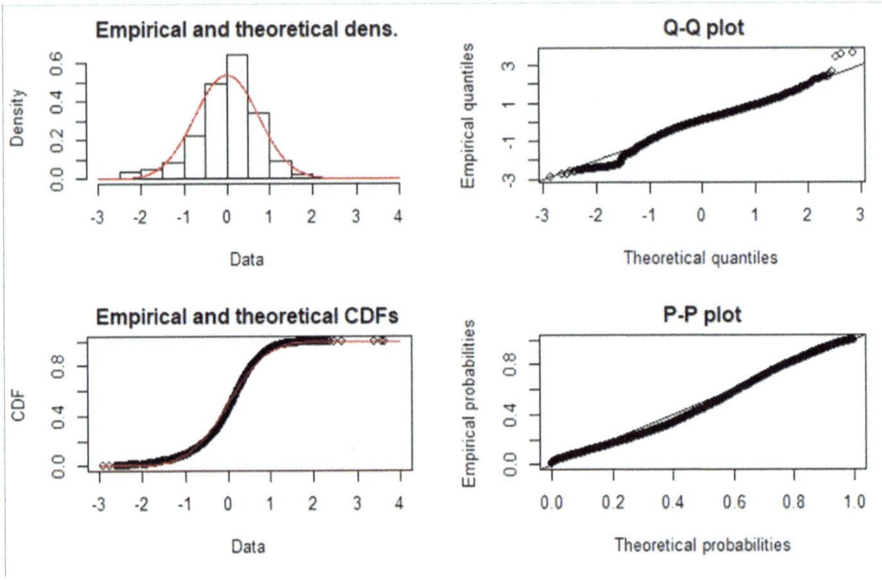

Fig. 6. Normal fit of the residuals of the log-linear model. From this we see that the residuals indeed follow a normal distribution.

We have plotted the residuals of the log-linear plot (not shown here), and from this plot it is clear that the residuals are nicely centered around zero. To investigate whether the residuals are normally distributed or not (as it should be according to the assumptions of this model), a normal fit is done and the results are shown in Fig. 6. From this, it is obvious that the residuals are most likely normally distributed. This property indicates that indeed this statistical model is a good fit. We note that the outliers may be disregarded.

6 Discussion and Conclusions

We have proposed a mathematical framework to simulate the spread of the novel coronavirus epidemic in a community based on a spatial Markov Chain network model, where the network topology can be made dynamic over time. The model is equipped with uncertainty assessment, in the sense that the probability of different events can be estimated. Different lockdown scenarios have been simulated to see the impact of the severity of lockdown policies. Mild lockdowns were not really effective in reducing the spread of the virus, while heavy lockdowns caused the number of infected cases to be approximately constant over the lockdown period and severe lockdown protocols could eradicate the virus (under the assumptions of the model). Lifting the lockdown rules caused multiple peaks in the number of active cases as the virus could spread more rapidly when the lockdown rules were relaxed. It is recommended for governments to strictly monitor what is happening when lifting lockdown rules. It might be wise to implement some more stricter rules even if the rules have been relaxed previously, when the spread of the virus is flaming up again. In reality, we see that many governments have indeed taken this approach. The proposed model is different from most of the models that have already been introduced, which are based on the general S(E)IR-models, like in [15–19]. There are two unique features in the presented model: (1) it is based on cellular automata (2) it uses exponentially distributed times between different states, which makes the model stochastic of nature. The model is able to predict the dynamics of the disease under various lockdown scenarios with different infection and recovery rates. However, we also note some limitations of the proposed model. The proposed model is still general as it has not yet been adjusted to specific countries or regions. It is recommended that the simple model is extended and adjusted to specific areas in the world as well as more features regarding the disease are added. This can be done by adding more compartments. To set up the mathematical model, a constant population size of n people is assumed. We assumed that the virus is only transferred from infected people to susceptible people, while in reality the virus is also able to spread through surfaces or objects. People who just recovered from the COVID-19 virus, are able spread the disease too. According to a study published in the journal JAMA, patients who recovered from the COVID-19 virus had been tested positive for the virus in every test between days 5 and 13 post-recovery [20].

The severity of the symptoms caused by the coronavirus vary from person to person. Some people only experience very mild symptoms like fever and dry cough, while others have difficulty breathing, chest pain and might even lose their speech or mobility and must be hospitalised. Elderly (above 70) also have a higher chance of dying from the coronavirus compared to younger individuals. Certain risk groups including individuals with chronic respiratory or pulmonary problems, heart patients or diabetics also have a higher chance of dying as a result of the coronavirus compared to 'normal people' of their age. Hence it is important for further research to investigate the infection rate for the different age-groups as well as the probability of getting heavy or mild symptoms.

Moreover, there is the possibility of reinfection. In this model, the recovery rate μ is assumed to constant for every individual, while later it is added in the extended model. Research shows that over a period of about three months, the number of antibodies in recovered people rapidly decreases. Recent studies found that there is a high chance of losing immunity to the COVID-19 virus after a period of time [21]. Death is assumed when a person has been infected for a certain period of time. In practice, some people might have mild symptoms for a very long time and recover or have very heavy symptoms and die within a few days. Variations are large among different individuals. Further research should be carried out to investigate the actual probability of reinfection (which is now only taken as a hypothetical parameter in the simulations) as well as introducing different sub-groups of infected people based on their symptoms, age and other health conditions (like asthma). To better quantify these transmission rates, one could assume that they are functions of time, since the rate of recovery does not only change with age, but also with the circumstances at that time (for instance IC capacity).

The different lockdown simulations showed that a severe lockdown is able to extinguish the outbreak in a limited amount of time, while less severe lockdown policies mainly cause a steady amount of infections as well as recovering and dying people over time. The term *flattening the curve* is often used in the media to describe: (i) reduction in the peak number of infections, to prevent the health care system from being overloaded and (ii) increasing the duration of the pandemic over a larger time interval, but with the same number of cases at the end. This phenomenon has been seen back in the simulations for various lockdown scenarios. The time of the epidemic was stretched over a longer period when a lockdown was implemented and the peak number of infections decreased. However, the total number of infections was the same as well as the total number of people who recovered and died, as can be seen in the cumulative graphs. Lifting the lockdown rules resulted into several peaks in the number of infections. The second and possibly third peak were a lot lower compared to the first one, due to a lower number of susceptible people. Nonetheless, governments are encouraged to impose appropriate measures in their lockdown policies to reduce the peak in the number of infections when they are relaxing the lockdown rules.

The model provides a theoretical framework to investigate the spread of the COVID-19 virus. The variability in the results due to the randomness in the model, makes the simulations more realistic as similar lockdown protocols in different countries have different effects on the number of casualties. Culture, population density and public health care are examples of variables that have a major impact on the number of casualties. In the paper by Cooper, Mondal and Antonopoulos in [22] a SIR-model is developed for various different communities. The paper by Cooper et al. present a study of the time evolution of different populations and the diversity in the parameters for the spread of the disease. Future research on this topic could help modify this model to a specific country or region. We realise that the presented model is still very basic and more simulations need to be carried out using extensions of the basic model. Perhaps it will be possible to find a pair of values for (λ_g, μ) and a lockdown strategy to eradicate the virus. It is also advisable to consider larger population samples or more simulation runs, since in this case we have only performed simulations for a constant population size of 10000 people and we have only done 100 simulation runs per pair of (λ_g, μ).

As the presented model is not yet adjusted to a specific country or region neither is it as extensive as many proposed S(E)IR-models. The results cannot be directly compared to previous studies. In future studies, it might be possible to relate the outcomes of this Spatial Cellular Automata Markov Chain model to the S(E)IR-models presented in for instance [15–19]. A downside of the current model, compared to the classical deterministic models like the S(E)IR-models, is that it is relatively expensive to execute. One could possibly optimise the model and use more computational resources within the Monte Carlo framework.

Estimation of the infection rate parameter for each individual is difficult based on the observed data. This is because the probability of getting infected by COVID-19 is dependent on many factors including the number of contacts that a person has (which in simulations must be randomised), the incubation period, lockdown policies, vaccination and many other conditions. In addition, the data is subject to under-reporting, due to the fact that in the beginning of the pandemic there were not that many tests performed and not everybody who becomes infected with COVID-19 reports it. This can be due to having had mild or almost no symptoms. The effect of lockdown is seen back in the number of active cases and in the Netherlands almost every two weeks the lockdown policy has changed in the first year of the pandemic. In addition, the number of tests increased over time, which resulted into more people who were tested positive for the virus compared to earlier. The rise of the different variants of COVID-19 as well as the increase in vaccinated people need to be taken in consideration as well. Further studies, which take these variables into account, will need to be carried out.

Currently, almost all the parameter estimation has been done based on statistical regression models such as the logistic regression, linear regression, log-linear regression and log-Poisson regression for the probability of recovery and duration of the disease given that the patient dies as a function of age and gender. In order to have a direct correspondence between the Markov Chain model and the data, more research is needed. For instance in [23] a physics-informed neural network is used to estimate the time-dependent contact rate and in [24] a Bayesian framework is used to estimate the infection parameter. Future research is encouraged to try to use any statistical method to estimate the parameter based on the mathematical model presented in this paper. Additionally, the estimation of the recovery rate parameter μ based on the actual data of the RIVM is also biased due to the assumptions made in the model. Further research should be undertaken to investigate this parameter under preferably less strong assumptions. All together, the proposed mathematical model is different from the regular S(E)IR-models used by many countries at this moment. It might provide a different way of modelling the pandemic and potentially lead to more accurate or different predictions as it looks at each individual individually. However, the proposed model is still simplistic and has to be extended and applied to real data in order to make predictions and give conclusions on the evolution of this disease.

References

1. worldometer. COVID-19 CORONAVIRUS PANDEMIC. https://www.worldometers.info/coronavirus/?utm_campaign=homeAdvegas1?%22%20%5Cl%20%22countries

2. WHO. Naming the coronavirus disease (COVID-19) and the virus that causes it (2020). https://www.who.int/emergencies/diseases/novel-coronavirus-2019/technical-guidance/naming-the-coronavirus-disease-(covid-2019)-and-the-virus-that-causes-it
3. Staff, M.C.: Covid-19 (coronavirus): Long-term effects, May 2021. https://www.mayoclinic.org/diseases-conditions/coronavirus/in-depth/coronavirus-long-term-effects/art-20490351
4. Kermack, W.O., McKendrick, A.G.: A contribution to the mathematical theory of epidemics. In: Proceedings of the royal society of London. Series A, Containing Papers of a Mathematical and Physical Character, vol. 115, no. 772, pp. 700–721 (1927)
5. Getz, W.M., Salter, R., Muellerklein, O., Yoon, H.S., Tallam, K.: Modeling epidemics: a primer and numerus model builder implementation. Epidemics **25**, 9–19 (2018)
6. Alleman, T.W., Vergeynst, J., De Visscher, L., Rollier, M., Torfs, E., Nopens, I., Baetens, J.: Assessing the effects of non-pharmaceutical interventions on SARS-CoV-2 transmission in Belgium by means of an extended SEIQRD model and public mobility data. Epidemics **37**, 100505 (2021)
7. Allen, L.J.: A primer on stochastic epidemic models: formulation, numerical simulation, and analysis. Infect. Dis. Model. **2**(2), 128–142 (2017)
8. Pellis, L., et al.: Eight challenges for network epidemic models. Epidemics **10**, 58–62 (2015)
9. Walters, C.E., Meslé, M.M., Hall, I.M.: Modelling the global spread of diseases: a review of current practice and capability. Epidemics **25**, 1–8 (2018)
10. Duan, W., Fan, Z., Zhang, P., Guo, G., Qiu, X.: Mathematical and computational approaches to epidemic modeling: a comprehensive review. Front. Comput. Sci. **9**(5), 806–826 (2014). https://doi.org/10.1007/s11704-014-3369-2
11. Britton, T.: Stochastic epidemic models: a survey. Math. Biosci. **225**(1), 24–35 (2010)
12. O'Neill, P.D.: A tutorial introduction to Bayesian inference for stochastic epidemic models using Markov chain monte Carlo methods. Math. Biosci. **180**(1–2), 103–114 (2002)
13. Vermolen, F., Pölönen, I.: Uncertainty quantification on a spatial markov-chain model for the progression of skin cancer. J. Math. Biol. **80**(3), 545–573 (2020)
14. Chen, F.: Better modelling of infectious diseases: lessons from covid-19 in china. BMJ **375**, 2363 (2021)
15. Fanelli, D., Piazza, F.: Analysis and forecast of covid-19 spreading in china, Italy and France. Chaos, Solitons Fractals **134**, 109761 (2020)
16. Yang, C., Wang, J.: A mathematical model for the novel coronavirus epidemic in Wuhan, china. Math. Biosci. Eng. **17**(3), 2708–2724 (2020)
17. Caccavo, D.: Chinese and italian covid-19 outbreaks can be correctly described by a modified sird model, *medRxiv* (2020)
18. Al-Raeei, M.: The forecasting of covid-19 with mortality using SIRD epidemic model for the united states, Russia, China, and the Syrian Arab republic. AIP Adv. **10**(6), 065325 (2020)
19. Rajagopal, K., Hasanzadeh, N., Parastesh, F., Hamarash, I.I., Jafari, S., Hussain, I.: A fractional-order model for the novel coronavirus (covid-19) outbreak. Nonlinear Dyn. **101**(1), 711–718 (2020)
20. Lan, L., et al.: Positive RT-PCR test results in patients recovered from COVID-19. JAMA **323**(15), 1502–1503 (2020). https://doi.org/10.1001/jama.2020.2783

21. Agel, F.: Antibodies, immunity low after COVID-19 recovery. https://www.dw.com/en/coronavirus-antibodies-immunity/a-54159332
22. Cooper, I., Mondal, A., Antonopoulos, C.G.: A sir model assumption for the spread of covid-19 in different communities. Chaos, Solitons Fractals **139**, 110057 (2020)
23. Grimm, V., Heinlein, A., Klawonn, A., Lanser, M., Weber, J.: Estimating the time-dependent contact rate of sir and seir models in mathematical epidemiology using physics-informed neural networks," Universität zu Köln, Technical Report, September 2020. https://kups.ub.uni-koeln.de/12159/
24. Irons, N.J., Raftery, A.E.: Estimating sars-cov-2 infections from deaths, confirmed cases, tests, and random surveys. In: Proceedings of the National Academy of Sciences, vol. 118, no. 31 (2021). https://www.pnas.org/content/118/31/e2103272118

A Novel Review of Temporomandibular Joint Replacement Options

Christine Walck(✉), Yeram Lim, and Seth Rosenstein

Embry-Riddle Aeronautical University, 1 Aerospace Boulevard, Daytona Beach, FL 32114, USA
Christine.walck@erau.edu

Abstract. The temporomandibular joint (TMJ) is one of the most complex joints in the human body with its ability to rotate and translate during jaw opening and closing movements. The motion during jaw-closing has two phases: 1) rotation and translation along the mandibular fossa from maximum mouth open (MMO), and 2) pure rotation from 22° until mouth closed position. However, this motion pattern is altered after the jaw undergoes open joint arthroplasty. This study analyzes the motion and torque capabilities associated with the Fibula Free Flap (FFF) procedure, and two types of TMJ implants: 1) a generic implant constrained to pure rotation, and 2) a patient-specific (PS) implant capable of both rotation and translation. Results show the PS implant most resembles the healthy jaw showing a 34.4 mm inferior displacement at MMO; a 1.15% decrease compared to the healthy jaw's (34.8 mm), and a maximum net torque of 18,600 N-mm; a 29.3% decrease compared to 26,300 N-mm. The rotation-constrained FFF procedure and generic implant limits the jaw to 70% of MC rotation decreasing the inferior displacement at MMO (FFF/Generic: 22.5 mm) by 35.3% compared to the healthy jaw and prohibits net torque about the MC after the pure rotation phase.

Keywords: Temporomandibular joint · Total open joint arthroplasty · Arthroplasty implants

Abbreviations

TMJ	Temporomandibular Joint
MMO	Maximum Mouth Open
FFF	Fibula Free Flap
PS	Patient Specific
ICR	Instantaneous Center of Rotation
CATIA	Computer-Aided Three-Dimensional Interactive Application
Φ_1	Angle from Axis to Ground Link
Φ_2	Angle from Axis to Arbitrary Link
Φ_3	Angle from Axis to Mandibular Link
l_m	Mandibular Link
l_g	Ground Link
l_b	Arbitrary Link

J. M. R. S. Tavares et al. (Eds.): CMBBE 2021, LNCVB 38, pp. 24–35, 2023.
https://doi.org/10.1007/978-3-031-10015-4_2

1 Introduction

The temporomandibular joint (TMJ) is one of the most complex joints in the human body due to its ability to rotate and translate during jaw-opening and jaw-closing movements. The motion of a healthy mandibular condyle during jaw-closing can be presented as two phases: 1) rotation and translation along the mandibular fossa from maximum mouth open (MMO) position (32° jaw rotation) to 22°of jaw rotation, and 2) pure rotation about the mandibular condyle from 22° of jaw rotation until mouth closed position (0° jaw rotation) [1]. The rate of rotation and translation of the mandibular condyle is assumed linear from the MMO position to the mouth closed position [1]. However, this motion pattern alters after the jaw undergoes a total open joint arthroplasty where the orthopedic surgeon removes damaged portions of the joint and, in some cases, replaces the joint with artificial components (implants). This study analyzes the altered motion pattern and the torque capabilities associated with the common Fibula Free Flap (FFF) procedure, and two types of TMJ implants: 1) a generic implant made of metal and polymer constrained to pure rotation, and 2) a patient-specific (*PS*) implant capable of both rotation and translation. Each has unique designs requiring different surgical procedures [2, 3].

The TMJ musculature involved in the jaw-closing movement of the healthy human is listed in Table 1. However, the FFF procedure requires the removal of the ramus and most of the mandible, including the temporalis, masseter, and pterygoid muscles [2]. The generic and *PS* implants require the mandibular condyle and 30% of the ramus to be cut, resulting in total removal of the deep masseter and the lateral pterygoid muscles

Table 1. Origin and insertion points of muscle line of action in a healthy human jaw. The YZ, XZ, and XY axes form the sagittal, frontal, and transverse planes, respectively [5, 6].

Muscles	Origin point (mm)			Insertion point (mm)		
	x-coordinate	y-coordinate	z-coordinate	x-coordinate	y-coordinate	z-coordinate
Temporalis						
Posterior temporalis	−13.1	50.0	−50.0	3.8	3.6	21.9
Middle temporalis	−13.1	70.6	−20.9	3.8	3.6	21.9
Anterior temporalis	−13.1	70.6	−20.9	3.8	3.6	21.9
Masseter						
Deep masseter	7.2	16.5	26.6	1.3	−22.3	18.6
Superficial masseter	7.2	16.5	45.5	1.3	−22.3	18.6
Pterygoid						
Medial pterygoid	−22.0	−1.2	35.4	2.7	−38.2	14.4

as well as 20% of the superficial masseter while 100% of the temporalis and medial pterygoid stay intact [2].

The FFF procedure and the generic implant resemble a mechanical revolute joint constraining movement to pure rotation on the defective side limiting jaw rotation to 22° (31.3% reduction compared to the healthy jaw) [1]. Whereas, the *PS* implant resembles a mechanical pin-in-a-slot joint allowing the condyle to rotate as well as translate along the natural path of the mandibular fossa, all of which influence the jaw's MMO position. The MMO position for the healthy jaw model aligns with the reported MMO ranges of 38–71 mm and 35–63 mm for adult (18–35 years of age) males and females respectively, worldwide [4].

2 Material and Methods

The subject-specific anthropometric measurements obtained from a model of a 50–65-year-old male of European descent acquired from the National Institute of Health [7] were imported into solid modeling computer-aided design (CAD) software to develop the kinematic and kinetic models shown in Fig. 1.

Accounting for the articular disc and retrodiscal pad relative to the condylar head during jaw-closing, points placed along the mandibular fossa defines its geometry using Eq. (1) while Eq. (2) defines the geometry of the mandibular condyle [9]. These geometries are depicted in Fig. 2. Furthermore, the instantaneous center of rotation (ICR) of mandibular condyle when the mouth is at its closed position defines the origin point of rotation [3, 5, 6, 9].

The transitional path for the healthy jaw condyle defined by Eq. (3) tracks the ICR and is constrained by the boundary of the mandibular fossa [3, 5, 9, 10]. Whereas, the *PS* implant's translational path is defined using Eq. (4) [5, 8]. The rotation constraint of the FFF procedure and the generic implant limits the mandibular condyle movement away from the point of origin negating a translational path and the use of any best-fit equations.

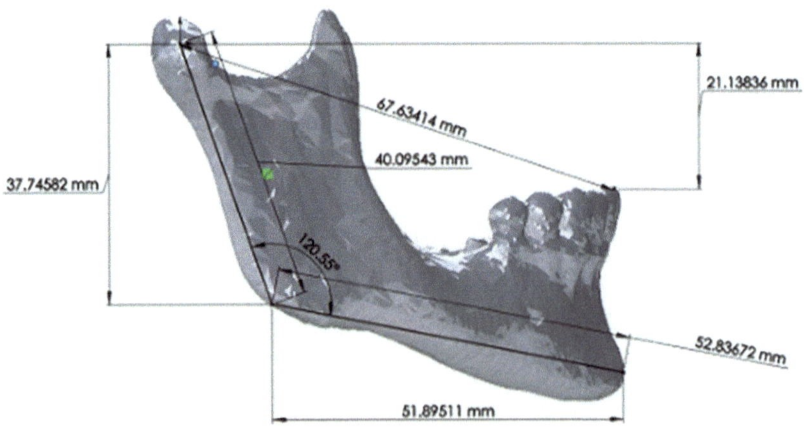

Fig. 1. Measurements were taken from the healthy model of the jaw for the kinematic analysis.

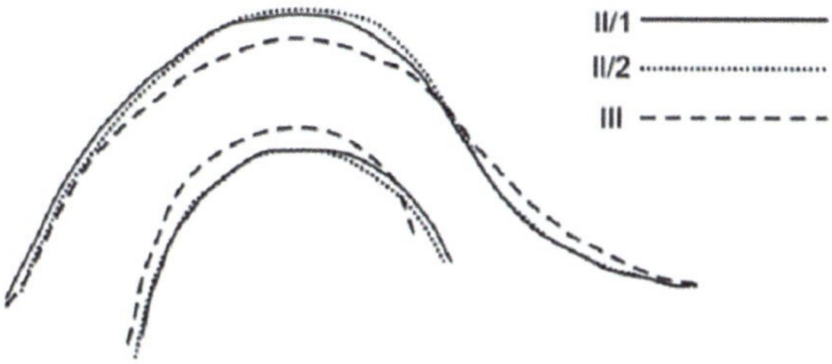

Fig. 2. Mandibular fossa and mandibular condyle average geometry of 189 patients with different classifications of mandibular condyle shape [11].

Best-Fit Equations

$$y = -12.3x^6 + 8.5x^5 + 6.6x^4 - 2.1x^3 - 2.8x^2 + 0.096x + 0.020 \quad (1)$$

$$y = 43.1x^6 + 51.5x^5 - 24.5x^4 - 2.3x^3 - 1.8x^2 + 0.15x - 0.24 \quad (2)$$

$$y = -0.0002x^4 + 0.0076x^3 - 0.1147x^2 + 0.3044x - 0.0664 \quad (3)$$

Coefficient of Determination $(R^2) = 0.997$

$$y = 6e^{-5}x^4 - 0.0027x^3 + 0.0208x^2 - 0.0941x + 0.0417 \quad (4)$$

Coefficient of Determination $(R^2) = 0.996$

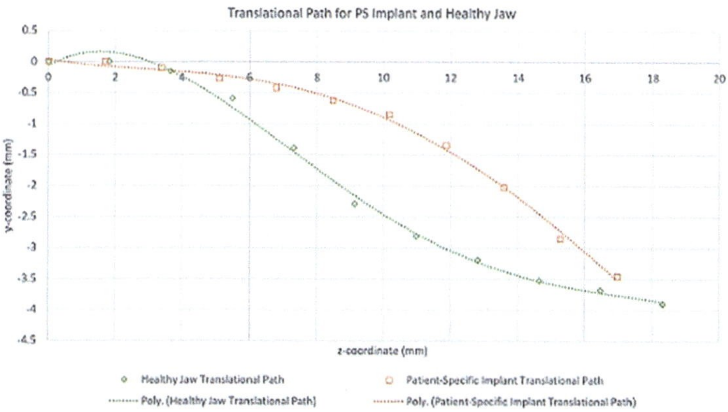

Fig. 3. Translational path of the mandibular condyle for the healthy jaw and patient-specific implant generated by CATIA models in the sagittal (yz) plane. The dotted lines represent the best-fit line from Eq.'s 3 and 4.

The kinetic models consist of eight muscles: the posterior temporalis, middle temporalis, anterior temporalis, deep masseter, superficial masseter, medial pterygoid, inferior lateral pterygoid, and superior lateral pterygoid. These muscles are projected onto the sagittal plane creating a 2-D muscle line of action model within the CAD software. Each line of muscle begins and ends using the muscle's insertion and origin points. The maximum muscle force, which relies on the procedure and the subsequent cutting of certain muscles (labeled as cut) are listed in Table 2 [3] (Fig. 3).

Table 2. Maximum muscle force, measured in Newton, for the muscles considered within this study [3].

Muscle	Healthy jaw	*PS* implant	Generic implant	FFF procedure
Posterior temporalis	75.6	75.6	75.6	0.0 (cut)
Middle temporalis	95.6	95.6	95.6	0.0 (cut)
Anterior temporalis	158	158	158	0.0 (cut)
Deep masseter	81.6	0.0 (cut)	0.0 (cut)	0.0 (cut)
Superficial masseter	190.4	38.1 (20% cut)	38.1 (cut)	0.0 (cut)
Medial pterygoid	174.8	174.8	174.8	0.0 (cut)
Inferior lateral pterygoid	66.9	0.0 (cut)	0.0 (cut)	0.0 (cut)
Superior lateral pterygoid	28.7	0.0 (cut)	0.0 (cut)	0.0 (cut)

2.1 Kinematic Analysis

Healthy Model

The length of the mandible (l_m) is defined from the origin to the end of the mandible [3, 10]. The mean mandible length for the healthy jaw is reported to be 100.2 mm and 92.5 mm for adult (18–31 years of age) males and females respectively [12]. The initial mandible angle at MMO is found by subtracting the rotation angle of the jaw (Φ_2) at each position from the maximum opening angle (32°) of the mandible. The ground link (l_g) represents the distance from the origin to the mandibular condyle at the MMO. To complete the closed-loop kinematic diagram, an arbitrary link exists between the endpoint of the mandible at maximum rotation angle to the origin of the translational path of the mandibular condyle (l_b).

The translational path found in the healthy jaw was modeled as the shape of the mandibular fossa, then scaled to achieve the mean condylar transition length of 3.9 mm inferior translation and 18.3 mm anterior translation from the mandibular condyle's origin [1, 11]. This same scale factor was applied to the translational path of the *PS* implant for consistency.

Four snapshot positions defined by the position of the mandibular condyle at the respective jaw rotation and appropriate translation using the CAD model are shown in Fig. 4. Position A depicts the jaw at MMO while position B depicts the jaw rotated 85% of the maximum rotation angle (27.2°) and at the midway point along the translational path. Both A and B closed-loop models consist of two slider joints and one revolute joint allowing for translational and rotational movement. Position C represents the pure rotation phase occurring at 70% of the maximum rotation angle to the mouth closed position. Position D illustrates the jaw at its closed position. The closed-loop models for positions C and D consist of one slider joint and two revolute joints.

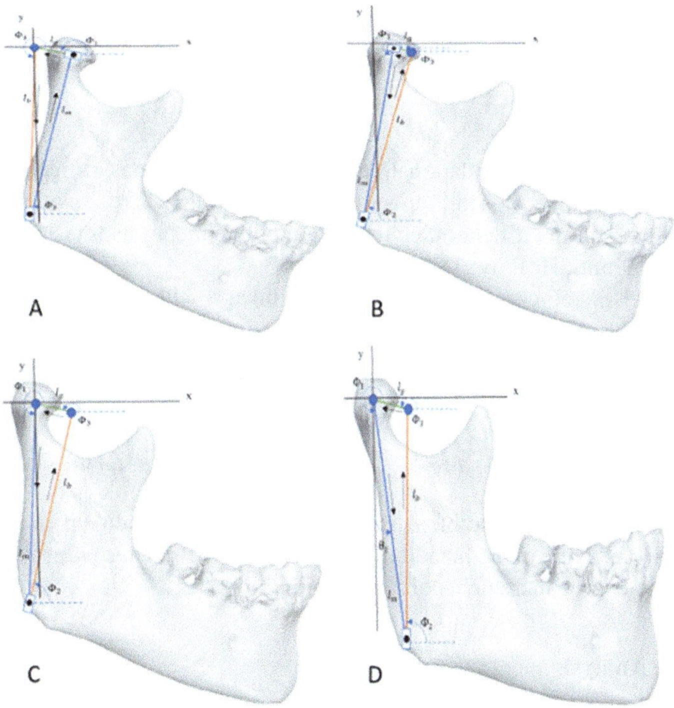

Fig. 4. A-B represents a two-degree-of-freedom kinematic model of the healthy jaw at A) MMO and B) 85% MMO. C-D represents a one degree-of-freedom at C) 70% MMO and D) mouth closed.

Fibula Free Flap (FFF) Procedure and Generic Implant
Like the healthy jaw, the closed-loop systems for the FFF procedure and the generic implant utilized two revolute joints and a slider joint, which slides along the length of the mandible as it opens and closes. However, since the FFF procedure and the generic implant restrict motion to pure rotation the kinematic models are positioned at MMO (Fig. 5A) and mouth closed position (Fig. 5B) for an equal assessment against the healthy jaw.

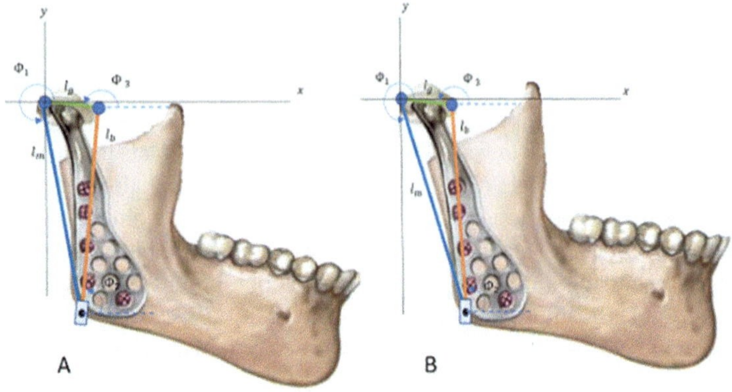

Fig. 5. Closed-loop diagram for the FFF procedure and generic implant at A) at MMO and B) mouth closed overlaid on a generic implant image [4].

Patient-Specific (PS) Implant

The *PS* implant has both translational and rotational capabilities mimicking the links used in the healthy jaw's closed-loop system. However, the *PS* kinematic model has a translational path of an artificial mandibular condyle reducing its translational path by 7.65% (*PS*: 16.9 mm; Healthy: 18.3 mm).

Each model is represented as a three-bar linkage within a closed-loop system resulting in the identical position and velocity equations, shown in Eq. (5) and (6).

$$l_m e^{j\Phi_1} + l_b e^{j\Phi_2} + l_g e^{j\Phi_3} = 0 \tag{5}$$

$$l_m\, e^{j\Phi_1} + l_m j\, \Phi_1\, e^{j\Phi_1} + l_b\, e^{j\Phi_2} + l_b j\, \Phi_2\, e^{j\Phi_2} + l_g\, e^{j\Phi_3} + l_g j\, \Phi_3\, e^{j\Phi_3} = 0 \tag{6}$$

However, to simplify the problem, inferior displacement was directly measured from a point placed on the top of the mandibular incisor teeth to the same point at its displaced location at one-degree increments of jaw rotation.

2.2 Kinetic Analysis

The net torque for the healthy jaw and the *PS* implant was taken from MMO (32°) to mouth closed position about the origin at one-degree increments of jaw rotation and respective translation, and calculated by multiplying muscle force by its perpendicular distance between its insertion point and the origin of the mandibular condyle. Figure 6A and Fig. 6B show the translational path for both the mandibular condyle of the healthy jaw and the *PS* implant. Whereas, the net torque for the FFF procedure and generic implant was calculated from MMO position (22°) to mouth closed position since these restrict translational movement.

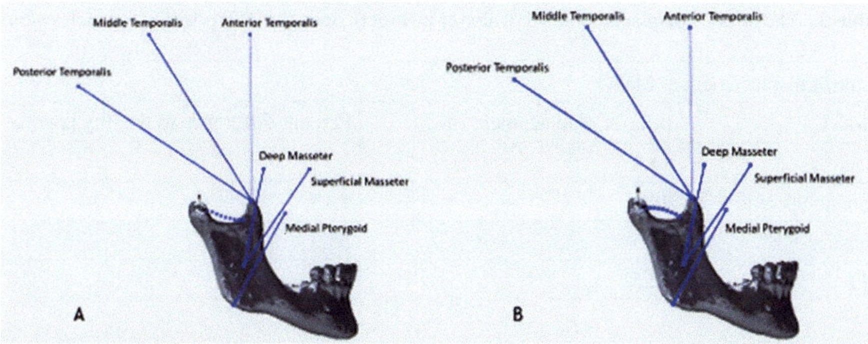

Fig. 6. Path of the mandibular condyle's movement in a A) healthy jaw and B) patient-specific implant with respective muscle line of action.

3 Results

During the rotation phase of jaw-closing, the inferior displacement of the FFF procedure and generic implant are identical to the healthy jaw as depicted in Fig. 7. However, at MMO a 35.3% decrease (FFF/Generic: 22.5 mm; Healthy: 34.8 mm) in inferior displacement is seen. Whereas the *PS* implant showed a maximum of 6.21% decrease (*PS*: 28.7 mm; Healthy: 30.6 mm) in inferior displacement at 28° jaw rotation and a 1.15% decrease (*PS*: 34.4 mm; Healthy: 34.8 mm) difference in inferior displacement at MMO (Table 3).

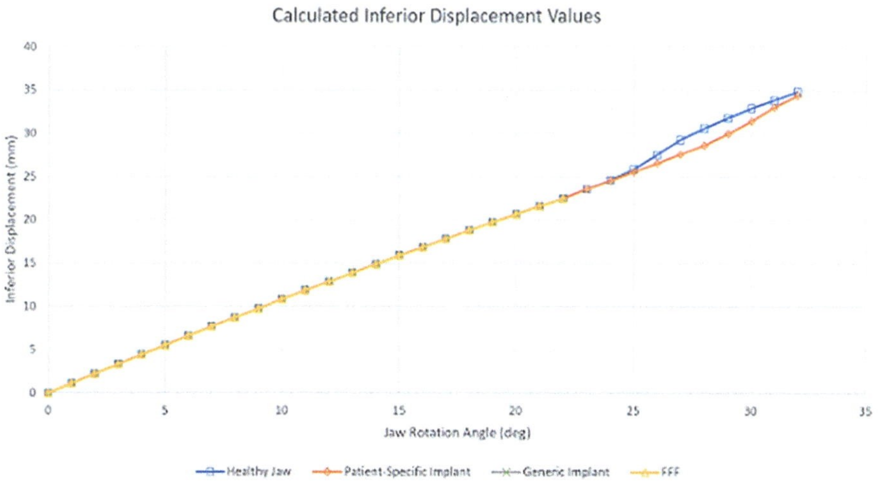

Fig. 7. Inferior displacement of the jaw for each system. The FFF procedure and generic implant reach MMO at 22° while the PS implant reaches MMO at 32°; a 31.3% and 0% difference, respectively, compared to that of the healthy jaw.

Table 3. The inferior displacement at the maximum mouth open (MMO) position for each system.

Inferior displacement at MMO		
Model	Inferior displacement (mm)	Percent difference to healthy jaw
Healthy	34.8	–
Patient-Specific	34.4	1.15
Generic	22.5	35.3
FFF	22.5	35.3

The net torque about the origin (ICR of the mandibular condyle) is equal to the muscle force and its respective moment arm found from the CAD model. The contributing muscles include the temporalis (anterior, middle, and posterior), masseter (superficial and deep), and the medial pterygoid muscle. However, the inferior lateral and superior lateral ptèrygoid muscles are neglected since their insertion points are at the origin [5, 9].

The healthy jaw produces a maximum net torque of 26,300 N-mm while the FFF procedure, generic implant, and *PS* implant produce 0.00 N-mm, 13,000 N-mm, and 18,600 N-mm, respectively. Investigating these numbers and the graph shown in Fig. 8, the net torque of the *PS* implant most closely resembles the healthy jaw with a maximum difference in magnitude of 29.3% at MMO (32°). The generic implant's net torque is identical to the *PS* implant up until the point of translation. At this point (22°) the generic implant's rotation-only capability causes the progression of torque to truncate after the maximum angle of rotation is reached. Given the removal of the muscles, the FFF procedure is incapable of producing torque about the origin leaving the torque at zero throughout its allowable translational path.

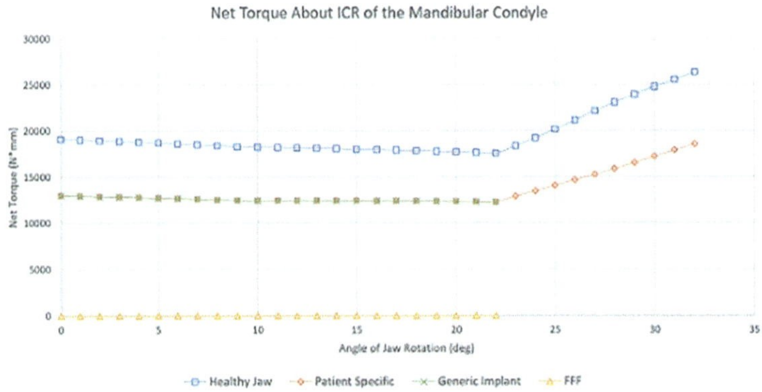

Fig. 8. Net torque about the origin (counterclockwise direction about the ICR of the mandibular condyle) for each system considering the muscles listed in Table 2.

4 Conclusions

Muscle force and net torque are influenced by muscle activation, subsequent muscle force, and moment arms and are observed at the TMJ. Anatomical features such as the positions of the mandibular condyle during the closing motion of the jaw, and the angle of the masseter, temporalis, and the medial and lateral pterygoid are all determinates for the mechanics of the closing of the jaw, thus the surgical procedure for the implants affect the outcome of the net moment calculations. The results from this study show that implants or procedures that constrain the TMJ joint to one degree of freedom (restricting translation along the sagittal plane) decreases the torque about the ICR of the mandibular condyle inhibiting how wide the mouth can open. The results also show that a *PS* implant is superior to both the FFF procedure and a generic implant in terms of torque about the ICR of the mandibular condyle and maximum inferior displacement during the jaw-closing movement in the sagittal plane (Fig. 9 and 10). This is because the *PS* implant allows both rotation and translation mimicking the function of a healthy jaw.

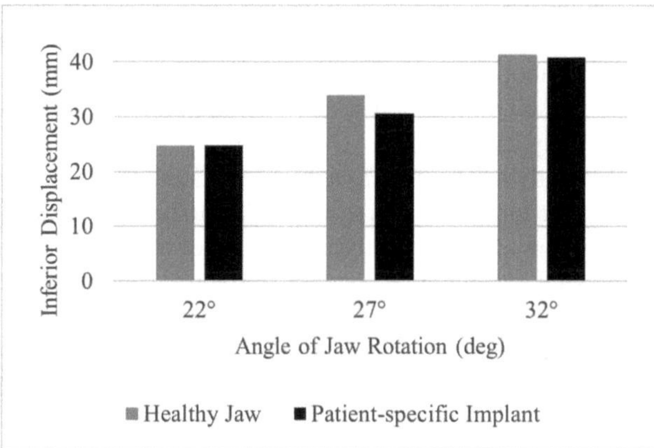

Fig. 9. Inferior displacement comparison for the healthy jaw and *PS* implant at jaw angle rotation of 1) beginning of translation (22°), 2) halfway to MMO (27°) and 3) at MMO (32°).

This study does have limitations. The healthy side of the jaw contralateral to the implants was ignored. If this was considered, it could very well affect the FFF procedure analysis in terms of the net torque and inferior displacement due to the severance of all the muscles considered. It may be beneficial to conduct a study without this limitation and to include other motions of the human jaw such as mandibular depression, elevation, lateral deviation, and retrusion and protrusion allowing for a more complete kinematic and kinetic analysis of each implant.

5 Discussion

The results from this study can be used to provide better insight into the outcomes of a TMJ implant and provide surgeons a better idea of the results of the procedure as well

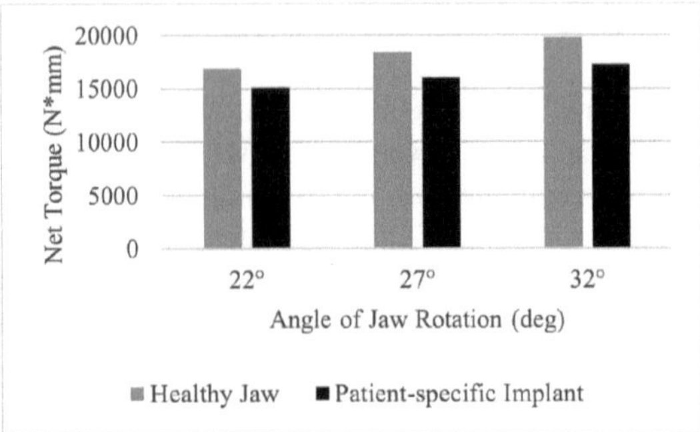

Fig. 10. Maximum net torque at mandibular condyle comparison for the healthy jaw and *PS* implant at jaw angle rotation of 1) beginning of translation (22°), 2) halfway to MMO (27°) and 3) at MMO (32°).

as provide a basis on future improvements of the surgical methods used to provide a higher torque seen at the TMJ. The surgical methods determine the muscles cut, which then influences the muscle activation, muscle force, and the moment arm of the muscle used to generate torque about the mandibular condyle. With this method of analysis, an improvement to the method of insertion of any TMJ implants can be improved to better resemble that of a healthy jaw.

Furthermore, this study can be used as a method of studying the *PS* implant, as well as determining any future improvements to the implant to better resemble the kinematics and kinetics of a healthy jaw. As seen in the calculation of both the inferior displacement and the net torque about the mandibular condyle, the path of translation of the *PS* implant plays a significant role. These results can be used as a baseline to improve the translational path of an implant that not only improves the inferior displacement, but also the torque observed throughout the closing motion of the jaw of all future implants to be developed.

Acknowledgments. We would like to give special thanks to Embry-Riddle Aeronautical University's Advanced Kinematics graduate course (ME 530). This research stemmed from the course project. All authors have viewed and agreed to the submission.

Declarations. Conflicting interests: None
Funding: None

References

1. Bermejo, E., Campomanes-Alvarez, C., Valsecchi, A., Ibanez, O., Damas, S., Cordon, O.: Genetic algorithms for skull-face overlay including mandible articulation. Inf. Sci. **420**, 200–217 (2017). https://doi.org/10.1016/j.ins.2017.08.029
2. Leclère, M.F., et al.: Reconstruction of large mandibulofacial defects with the composed double skin paddle fibula free flap: a review of 32 procedures. Laryngoscope **124**(6), 1336–1343 (2013). https://doi.org/10.1002/lary.24452

3. Ahn, S.J., Tsou, L., Sanchez, A.C., Fels, S., Kwon, H.B.: Analyzing center of rotation during opening and closing movements of the mandible using computer simulations. J. Biomech. **48**(4), 666–671 (2014). https://doi.org/10.1016/j.jbiomech.2014.12.041
4. Chaudhary, F., Ahmad, B., Butt, D., Hameed, S., Bashir, U.: Normal range of maximum mouth opening in pakistani population: a cross-sectional study. J. Int. Oral Health **11**(6), 353 (2019)
5. Koolstra, J.H., Van Eijden, T.M.G.J.: The jaw open-close movements predicted by biomechanical modelling. J. Biomech. **30**(9), 943–950 (1997). https://doi.org/10.1016/S0021-9290(97)00058-4
6. She, X., Feng, W., Damon, B.J., Coombs, M.C., Lee, D.G.: Three-dimensional temporomandibular joint muscle attachment morphometry and its impacts on musculoskeletal modeling. J. Biomech. **79**, 119–128 (2018). https://doi.org/10.1016/j.jbiomech.2018.08.010
7. NIH 3D Print Exchange, "Male Mandible, European Descent, 50–65 Years of Age (2016). https://3dprint.nih.gov/discover/3DPX-003381
8. Katsavrias, E.G., Halazonetis, D.J.: Condyle and fossa shape in class II and class III skeletal patterns: a morphometric tomographic study. Am. J. Orthod. Dentofac. Orthop. **128**(3), 337–346 (2005). https://doi.org/10.1016/j.ajodo.2004.05.024
9. Koolstra, J.H., Van Eijden, T.M.G.J.: Combined finite-element and rigid-body analysis of human jaw joint dynamics. J. Biomech. **38**(12), 2431–2439 (2005). https://doi.org/10.1016/j.jbiomech.2004.10.014
10. Quinn, P.D., Dolwick, F., Sinn, D.P.: Total Mandibular Joint Replacement Surgical Guidelines, Powerpoint, Accessed 19 May 2020. https://www.zimmerbiomet.com/content/dam/zimmer-biomet/medical-professionals/000-surgical-techniques/cmf/total-mandibular-joint-replacement-surgical-guidelines.pdf
11. Xingbin, C.: The instantaneous center of rotation during human jaw opening and its significance in interpreting the functional meaning of condylar translation. Am. J. Phys. Anthropol. **106**(1), 35–46 (1998). https://doi.org/10.1002/(SICI)1096-8644(199805)106:1%3c35::AID-AJPA3%3e3.0.CO;2-C
12. Naeije, M.: Local kinematic and anthropometric factors related to the maximum mouth opening in healthy individuals. J. Oral Rehabil. **29**, 534–539 (2002)

Could an Exoskeleton-Driven Rehabilitation Treatment Improve Muscle Forces Generation in PD? - a Pilot Study

Marco Romanato[1](✉), Fulvia Fichera[2], Fabiola Spolaor[1], Daniele Volpe[2], and Zimi Sawacha[1,3]

[1] Department of Information Engineering, University of Padova, Padova, Italy
marco.romanato.2@phd.unipd.it
[2] Fresco Parkinson Center, Villa Margherita, S. Stefano, Vicenza, Italy
[3] Department of Medicine, University of Padova, Padova, Italy

Abstract. Research focusing on optimal rehabilitation methods has been directed towards powered lower-limb exoskeletons which combines the advantages delivered from the grounded robotic devices with the ability to train the patient in a real-world environment. In this context literature has highlighted the benefit of coupling gait analysis and musculoskeletal modeling for treatment planning. Recently, this combined approach has been successfully applied to detect the alterations in motor control related to Parkinson's Disease (PD). However, no study has reported about the effects of an overground wearable exoskeleton in terms of both gait analysis and musculoskeletal modeling-derived parameters in people with PD. The aim of this study was to quantitatively assess the effect of an overground rehabilitation treatment on a PD participant both in terms of gait parameters and muscle forces. One people with PD has been enrolled and gait analysis was performed before and after a 4-weeks gait training intervention with an overground exoskeleton. Inverse kinematics, inverse dynamics, and static optimization were performed in OpenSim. Results from joint moments and muscle forces were compared with a group of healthy controls. Preliminary results showed that after the therapy joint loads in both ankle and knee joints were reduced during the stance phase and muscle forces displayed an increased magnitude in their peak after the treatment. For the best of our knowledge, the presented case study is the first attempt to track rehabilitation improvement via muscle forces assessment. Further studies should focus on increasing the sample size to generalize the outcomes.

Keywords: Neuro rehabilitation · Clinical gait analysis · Musculoskeletal modeling

1 Introduction

The ability to walk independently is a primary goal when rehabilitating an individual with Parkinson's disease (PD). Indeed, people affected by PD frequently display a flexed posture that coupled with an excessive joint stiffness leads to poor walking mechanics,

J. M. R. S. Tavares et al. (Eds.): CMBBE 2021, LNCVB 38, pp. 36–49, 2023.
https://doi.org/10.1007/978-3-031-10015-4_3

which increases their risk of falls. Literature reported that locomotor functions are positively recovered by robotic gait training. Studies involving post-stroke participants, who underwent such a rehabilitation program, have shown to be more likely to achieve independent walking than people who did not receive the treatment [1]. Although studies have already shown the many benefits of robotic-assisted gait training in PD participants on spatiotemporal parameters [2–4], this walking-environment showed less control over the gait initiation by the patient and lacked in variability of visuospatial flow [5]. Therefore, research focusing on optimal rehabilitation methods has been directed towards powered lower-limb exoskeleton. In post-stroke patients the effect of this treatment enhanced the potential for patient-specific rehabilitation [5, 6], showing improvements in spatiotemporal parameters and joint kinematics [2]. As gait involves both central and peripheral nervous systems, targeted rehabilitation must restore not only mechanics but also neurophysiological gait patterns. Hence, this requires improvements at the level of both balance and lower limb joint motion [2]. In this direction, wearable lower-limb powered exoskeletons promote functional training in a realistic walking-environment combined with a greater patient's engagement than in grounded devices [7]. These systems provide a greater level of subject participation for maintaining trunk and balance control, as well as navigating their path over different surfaces and increasing the subject's functional abilities while the wearable robotic system guarantees less support. Furthermore, the stability that the exoskeleton addresses to the patient, allows a hands-free walking trial which represents an integral part of a physiological locomotion restoration [7].

Gait in people with PD has been studied thoroughly with 3-dimensional gait analysis systems in recent years, documenting a typical hypokinetic gait with an increment of cadence, stance and double support phases, which compensates for the reduced stride length, step length and velocity [8]. Recently it has been shown that combining gait analysis and musculoskeletal modeling (MSM) enables to track disease progression with enhanced precision [9, 10]. For each individual, a MSM is created to track their biomechanics [11, 12]. This allows to provide new biomarkers to assess and track PD motor impairment. Therefore, the present contribution proposes a framework to quantitatively assess the individual's gait and neuromuscular functions recovery of the sensorimotor alterations at the peripheral nervous system after an over-ground wearable exoskeleton training (OWET). Hence, the current study investigates the effects of a rehabilitation program involving an overground and wearable robotic gait trainer on one PD participant. Comprehensive gait analysis is enriched by MSM to determine modifications induced by the active intervention. As health care professionals and researchers need objective, reliable, and valid tools to plan subject-specific interventions, quantify therapeutic outcomes, and monitor changes over time, the proposed investigation includes coupling of neuromechanical variables with clinical scales.

2 Material and Methods

2.1 Participant

One participant (age = 69 years, BMI = 22.8 kg/m2) with clinically established diagnosis of PD according to the United Kingdom Parkinson's Disease Society Brain Bank Diagnostic criteria has been recruited. The diagnosis has been reviewed by a neurologist specialized in movement disorders. Local Ethic Committee approval was obtained (Comitato Etico per le Sperimentazioni Cliniche (CESĊ) della Provincia di Vicenza – CE/PROG. 55/20, July 14th, 2020). The inclusion/exclusion criteria were:

Inclusion Criteria:

- Aged between 20 and 90 years old
- Patient with rigid-akinetic bilateral PD form
- Hoehn-Yahr between 3–4
- At least 4 years of disease history
- Stable drug therapy response without any change performed in the 3 months before the study
- Presence of freezing of gait and of postural instability not responding to parkinsonian therapy
- Mini Mental State Evaluation > 24/30

Exclusion Criteria:

- Systemic illness
- Presence of cardiac pacemaker
- Postural abnormalities, orthopedic comorbidities that do not match the active physiotherapy treatment and the use of the proposed robotic device
- Presence of deep brain stimulation
- Presence of severe dysautonomia with marked hypotension
- Obsessive-Compulsive disorder
- Major depression
- Dementia and psychosis
- History or active neoplasia
- Pregnancy
- Other criteria that do not respect the device counterindications

Additionally, this trial was registered in ClinicalTrials.gov (a service of U.S. National Institutes of Health) Identifier NCT04778852 (March 3, 2021) named "Quantitative Assessment of training Effects Using a Wearable Exoskeleton in Parkinson Disease Patients". The participant was asked for written informed consent according to the standard forms. This study is currently recruiting patients (study start date: May 2020 – due to pandemic COVID-19 enrolment start date: January 2021).

2.2 Rehabilitative Intervention Design

The rehabilitative intervention was delivered with EksoGT®, a robotic device developed by EKSOBIONICS. The OWET was delivered in 12 sessions, spanned in 4 weeks by trained physical therapists, in possession of the patent to deliver the robotic training and under the supervision of medical mentors. Each session dosage was 45 min long. Instrumented clinical gait analysis evaluation was made within 3 days prior the start of the training (T0) and within 3 days after (T1). Schematic representation of the study design is provided in Fig. 1.

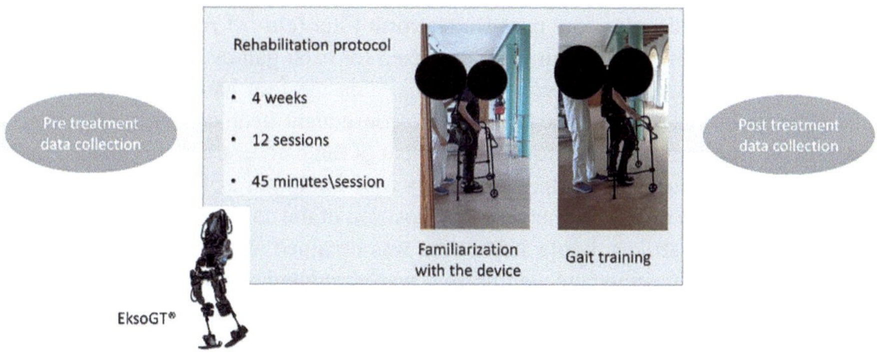

Fig. 1. Study design for the rehabilitative intervention.

2.3 Clinical Evaluation

For the clinical assessment, the Unified Parkinson's Disease Rating Scale (UPDRS) was delivered by trained physical therapists at each time frame. The UPDRS consists of four parts: Part I investigates non-motor experiences of daily living, Part II follows motor experiences of daily living up. Part III includes instructions that the researcher can either provide or show to the patient for the motor assessment. Part IV is completed by the researcher investigating motor complications. Higher scores show increased severity in a 0 to 4 scale. The numerical score obtained from the UPDRS scale offers the opportunity for repeated comparisons over time of the same patient obtaining information on the progress of the disease over time [13].

2.4 Instrumental Evaluation

Gait analysis was carried out at the Motion Analysis Laboratory (10m walkway) of the Fresco Parkinson Center of Villa Margherita (Arcugnano, Vicenza, Italy). Several gait cycles were collected with an 8-cameras optoelectronic system (120–240 Hz, Vicon, UK), synchronized with two force plates (1080 Hz, AMTI, USA) and an 8-channel electromyographic system (2000 Hz, Cometa, Italy) which recorded the activity of four lower limb muscles bilaterally, namely Tibialis Anterior, Gastrocnemius Lateralis, Rectus Femoris, Biceps Femoris Long Head [14]. Surface EMG electrodes were placed following the minimal crosstalk area guidelines to maximize the measured selective activity per muscle [15]. A modified version of the IOR-Gait protocol was adopted both for anatomical landmarks identification and joint angles calculation [16]. Multimodal signals were acquired within the Nexus (version 2.8, Vicon, UK) environment (Fig. 2).

2.5 Data Processing

Gait analysis data was processed in Matlab (version R2019b, MathWorks, MA, USD). Markers' trajectories were filtered with a cutoff frequency of 6 Hz, force platform data were filtered by a 3rd order, low-pass Butterworth filter (cut-off frequency 5 Hz) [17]. Spatiotemporal parameters of gait were extracted for each gait cycle. Gait cycles were identified as two consecutive heel strikes, *i.e.*, when the foot naturally landed on the force plate and force vector appeared and the consequent heel strike was determined detecting the minimum of the vertical component of the marker trajectory placed in the correspondent calcaneus. Extracted parameters were computes as follows: stride length was computed as the difference between the position of the calcaneus marker in second and in the first heel strikes, while stride time was obtained as the temporal difference between the two gait events; stance duration was computed as the difference between the temporal instant correspondent to the heel strike and the toe off; stance period was then reported in percentage of the total gait cycle [18].

sEMG signals were processed to extract the linear envelope and a low pass filter with a 4th-order Butterworth filter (cutoff frequency of 5 Hz) was adopted. The envelope was normalized with the respect of the value of the envelope peak obtained during each walking trial [19].

To extract the neuromuscular parameters, the pipeline for MSM described in the following was applied. MOtoNMS (SimTK, https://simtk.org/projects/motonms) [20] was adopted to export the motion data used in OpenSim (version 3.3, SimTK, https://simtk.org/projects/opensim) framework. A generic musculoskeletal model (gait2392 [21]) was used to linearly scale each subjects' geometry in OpenSim, matching the virtual markers of the model with the experimental ones of the subjects acquired during the static pose trial. Then a muscle optimizer tool was considered to adjust musculotendon parameters [22]. Inverse kinematics, inverse dynamics and static optimization tools were performed to obtain joint angles and moments and muscle forces.

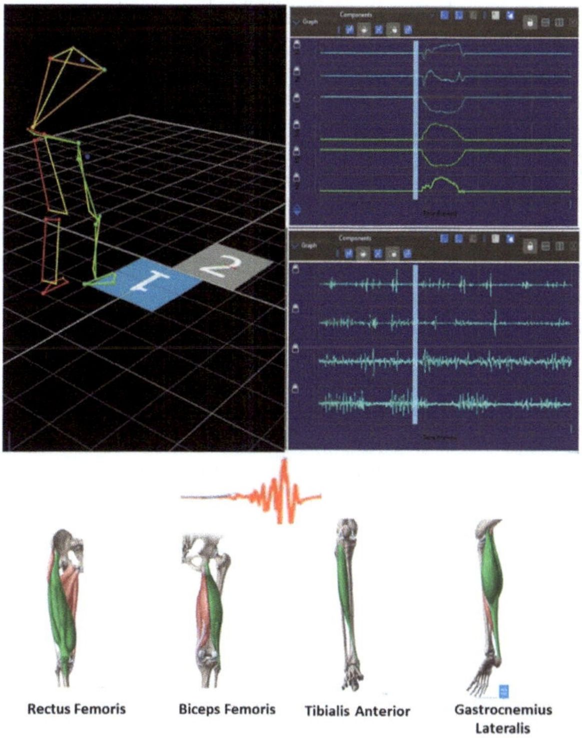

Fig. 2. Experimental setup. In the upper part, a snapshot of the experimental setup reconstructed in the Nexus environment (https://www.vicon.com/software/nexus/). On the left side the stick diagram of one of the subject's walking trials. On the right side, synchronized force components are displayed along with electromyographic signals. In the bottom, the considered muscles are reported.

Data retrieved at each time frame were presented as the mean ± 1 time the standard deviation of 6 representative trials (3 left and 3 right) after checking for gait asymmetry. Asymmetrical gait was assessed testing the difference in stride time between the right and the left leg (Kruskal-Wallis test) since normality distribution was not verified (Shapiro-Wilk test). An overall report is then provided. All the data were compared with normative bands of 10 healthy subjects (HS) (age 65.5 ± 7 years, BMI 28.0 ± 3 kg/m^2) enrolled and computed in a previous study [23].

3 Results

3.1 Clinical Results – UPDRS

UPDRS scale motor examination (Table 1) reports moderate improvements in lower limb stiffness with increased agility in sit-to-stand gesture. The postural stability item reported the greatest change after treatment with the exoskeleton. Through the retropulsion test, the response to a sudden movement of the body produced by a rapid and vigorous traction on the shoulders was evaluated while the patient is in an upright position, with the eyes open and the feet parallel to each other. Given that a recovery of balance within two steps is to be considered normal, the patient reports a clear improvement.

Table 1. UPDRS subtotal scores.

	UPDRS-I	UPDRS-II	UPDRS-III	UPDRS-IV
T0	9	12	26	0
T1	6	5	15	0

3.2 Spatiotemporal Parameters

Spatiotemporal parameters of gait were reported in the Table 2. Between the two time frames no changes were appreciated. Even after the treatment the walking velocity was noticeably lower than the one of the healthy controls.

Table 2. Spatiotemporal parameters of gait.

	Space (m)	Time(s)	Velocity (m/s)	Stance (%)	Swing (%)	Cadence (step/min)
T0	1.00 ± 0.07	1.36 ± 0.08	0.74 ± 0.08	65.55 ± 1.67	34.45 ± 1.67	44.38 ± 2.39
T1	0.99 ± 0.02	1.32 ± 0.02	0.75 ± 0.02	65.97 ± 0.64	34.03 ± 0.64	45.39 ± 0.80
HS	1.20 ± 0.10	1.22 ± 0.12	0.99 ± 0.13	61.57 ± 1.05	38.43 ± 1.05	49.67 ± 4.67

3.3 Joint Angles and Moments

Joint angles and joint moments results extracted from the IOR-protocol protocol for angular calculation were reported in the Figs. 3 and 4 respectively.

Before therapy, the subject denotes an increased external rotation at the hip during both initial contact and loading response, accompanied by a reduced hip flexion angle throughout the gait cycle. At the knee level an increased flexion angle can be observed throughout the gait cycle, while the ankle presents a reduced flexion. After therapy, hip, knee, and ankle joint patterns on the sagittal plane show improvements as their values are closer to controls, meanwhile, on the coronal plane a more altered hip joint pattern is revealed. Knee flexion extension moments show a pattern closer to controls after therapy, which finds agreement with the improvements in the joint flexion-extension angle.

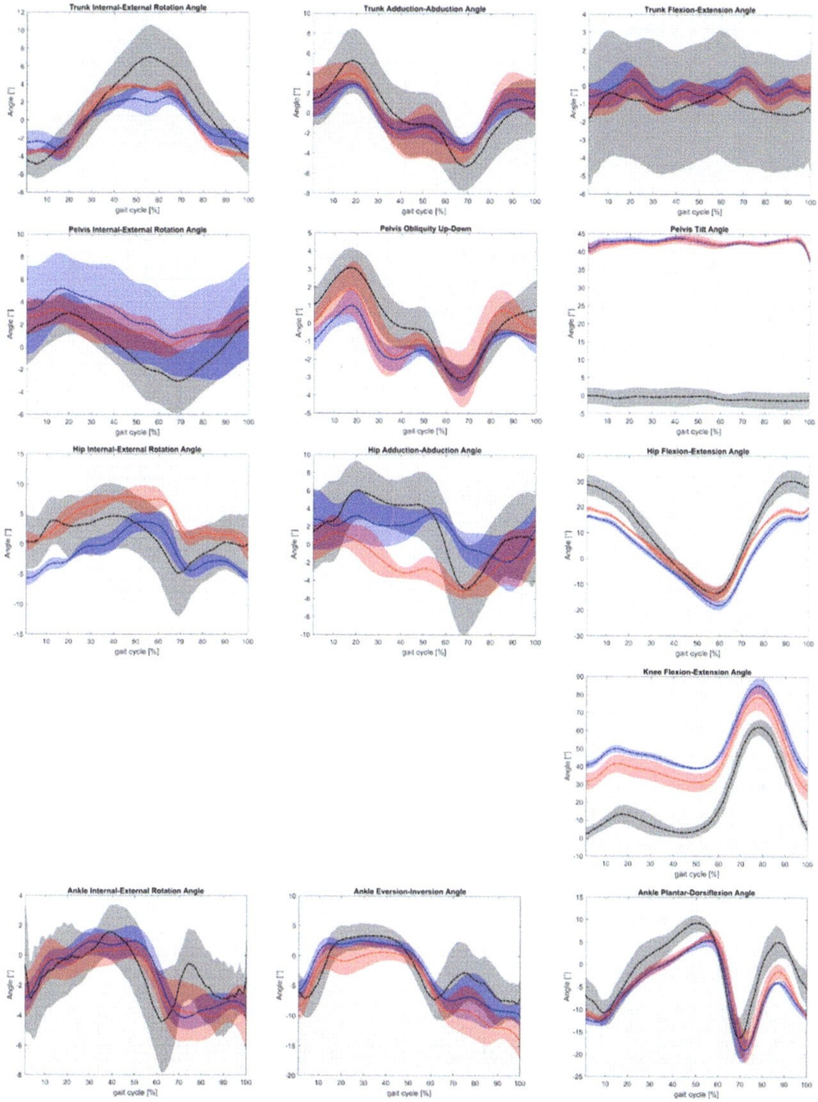

Fig. 3. Joint angles for the trunk, pelvis, hip, knee and ankle joints for the sagittal, coronal and transversal planes. Mean (dashed lines) and standard deviation clouds (shaded areas). Blue color is for the participant before the treatment (T0), red color is for the participant after the treatment (T1), black color is for the normative bands.

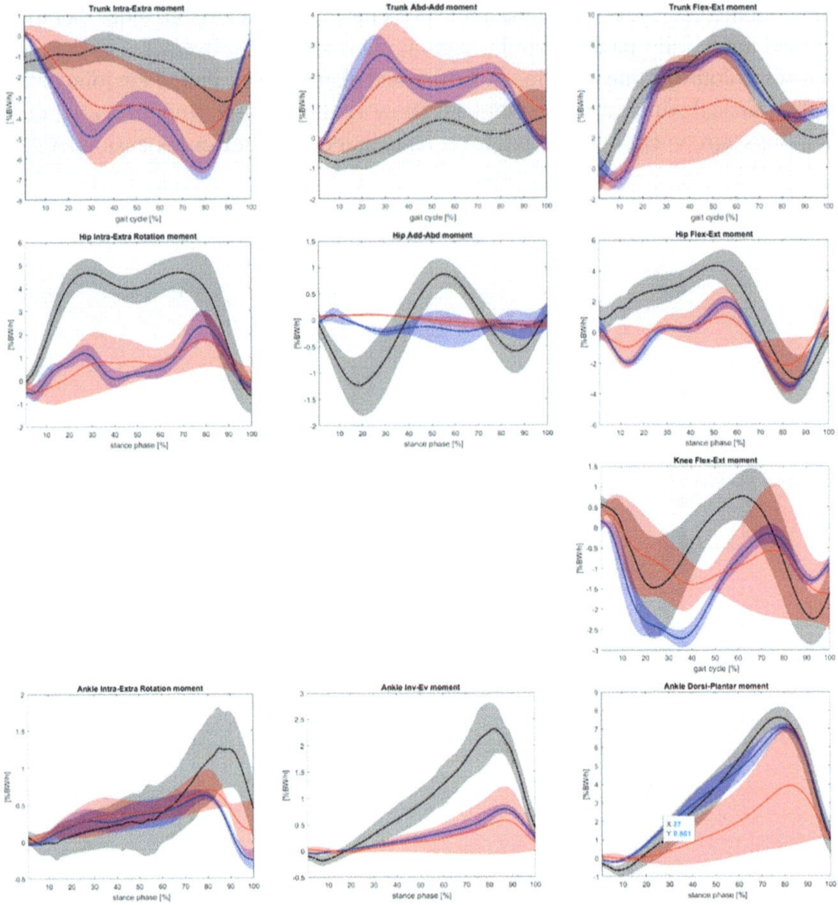

Fig. 4. Joint moments for the trunk, hip, knee and ankle joints for the sagittal, coronal and transversal planes. The normalization was performed in percentage of the body weight × height (BW × h %). Mean (dashed lines) and standard deviation clouds (shaded areas). Blue color is for the participant before the treatment (T0), red color is for the participant after the treatment (T1), black color is for the normative bands.

3.4 Electromyography

Surface electromyography shows noticeable changes in the timing of muscles activation: Rectus Femoris, Gastrocnemius Lateralis and Biceps Femoris show values closer to control during swing phase, the terminal stance phase and between 20 to 80% of gait cycle respectively (Fig. 5).

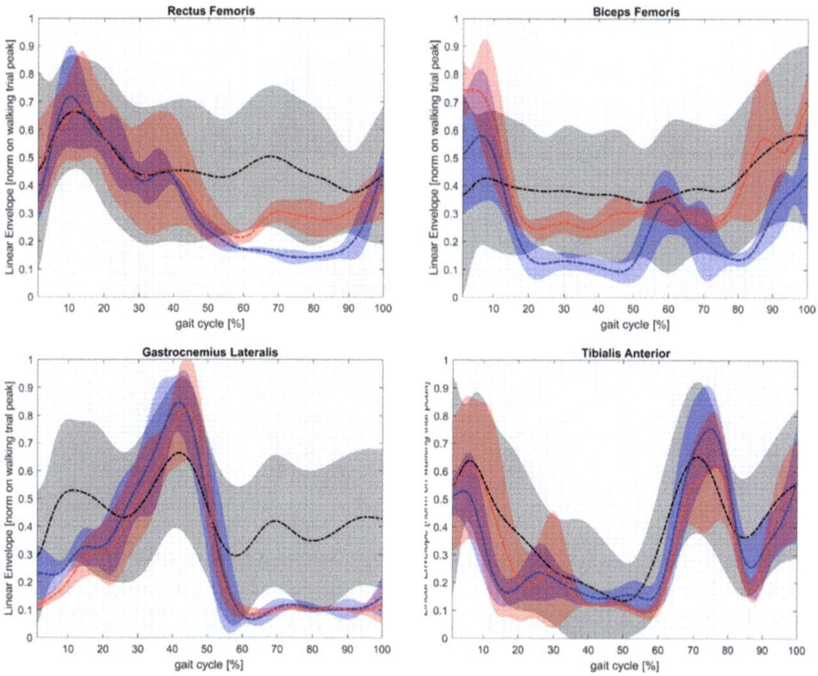

Fig. 5. Envelope profiles of the muscles Rectus Femoris, Biceps Femoris Caput Longus, Gastrocnemius Lateralis and Tibialis Anterior. The normalization was performed on the maximum value within the walking trials ([0, 1] range). Mean (dashed lines) and standard deviation clouds (shaded areas). Blue color is for the participant before the treatment (T0), red color is for the participant after the treatment (T1), black color is for the normative bands.

3.5 Muscle Forces (MSM Output)

In terms of muscle forces, reported in Figure 6, it is worth mentioning the presence of an increase force production in knee extensors as Biceps Femoris and Semimembranosus muscles during the stance phase. In the same terms, for the stance phase can be noticed a reduction of the magnitude in production of force for the knee flexors as the Vastii and the Rectus Femoris. However, the latter displayed an increased mean peak in correspondence of the toe-off instant. Considering the ankle dorsiflexors a decreased magnitude in the Soleus force production has been noticed, couple with an increased peak of force for the two Gastrocnemii showed in the late stance phase. Finally, an appreciable increasing of the Tibialis Anterior within all the first half of the gait cycle could be highlighted.

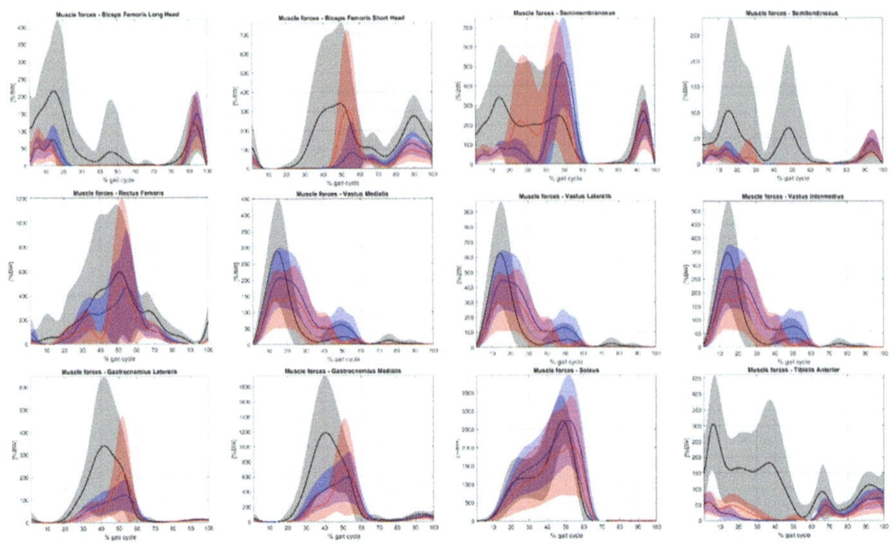

Fig. 6. The normalization was performed in percentage of the body weight (BW %). Mean (dashed lines) and standard deviation clouds (shaded areas). Blue color is for the participant before the treatment (T0), red color is for the participant after the treatment (T1), black color is for the normative bands.

4 Discussion

The purpose of the presented work was to evaluate the effects of an OWET on gait impairments in one individual affected by PD throughout clinical gait analysis assessment and a feasible MSM pipeline. As gait is a complex task that involves both central and peripheral nervous systems, targeted rehabilitation is addressed to restore not only gait mechanics (e.g., spatiotemporal parameters of gait) but also neurophysiological gait patterns (*e.g.*, joint kinematics and dynamics, muscle activity and neural-to-mechanical activation dynamics). To link the brain activity and the motor behavior in a quantitative fashion, in the presented work, standard methods and innovative analysis have been applied for data acquisition and processing. Hence, gait analysis, coupled with MSM, provided a great amount of valuable information to evaluate treatment outcomes. Furthermore, measurement obtained from the adopted modeling technique allowed the authors to consider for the muscle forces influence on the biomechanics of gait. These variables were proposed as compact biomarkers to link the causal effect from the neural command to the joint dynamics and might be able to portray relevant information on the disease progression.

In particular, the participant who underwent the proposed rehabilitation protocol denoted an overall improvement when comparing the clinical scales before and after the enrollment. Indeed, UPDRS reported a sensitive enhancement at the motor and postural level. The positive gain that the treatment has driven in postural stability might explain the different locomotion dynamic strategies detected between the two timeframes of acquisition (*i.e.*, T0 and T1). For instance, a deficiency in the force exerted by dorsiflexors muscles, as the Soleus muscle, has been revealed after the treatment. This loss though, has been accompanied by a gain in propulsion forces during the heel rising phase, where both the Gastrocnemii increased their force magnitude. In this phase, the activity of the dorsiflexors group is crucial for ankle joint stability; particularly, Medial and Lateral Gastrocnemii are responsible for the heel rising [24]. Indeed, decreasing the ankle joint flexor moment, those muscles promote walking progression and stability during the stance. Moreover, the Soleus provides both a synergic support to the other plantarflexors and the required forefoot stability in the final stance [24]. Results reported in Fig. 6 showed how, after the treatment, the Gastrocnemii are expressing a higher intensity of force magnitude with the respect of T0. Although the Soleus has a reduced force magnitude after the treatment, when considering the joint level we might notice its involvement on ankle joint inversion, which resulted modified from T0. The active intervention did not account for an increased muscle force in plantarflexors during the heel strike phase (*i.e.,* Tibialis Anterior), as the magnitude resulted unchanged between the two timeframes. This was reflected in the ankle joint kinematics on the sagittal plane, which displayed a more plantarflexed gait initiation with the respect of the normative bands of healthy controls. Overall, an increase of force on the posterior kinematic chain has been revealed. As in normal gait the toe off phase involves the activation of hamstrings muscles [24], the participant of the presented study confirmed this hypothesis in T0, although with a reduced force production with the respect of normative bands. After the treatment, it was interesting to notice that this synergistic contribution was not respected. Hence, Biceps Femoris and Semimembranosus muscles reported some increasing in the magnitude. Moreover, the latter muscle presented an augmented similarity in the overall force pattern with respect to healthy peers. Authors might speculate that these results could be interpreted as a tentative to gain the functional role of this muscle group during the terminal stance in supporting the knee flexion.

To conclude, the proposed methodology allowed to retrieve an insight on the participant's kinematics, kinetics, and on his musculotendon dynamics and force during locomotion. The changes we observed after the treatment were linked with the ones accounted in the clinical scales, suggesting the reliability of the approach. Moreover, the adoption of musculoskeletal modeling was pivotal on the evaluation and discussion of the muscle functions, even in those muscles that were not recorded with experimental EMGs.

This work comes with several limitations that must be acknowledged. The main drawback of the work is the sample size. Participant number must be increased to generalize the outcomes both in terms of therapy effectiveness over the considered population and in terms of the methodological assessment approach. However, in authors opinion, those preliminary results were worth to be reported as, for the best of their knowledge, it was the first time that an OWET was proposed in PD population and the effectiveness of the intervention was evaluated with such a pipeline. Furthermore, with an increased enrollment of participants, findings linked with the proposed study will likely give substantial solutions to the management of gait and postural disorders (posture, balance, and gait) in people with PD where an agreement on the optimal intervention (*e.g.*, pharmacological, neurosurgery, traditional physiotherapy) is still lacking [25]. Moreover, a MSM that identifies patient-specific variables for rehabilitative intervention could be used to assess treatment outcomes but also to conduct on-line rehabilitation therapy by means of remote control of the assistive robotic device. This would provide several advantages over conventional approaches, proposing an active treatment that is personalized and scalable to a larger population and including a standardized training environment and an adaptable support that can increase the treatment intensity and dose, without being a burden for the therapists.

Acknowledgement. This study was founded by the Marlene and Paolo Fresco Institute for Parkinson's and Movement Disorders at NYU Langone Health grant "Quantitative assessment of training effects using a wearable exoskeleton in Parkinson's disease patients". Marco Romanato's PhD scholarship is supported by Fresco Parkinson Institute Italia. Fulvia Fichera and Fabiola Spolaor scholarships are supported by Marlene and Paolo Fresco Institute for Parkinson's and Movement Disorders at NYU Langone Health grant.

References

1. Jayaraman A., et al.: Stride management assist exoskeleton vs functional gait training in stroke. A randomized trial. Neurology. **92**(3) e263–e273 (2019)
2. Galli, M., et al.: Robot-assisted gait training versus treadmill training in patients with Parkinson's disease: a kinematic evaluation with gait profile score. Funct. Neurol. **31**(3), 163–170 (2016)
3. Sale, P., et al.: Robot-assisted walking training for individuals with Parkinson's disease: a pilot randomized controlled trial. BMC Neurol. **13**, 50 (2013)
4. Picelli, A., et al.: Robot-assisted gait training in patients with Parkinson disease: a randomized controlled trial. Neurorehabil. Neural Repair. **26**(4), 353–361 (2012)
5. Louie, D.R., Eng, J.J.: Powered robotic exoskeletons in post-stroke rehabilitation of gait: a scoping review. J. NeuroEng. Rehabil. **13**, 53 (2016)
6. Bortole, M., et al.: The H2 robotic exoskeleton for gait rehabilitation after stroke: early findings from a clinical study. J. NeuroEng. Rehabil. **12**, 54 (2015)
7. De Luca, A., et al.: Exoskeleton for gait rehabilitation: effects of assistance, mechanical structure, and walking aids on muscle activations. Appl. Sci. **9**, 2868 (2019)
8. Volpe, D., Signorini, M., Marchetto, A., Lynch, T., Morris, M.E.: A comparison of Irish set dancing and exercises for people with Parkinson's disease: a phase ii feasibility study. BMC Geriatr. **13**, 54 (2013)
9. Scarton, A., et al.: Comparison of lower limb muscle strength between diabetic neuropathic and healthy subjects using OpenSim. Gait Posture. **58**, 194–200 (2017)
10. Varotto, C., Sawacha, Z., Gizzi, L., Farina, D., Sartori, M.: Closed-loop EMG-informed model-based analysis of human musculoskeletal mechanics on rough terrains. In: International Conference on Rehabilitation Robotics, pp. 364–368 (2017)
11. Sartori, M., et al.: Toward modeling locomotion using electromyography-informed 3D models: application to cerebral palsy. WIREs Syst. Biol. Med. **9**(2), e1368 (2017)
12. Sartori, M., Lloyd, D.G., Farina, D.: Neural data-driven musculoskeletal modeling for personalized neurorehabilitation technologies. IEEE Trans Biomed Eng. **63**(5), 879–893 (2016)
13. Goetz, C.G., Poewe, W., Rascol, O., Sampaio, C., Stebbins, G.T. and Fahn, S.: Movement disorder society task force on rating scales for Parkinson's disease. Mov. Disord. **18**(7), 738–50 (2003)
14. Spolaor, F., et al.: Relationship between muscular activity and postural control changes after proprioceptive focal stimulation (Equistasi ®) in middle-moderate Parkinson's disease patients: an explorative Study. Sensors. **14**, 21(2), 560 (2021)

15. Blanc, Y., Dimanico, U.: Electrode placement in surface electromyography (sEMG) "Minimal Crosstalk Area" (MCA). TOREHJ. **3**(1), 110–126 (2010)
16. Sawacha, Z., Gabriella, G., Cristoferi, G., Guiotto, A., Avogaro, A., Cobelli, C.: Diabetic gait and posture abnormalities: a biomechanical investigation through three dimensional gait analysis. Clin. Biomech. **24**, 722–728 (2009)
17. Romanato, M., et al.: Changes of biomechanics induced by Equistasi® in Parkinson's disease: coupling between balance and lower limb joints kinematics. Med. Biol. Eng. Compu. **59**(7–8), 1403–1415 (2021). https://doi.org/10.1007/s11517-021-02373-3
18. Sawacha, Z., et al.: Biomechanical assessment of balance and posture in subjects with ankylosing spondylitis. J. NeuroEng. Rehab. **9**, 63 (2012)
19. Barrett, R.S., Besier, T.F., Loyd, D.G.: Individual muscle contributions to the swing phase of gait: an EMG-based forward dynamic modelling approach. Simul. Model. Pract. Theory **15**(9), 1146–1155 (2007)
20. Mantoan, A., Pizzolato, C., Sartori, M., Sawacha, Z., Cobelli, C., Reggiani, M.: MOtoNMS: a MATLAB toolbox to process motion data for neuromusculoskeletal modeling and simulation. Source Code Biol. Med. **10**, 12 (2015)
21. Delp, S.L., Loan, J.P., Hoy, M.G., Zajac, F.E., Topp, E.L., Rosen, J.M.: An interactive graphics-based model of the lower extremity to study orthopaedic surgical procedures. IEEE Trans. Biomed. Eng. **37**(8), 757–767 (1990)
22. Modenese, L., Ceseracciu, E., Reggiani, M., Lloyd, D.G.: Estimation of musculotendon parameters for scaled and subject specific musculoskeletal models using an optimization technique. J Biomech. **49**, 141–148 (2016)
23. Volpe, D., et al.: Muscular activation changes in lower limbs after underwater gait training in Parkinson's disease: a surface emg pilot study. Gait Posture **80**, 185–191 (2020)
24. Perry, J., Burnfield, J.M.: Gait analysis: normal and pathological function. J. Sports Sci. Med. **9**(2), 353 (2010)
25. Lang, A.E., Lees, A.: Management of Parkinson's disease: an evidence-based review. Mov. Disord. **17**(4), S1-S166 (2002)

EMG Signals as a Way to Control Soft Actuators

António Diogo André[1(✉)], Ana Margarida Teixeira[1], and Pedro Martins[2,3]

[1] INEGI, Porto, Portugal
{aandre,ateixeira}@inegi.up.pt
[2] INEGI-LAETA/FEUP, Porto, Portugal
palsm@fe.up.pt
[3] ARAID, I3A, Universidad de Zaragoza, Zaragoza, Spain

Abstract. Physical impairments have multiple causes, making them common. Locomotion disorders have been afflicting society for a long time, motivating researchers and engineers to mitigate their consequences. Nowadays, solutions such as exoskeletons and exosuits are in constant development and may become reliable options to help people in these circumstances. However, prospective solutions need a control system acting as a "bridge" between the external device (actuators) and the user. Among several possibilities, movement prediction is prioritized over movement reaction. This task may be done by capturing and processing biological signals from a user's body. Within this paradigm, muscle electromyographic (EMG) signals were acquired, processed and sent as input to piezoelectric soft actuators.

Keywords: Soft actuators · Piezo-based materials · EMG signals · Movement prediction · Exoskeleton

1 Introduction

Physical impairments are very common nowadays since they could essentially be caused by three main reasons [1]: the normal ageing process, also associated with higher life expectancy [2,3]; neurodegenerative diseases [4]; and finally, daily accidents [5].

Locomotion disorders caused by the normal ageing process are typically associated with older persons [6] and more accentuated in regions where life expectancy is higher. Some countries, such as Norway, Germany or Hong Kong (China) have a life expectancy over 80 years old, contrasting with some African nations (The Central African Republic or Nigeria have an expectancy of 54 years old). Statistics from United Nations (UN) reveal that elderly people older than 60 years old will be around 20% of the global population by 2050 [7].

Supplementary Information The online version contains supplementary material available at https://doi.org/10.1007/978-3-031-10015-4_4.

J. M. R. S. Tavares et al. (Eds.): CMBBE 2021, LNCVB 38, pp. 50–58, 2023.
https://doi.org/10.1007/978-3-031-10015-4_4

Neurological pathologies are also responsible for a fragile physical state. They cause a progressive loss of structure and function of the central nervous system (CNS). The CNS is the main responsible to coordinate muscle control and command. Nowadays, it is estimated that about 50 million people worldwide could have some kind of dementia [8], with Parkinson's disease representing more than 10% of this value [9].

Injuries, particularly in the spinal cord (SC) and brain, could also cause permanent or temporary changes in muscular strength or body functions. They are the result of external traumas (such as car accidents (39.3%), falls (31.8%), gunshots (13.5%), sports-related (8%) [10]) or due to tumours in these specific locations [11]. Every year, about 40 million people suffer SC or brain injuries [12].

These locomotion issues have a significant societal impact, which motivated the effort of researchers and engineers [13] to mitigate some of their consequences and impact. Despite the solutions already in used, such as wanderers, wheeled vehicles or wheelchairs [14], the current focus is on the development of exoskeletons and exosuits equipped with different types of actuators [15].

A control system, acting as a bridge between the final user and the external device, is a fundamental requirement that does not depend on the adopted actuation system. In this work, biological signals (bioelectrical) acquired from muscle activities, will be used as inputs to control soft actuators.

Bioelectrical signals are the result of chemical activities at a cellular level, resulting in electrical potentials (typically in the mV range) [16] detected by appropriate sensors, such as EMG sensors [17,18].

EMG sensors (intramuscular [19] or surface [20]) are able to measure small electrical signals [21] generated by muscle contraction. For this reason, they were successfully applied in the control of external devices [22,23]. Solutions based on this method bring the advantage of reading and capturing a user's intention and control actuation, even if a person can't produce sufficient joint torques [24] to move.

However, the signal might be biased due to several factors, such as muscle crosstalk [25], a skin condition (in case of surface sensors), muscle fatigue [26] or the inability to access deep muscle fibres [27].

In the current work, a setup was developed to capture muscle bioelectrical signals, and use them as inputs to piezoelectric soft actuators - materials capable of changing their shape and size when electrically stimulated.

2 Control and Actuation System/Experimental Setup

An experimental setup was developed to capture bioelectrical signals and use them as inputs to induce mechanical deformation on soft materials. The solution proposed consists of the following elements (Fig. 1): a BITalino (and EMG sensors) [28] connected to a personal computer (PC) and Matlab software [29] via Bluetooth connection; a function generator connected to a PC via USB;

and finally an oscilloscope connected to the function generator for display and analysis of the output signal.

The EMG sensors connected to a BITalino board [28] acquire electrical activity at muscle level, while sending the data to software implemented in Matlab. In parallel, an algorithm (Matlab software) processes the acquired information and if the signal fulfils pre-established criteria (such as a threshold value of signal intensity), a command is sent to the function generator to change the output signal. This signal constitutes the electrical stimulation input for the soft material, and is displayed on the oscilloscope. All data acquisition and processing happens in real-time (with maximum delays of approximately 1 s).

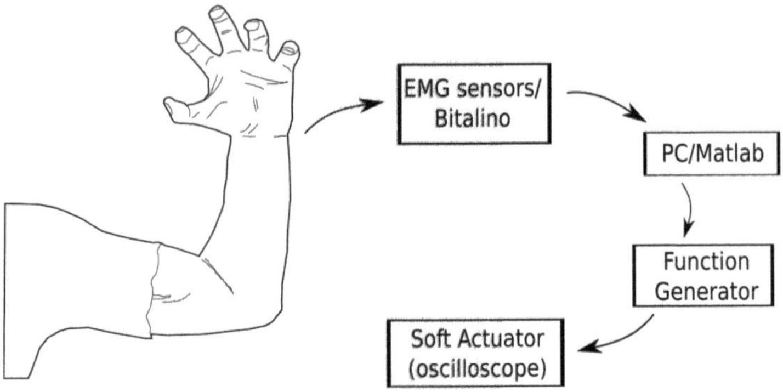

Fig. 1. Outline of the experimental setup.

2.1 Hardware

The hardware setup consists of a PC (Intel Core i7-2700k and 32GB RAM) running Windows 10 Enterprise OS (OS build 19042.928), an oscilloscope (UNI-T UTD2062C) for displaying the output signal, the EMG sensors and a function generator (Joy-it JDS6600).

The EMG sensors were acquired from BITalino (Plugged Kit BT, Bluetooth version with firmware v5.1) [28]. BITalino (Fig. 2) is a microcontroller based on the ATMEGA family, specifically developed for biometric applications with a maximum resolution of 10 bits [16].

BITalino is a modular system. The main board may be coupled with several modules. The energy module powers the device, has two energy outputs (3.3V and 1.65V) and a micro-USB connection to recharge a Li-ion battery. The central processing module hosts the microprocessor and the firmware. Finally the communication BLE module enables communication with other devices via Bluetooth. These components make the central unit of BITalino

Additionally, different sensors can be plugged into the main board, for electrocardiography (ECG) or electroencephalography (EEG) analysis [16]. In the current work, only EMG sensors were used. EMG sensor is an operational amplifier for biometric applications [28]. It has three electrodes - positive and negative placed at the muscle and a reference electrode, placed at a reference point such as an elbow.

The last component of the setup is function generator Joy-it JDS6600 [30]. The relevant characteristics of the equipment are the possibility to generate arbitrary waveforms (signals) as well as support for PC connection.

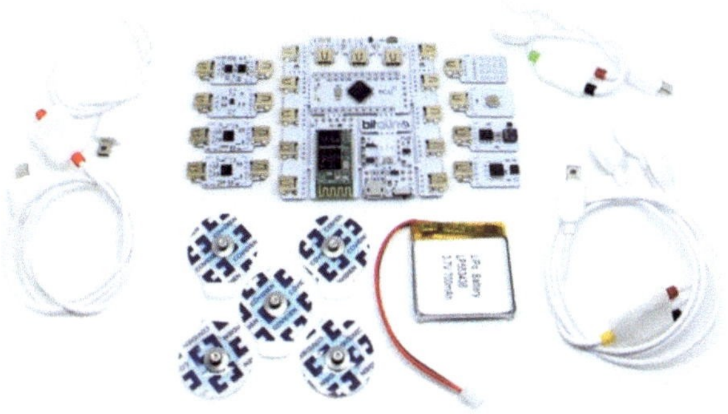

Fig. 2. BITalino hardware.

2.2 Software

The control and actuation of the soft material requires software purposely built for the task. Matlab 2021a [29] and Python (v3.7) [31], are the main requirements. Additionally, the BITalino toolbox for Matlab [32] and the python module 'pyserial' [33] are needed for communication and data transfer between the BITalino and the PC.

All signal processing tasks are addressed using Matlab script EMG_signal.m, the main script which controls all hardware. While EMG_signal.m is running, two additional Python scripts are called hierarchically. The first, BITwave.py, defines the signal parameters. The second, jds6600.py, is the bridge between the PC and the function generator.

The biolectrical signal acquired with BITalino's EMG sensors, requires to be processed with a sequence of mathematical operations [28,34,35]. The sequence depicted in Fig. 3 gives the reader a clear understanding of the signal processing operations, applied to the raw signal. In the first step, the raw signal is converted into an EMG signal using Eq. 1 [28], rectified and displayed in the last plot of Fig. 3.

$$EMG(mV) = \frac{(\frac{ADC}{2^n} - \frac{1}{2}) \times VCC}{G_{EMG}} \times 1000 \qquad (1)$$

wherein ADC is the value sampled from the channel, n the number of bits of the channel, VCC the operating voltage (3.3V) and, finally, G_{EMG} is the sensor gain (1009).

The last step involves an envelope peak function from Matlab libraries.

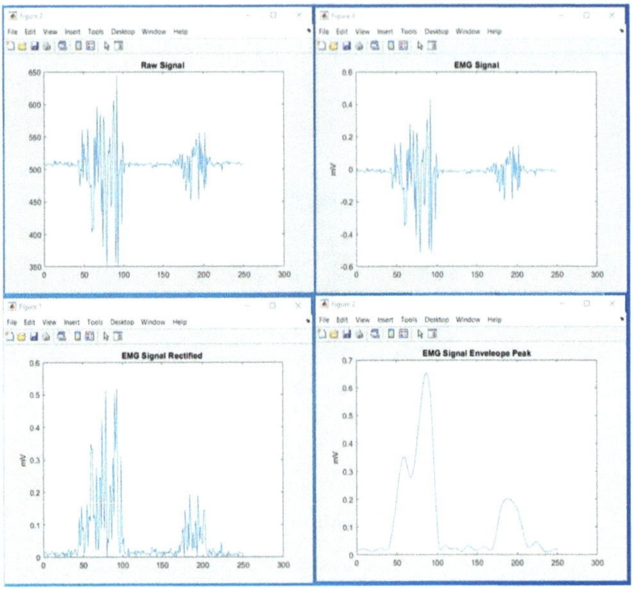

Fig. 3. Signal acquired at different stages (example).

3 Preliminary Results and Discussion

Using the setup described previously, a test was carried out on a healthy male volunteer (29 years old). The results are observed in Fig. 4.

Observing the first image, when no muscle contraction was detected, the output signal (oscilloscope) is 0V. As the muscle contraction starts, the EMG signal is acquired by EMG sensors of Bitalino and the algorithm detects the changes of the output signal from 0V to an arbitrary positive value (second image). While the signal contraction from muscles remains constant, the output state is held (third image). Moreover, when the muscles relaxation begins and is detected by the sensors, the control system does the opposite behaviour. The output value observed returns to 0V (fourth image).

All steps involved are processed (max delay ≈1s) in real-time from data acquisition to signal output (see supplementary video).

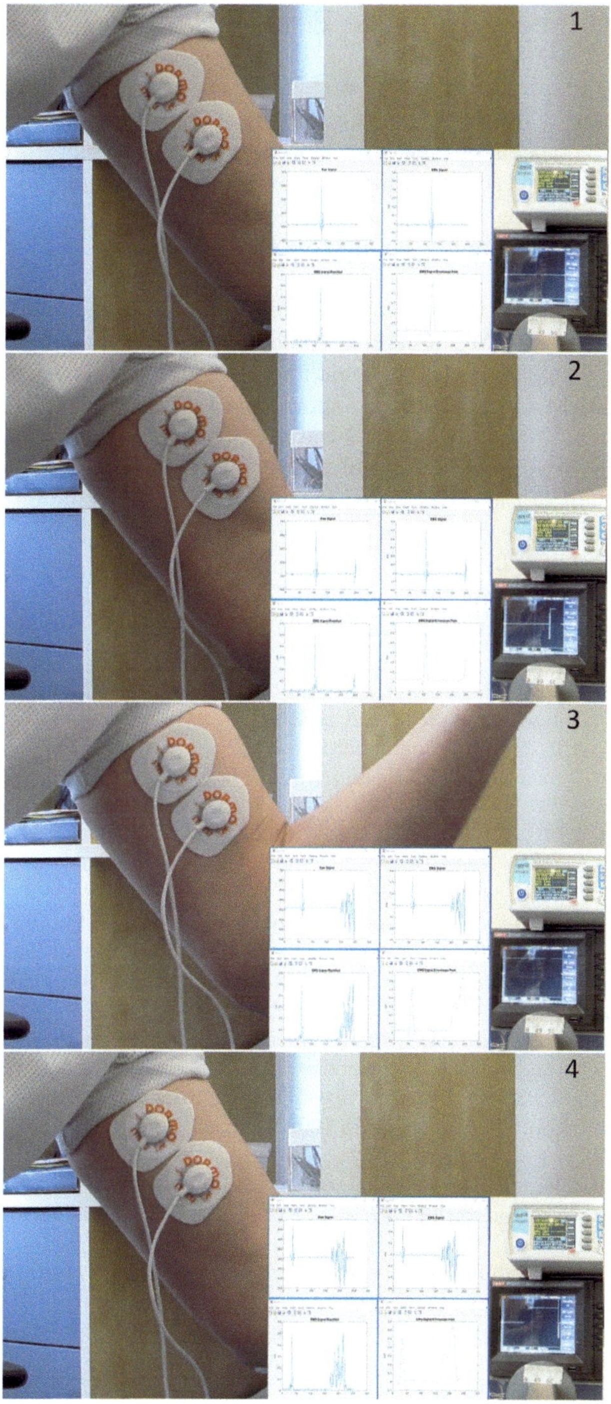

Fig. 4. Oscilloscope output signal for four sequential instances (different stages) during muscle contraction. 1. 0 V without muscle contraction; 2. positive tension value associated with contraction detection; 3. Stable tension while contraction is maintained; 4. muscle relaxation induces progressive tension drop to 0 V.

4 Conclusions and Future Perspectives

The procedure adopted in this work enables the use of an EMG signal as input for different use cases. The actuation of a piezoelectric soft actuator almost in real-time is a possible use of this method. Moreover, the most relevant characteristic of the proposed setup is the ability to provide adequate input to the actuator, even with slight muscle contraction. This behavior can be fully customised, since the algorithm uses sensibility parameters that can be changed according with usage requirements.

Despite its usability, the system has limitations. Since the bioelectrical signal comes from a surface EMG sensor, it could be biased due to skin conditions or muscle crosstalk signals. This fact could impair the applicability of the developed setup in real-world scenarios such as movement assistance or rehabilitation.

The research team aims to test the setup using a working soft actuator, and to address or minimize some of the technical issues already encountered. The improvement of the EMG signal via noise reduction, using filters or Fourier transforms will be one of the next iterations of the project. The creation of a portable solution will be another milestone for the project, since the team aims to combine this technology with augmented reality (AR) and virtual reality (VR) for assisted rehabilitation purposes.

Acknowledgements. The authors gratefully acknowledge funding from FCT, Portugal, MCTES, FSE and EU under grant SFRH/BD/147807/2019, and project MImBI - PTDC/EME-APL/29875/2017 financed through FEDER and FCT. This work was supported by FCT, through INEGI, under LAETA, project UIDB/50022/2020.

References

1. Miranda, A.B.W., et al.: Bioinspired mechanical design of an upper limb exoskeleton for rehabilitation and motor control assessment. In: 2012 4th IEEE RAS EMBS International Conference on Biomedical Robotics and Biomechatronics (BioRob), pp. 1776–1781 (2012)
2. Dima-Cozma, C., et al.: The importance of healthy lifestyle in modern society: a medical, social and spiritual perspective. Eur. J. Sci. Theol. **10**(3), 111–120 (2014)
3. Chen, B., et al.: State-of-the-art research in robotic hip exoskeletons: a general review. J. Orthop. Translation 4–13 (2020). https://doi.org/10.1016/j.jot.2019.09.006
4. Moreno, J.C., et al.: Hybrid wearable robotic exoskeletons for human walking. Wearable Robot. Syst. Appl. 347–364 (2020). https://doi.org/10.1016/B978-0-12-814659-0.00018-7
5. Nas, K., et al.: Rehabilitation of spinal cord injuries. World J. Orthop. 8–16 (2015). issn: 2218–5836. https://doi.org/10.5312/wjo.v6.i1.8
6. Levesque, L., Doumit, M.: Study of human-machine physical interface for wearable mobility assist devices. Med. Eng. Phys. 33–43 (2020). https://doi.org/10.1016/j.medengphy.2020.03.008
7. Chen, B., et al.: Recent developments and challanges of lower extremity exoskeletons. J. Orthop. Transl. **5**, 26–37 (2015). https://doi.org/10.1016/j.jot.2015.09.007

8. Bellou, E., Stevenson-Hoare, J., Escott-Price, V.: Polygenic risk and pleiotropy in neurodegenerative diseases. Neurobiol. Dis. (2020). https://doi.org/10.1016/j.nbd.2020.104953
9. Vos, T., et al.: Global, regional, and national incidence, prevalence, and years lived with disability for 310 diseases and injuries, 1990–2015: a systematic analysis for the global burden of disease study 2015. The Lancet 1545–1602 (2016). https://doi.org/10.1016/S0140-6736(16)31678-6
10. National Spinal Cord Injury Statistical Center. https://www.nscisc.uab.edu/
11. Kirshblum, S., et al.: Rehabilitation of persons with central nervous system tumors. Cancer (2001). https://doi.org/10.1002/1097-0142(20010815)92:4+⟨1029::AID-CNCR1416⟩3.0.CO;2-P
12. Yip, P.K., Malaspina, A.: Spinal cord trauma and the molecular point of no return. Mol. Neurodegeneration (2012). https://doi.org/10.1186/1750-1326-7-6
13. Kim, Y.S., et al.: A force reflected exoskeleton-type masterarm for human-robot interaction. IEEE Trans. Syst. Man Cybern Part A Syst. Hum. **35**(2), 198–212 (2005). issn: 10834427. https://doi.org/10.1109/TSMCA.2004.832836
14. van den Bogert A.J.: Exotendons for assistance of human locomotion. BioMed. Eng. Online (2003). https://doi.org/10.1186/1475-925X-2-17
15. Sado, F., et al.: Design and control of a wearable lower-body exoskeleton for squatting and walking assistance in manual handling works. Mechatronics (2019). https://doi.org/10.1016/j.mechatronics.2019.102272
16. Casanova, P.: Sensorização Espacial no Contexto da Reabilitação Humana (2020)
17. Muhammad Zahak Jamal: Signal acquisition using surface EMG and circuit design considerations for robotic prosthesis. Comput. Intell. Electromyogr. Anal-A Perspect. Current Appl. Future Challenges **18**, 427–448 (2012)
18. Jamal, M.Z., Dong-Hyun, L., Hyun, D.J.: Real time adaptive filter based EMG signal processing and instrumentation scheme for robust signal acquisition using dry EMG electrodes. In: 2019 16th International Conference on Ubiquitous Robots (UR), pp. 683–688 (2019). https://doi.org/10.1109/URAI.2019.8768662
19. Péter A., Arndt A., Hegyi, A., et al.: Effect of footwear on intramuscular EMG activity of plantar flexor muscles in walking. J. Electromyogr. Kinesiol. **55**, 102474 (2020). https://doi.org/10.1016/j.jelekin.2020.102474
20. Qi, J., Jiang, G., Li, G., Sun, Y., Tao, B.: Surface EMG hand gesture recognition system based on PCA and GRNN. Neural Comput. Appl. **32**(10), 6343–6351 (2019). https://doi.org/10.1007/s00521-019-04142-8
21. Asghari Oskoei, M., Hu, H.: Myoelectric control systems - a survey. Biomed. Sig. Process. Control **2**, 275–294 (2007). https://doi.org/10.1016/j.bspc.2007.07.009
22. Lenzi, T., et al.: Intention-based EMG control for powered exoskeletons. IEEE Trans. Biomed. Eng. **59**(8), 2180–2190 (2012). https://doi.org/10.1109/TBME.2012.2198821
23. Noda, T., et al.: An electromyogram based force control coordinated in assistive interaction. In: 2013 IEEE International Conference on Robotics and Automation, pp. 2657–2662 (2013). https://doi.org/10.1109/ICRA.2013.6630942
24. Peternel, L., et al.: Adaptive control of exoskeleton robots for periodic assistive behaviours based on EMG feedback minimisation. PLOS ONE (2016). https://doi.org/10.1371/journal.pone.014894
25. Park, J., et al.: Feasibility of propertional EMG control for a hand exoskeleton: a fitts. law approach. IFAC-PapersOnLine **51**, 214–219 (2018). https://doi.org/10.1016/j.ifacol.2018.11.544

26. Zhang, K., de Silva, C.W., Chenglong, F.: Sensor fusion for predictive control of human-prosthesis-environment dynamics in assistive walking: a survey. In: CoRR abs/1903.07674 (2019). arXiv: 1903.07674. http://arxiv.org/abs/1903.07674
27. Farina, D., et al.: Surface EMG crosstalk between knee extensor muscles: experimental and model results. Muscle Nerve **26**(5), 681–695 (2002). https://doi.org/10.1002/mus.10256
28. BITalino. BITalino. https://bitalino.com/
29. MathWorks. MathWorks. https://www.mathworks.com/products/matlab.html
30. Joy-it. Joy-it. https://joy-it.net/de/
31. Python Org. Python Og. https://www.python.org/
32. MathWorks Instrument Control Toolbox Team. BITalino Toolbox. https://www.mathworks.com/matlabcentral/fileexchange/53983-bitalino-toolbox (2021)
33. Python Modules. Python Modules. https://pypi.org/project/pyserial/
34. Reaz, M.B.I., Hussein, M.S., Mohd-Yasin, F.: Techniques of EMG signal analysis: detection, processing, classification and applications. Biol. Proced. **8**, 11–35 (2006). https://doi.org/10.1251/bpo115
35. Farinha, D.M.: Processamento de sinal EMG para dispositivos de reabilitação e assistênica motora (2018)

Movement Optimization Through Musculoskeletal Modeling and Multidimensional Surface Interpolation

Christine Walck(✉), Christopher Lamb, and Pablo Vilches

Embry Riddle Aeronautical University, Daytona Beach, FL 32114, USA
christine.walck@erau.edu, {lambc9,vilchesp}@my.erau.edu

Abstract. In an effort to enhance medical decision-making and precision treatment, biomechanical musculoskeletal modeling and computer simulations originating from observations of human movements are being implemented. Such a combination allows for the establishment of cause-and-effect relationships providing a unique, non-invasive method of evaluation into internal joint and muscle force function, as well as athletic performance and neuromuscular coordination. While these are beneficial, current musculoskeletal models lack the ability to give insight into the mechanical stimuli required to prevent physiological deconditioning (i.e., muscle atrophy, bone decalcification, and poor cardiovascular health). This is seen on the International Space Station (ISS) as astronauts still experience deconditioning symptoms even with strict exercise regimes. Therefore, this study overviews the development of a multidimensional surface interpolation, which uses a radial basis function derived from the proper orthogonal decomposition). Inputs to the interpolation scheme are biomechanical responses found by solving an inverse dynamic problem using the ISS's Advanced Resistive Exercise Device model and simulation within OpenSim. A comparison of 41 lower extremity muscle forces between the interpolation and the static optimization analysis in OpenSim was made. Results highlighted a 1.30% average deviation, with errors concentrated at fluctuating data regions in the minimum force production category. The interpolation scheme shows promise in aiding researchers in the development of an optimized singular exercise movement versus the three exercise machines currently used on the ISS.

Keywords: Machine learning · Space deconditioning countermeasure · Microgravity

1 Introduction

According to NASA, over five hundred astronauts have been launched into space since 1961, when Yuri Gagarin made the first journey. Consequently, many crewmembers endure long-term significant pain and functional limitations specifically deconditioning of the multifidus muscles, bone decalcification, and poor cardiovascular health. Studies have shown that crewmembers experience a reduction in muscle volume of up to a 20%

J. M. R. S. Tavares et al. (Eds.): CMBBE 2021, LNCVB 38, pp. 59–64, 2023.
https://doi.org/10.1007/978-3-031-10015-4_5

in the lower extremities [1] and that space-induced osteoporosis occurs at a rate of 1% to 2% per month compared to the average loss of approximately 1% per year on Earth [2, 3]. This suggests that the current countermeasures (i.e., exercise machines) used on the International Space Station (ISS) as well as the exercise regime of three hours per day, are insufficient in combating such negative effects [4–6].

In efforts to improve this, an optimized solution is needed to find the required mechanical stimuli needed to combat the negative effect microgravity imposes on the human body. This is because space-induced muscle atrophy and bone decalcification have been hypothesized to be caused by a reduction of mechanical stimuli on the musculoskeletal (MSK) system. Current methods best match the human strength curve to the machine's resistance curve to optimize work done by the user. This method is less optimal when applied in a microgravity environment (i.e., the absence of the gravitational force decreases the mechanical stimuli applied to the astronaut). To enhance current countermeasures, this study overviews the potential benefits of pairing biomechanical MSK modeling and computer simulations with multidimensional surface interpolation.

2 Methods

Biomechanical responses (i.e., joint angles and moments) found from an inverse dynamic problem solved using Advanced Resistive Exercise Device (ARED) model within OpenSim were paired with a multidimensional surface interpolation to investigate its potential in aiding in the development of an optimized singular exercise movement for space exercise. The interpolation uses a radial basis function (RBF) trained by the proper orthogonal decomposition (POD) and is designed to learn from progressive calculations outputting the ideal response that will then interpolate a set of new parameters [1]. In addition, for the algorithm to understand the biomechanical constraints and restrictions, boundary conditions were set for joint ranges of motion, foot offset, and ARED machine resistive loading. The workflow of this algorithm is shown in Fig. 1. Furthermore, the database set is characterized by a parameter matrix of [PxM] and a simulation matrix of [NxM], where P is the parameter in question, N is the simulation data and M is the simulation number or scenario.

2.1 Development of Interpolation

The POD filters data by creating a truncated basis, which is widely used in image compression, signal processing, and facial recognition [7]. Operatively, the POD process begins by finding a covariance matrix for the OpenSim biomechanical response matrix by multiplying it by a transposed version of itself. The matrix's eigenvectors are used to represent it on its orthogonal basis [7, 8], where the covariance can be manually inspected by plotting a descending order logarithmic scale plot, in which each eigenvalue represents the amount of energy or variance contained in a corresponding eigenvector. By observing a trend in the eigenvalue plot, a truncation point can be determined for the eigenvectors. Equation 1 displays this method.

$$\text{Covariance Matrix} = \text{INP}^{\text{T}} * \text{INP} \tag{1}$$

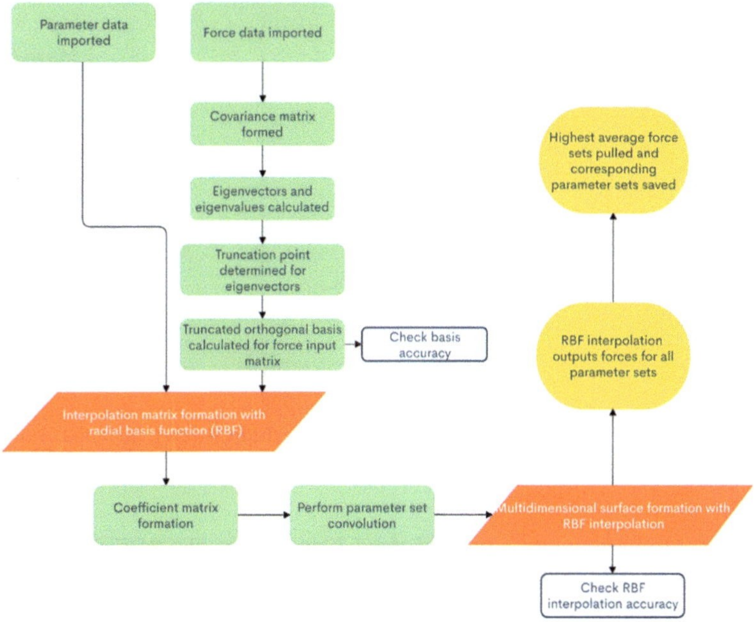

Fig. 1. Depiction of POD - RBF workflow used in this investigation

The INP is defined as the input matrix of simulation results. For the purposes, and as an example, Fig. 2 graphically describes this process of eigenvalue truncation, where the eigenvalues are sorted in descending order to show the energy or variance contained in the eigenvector matrix columns. The point in which truncation is prescribed can be found by experimenting with different values and observing reconstruction accuracy. In Fig. 2, the eigenvalues are plotted on a logarithmic scale, to see at which point the eigenvalues and corresponding eigenvectors, begin to contain less information. The point in which the function changes is a candidate for the truncation point. Only high-energy eigenvectors are kept. Equation 2 creates a filtered orthonormal basis for interpolation resulting in a truncated basis. [8].

$$\text{Truncated Basis Matrix} = \text{INP} * (\text{Truncated Eigenvector Matrix}) \qquad (2)$$

Once the Truncated Basis is calculated, it is fundamental to gauge whether the truncation point was applied in an effective manner, and confirm that the new basis accurately reconstructs the original dataset. This is carried out using Eq. 3, which reconstructs the simulation results.

$$W = (Truncated\ Basis) * (Truncated\ Basis)^T * INP \qquad (3)$$

A small truncation will produce an identity matrix when the truncated basis is multiplied with its transpose, due to the orthonormality of the basis formed. If the identity matrix criteria are not met, the reconstruction, W, is not the same as the input data. This data is then compared at every interval to find a standard deviation, which is used to gauge an average error. To produce zero error with the new basis, truncation would be small.

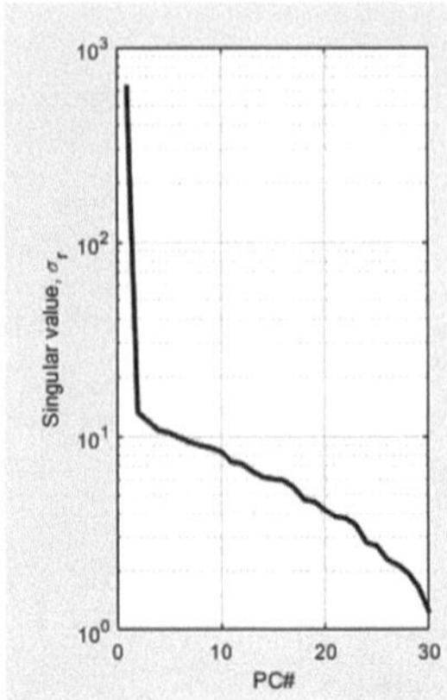

Fig. 2. Displays the eigenvalue (singular value) plot, where the PC# axis is bounded by the number of simulations performed.

Following the partial orthogonal decomposition process, the RBF algorithm utilizes the Hardy Inverse multi–quadratics equation to create the interpolation matrix. This formula is represented in Eq. 4, where P represents the matrix of parameters input from the simulation. Indexes i & j are placeholders for a looping algorithm, both being limited by the number of simulations, which set H to a matrix of [MxM] rows and columns.

$$H(i,j) = \left[\sum\left(\left(\frac{P(:,i) - P(:,j)}{Max(P(:,i)) - Min(P(:,j))}\right)^2 + C^2\right)\right]^{-1/2} \tag{4}$$

Calculating a coefficient matrix, developed from Eq. 5, consists of an amplitude matrix A. Equation 6 defines how this amplitude matrix A is obtained. The accuracy of the RBF can be tested by re-creating the input snapshot matrix.

$$Coeff.Matrix = (Truncated\ basis) * A * H^T \tag{5}$$

$$A = (Truncated\ Basis)^T * INP \tag{6}$$

To generate all possible parameter combinations within the bounds of the simulation parameters, a permutation algorithm is used. The bounds of each individual parameter has a list formed between the intervals at any resolution, making it possible to assign unique resolution to particular variables. With 100 values between the bounds, the generated parameter list can become immense, hence the fidelity of the response surface will be high. These values are then combined in every possible arrangement. The parameters of the expanded matrix can then be run through the RBF to produce a matrix. The RBF algorithm is shown below, in Eq. 7.

$$RBF = Coeff.Matrix(:,i) * \left(\left(\frac{Expanded\ matrix(:,a) - Par(:,i)}{\max(Par(:,i) - \min(Par(:,i)))}\right)^2 + C^2\right)^{-\frac{1}{2}} \tag{7}$$

Again indexes i & a are placeholders for loop variables and are defined by several input columns. The number of columns in the expanded matrix respectively. The smoothing factor, C, is a number chosen to help fit RBF results to original data. The RBF is re-generated every loop iteration, and the summed value is stored in Res for analysis as soon in Eq. 8.

$$Res(:, a) = Sum(RBF, 2) \tag{8}$$

To find the most desirable parameter sets, the interpolated snapshots (in the matrix of generated simulation outputs), with the highest average forces, are gathered and presented with their corresponding parameter set. The most viable parameter set will be chosen, from a batch of candidates, to model a new exercise movement applied in a space environment. For this study, preliminary input data to the POD – RBF was found from solving an inverse dynamic problem using the ARED model and simulation data provided by OpenSim.

3 Results

Interpolation results were compared to results from solving a static optimization problem within OpenSim for 41 lower extremity muscle forces during a single squat exercise ARED. Initial results show an average deviation of 1.30% between the interpolation and the OpenSim static optimization results, with larger errors seen in the datasets of minimum force production predictions. These readings with high error tended to fluctuate to zero throughout the simulation snapshots as depicted in Fig. 3.

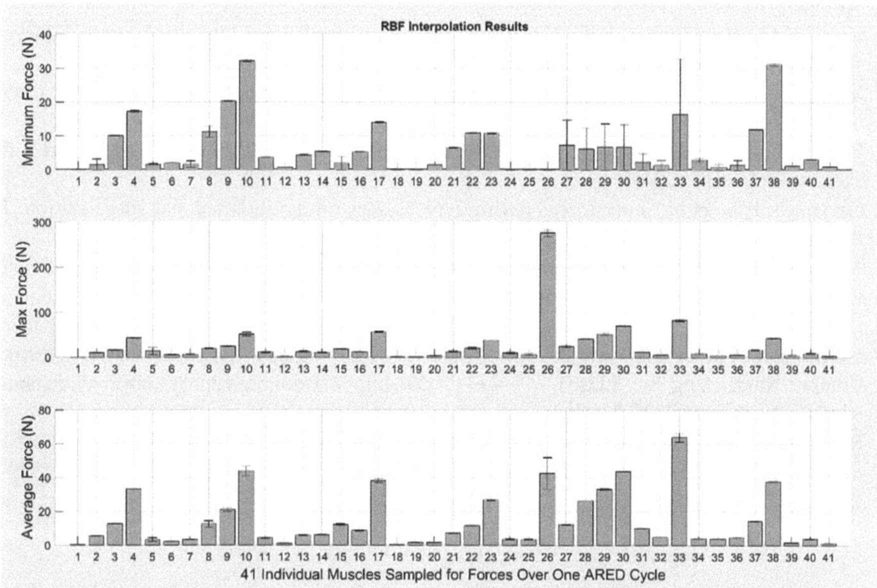

Fig. 3. POD-RBF interpolation results found in 41 lower extremity muscles during an ARED squat

4 Discussion

Joint angles from the ARED exercise within OpenSim allowed for a realistic simulation of space exercise without the gravitational acceleration variable. However, because today's MSK models are developed for specific tasks (i.e., movements), many models must be used in the development of a database of multiple exercises. Furthermore, to find muscle forces for each model and simulated exercise, a static optimization problem must be solved. This is computationally taxing and time-consuming. Furthermore, results give no insight into taking the benefits from each exercise to create one optimized exercise. However, by taking only the joint angles from multiple exercises as inputs into the developed interpolation scheme, results will give optimized parameters to develop a single exercise. With results showing an initial average deviation of 1.30% between the interpolation and the OpenSim static optimization results, such a method holds great promise in increased performance and predictive capability with much less time dedicated to computational analysis.

Future studies should expand the input database of simulated motions with more diversity including exercises such as walking, cycling, cutting maneuvers, and a variety of lunges, squats, and jumps to increase the predictive capability.

References

1. Trappe, S., et al.: Exercise in space: human skeletal muscle after 6 months aboard the International Space Station. J. Appl. Physiology. 1159–68. https://doi.org/10.1152/japplphysiol.91578.2008. Epub 2009 Jan 15. PMID: 19150852. Author, F., Author, S.: Title of a proceedings paper. In: Editor, F., Editor, S. (eds.) CONFERENCE 2016, LNCS, vol. 9999, pp. 1–13. Springer, Heidelberg (2016)
2. Genc, K.O., Humphreys, B.T., Cavanagh, P.R.: Enhanced daily load stimulus to bone in spaceflight and on earth. Aviat. Space Environ. Med. **80**(11), 919–926 (2009)
3. Myer, G.D., Ford, K.R., Hewett, T.E.: Rationale and clinical techniques for anterior cruciate ligament injury prevention among female athletes. J. Athl. Train **39**(4), 352–364 (2004)
4. Diaz, A.A., Heldt, T., Young, L.: Cardiovascular responses to artificial gravity combined with exercise. vol. 2015 (2015)
5. Cavanagh, P.R., et al.: Foot forces during typical days on the international space station. J. Biomech. **43**(11), 2182–2188 (2010)
6. Navasiolava, N., et al.: Vascular and microvascular dysfunction induced by microgravity and its analogs in humans: mechanisms and countermeasures. Front. Physiol. **11**, 952 (2020)
7. Das, A., Khoury, A., Divo, E., Huayamave, V., Ceballos, A., Eaglin, R., et al.: Real-time thermomechanical modeling of PV cell fabrication via a POD-trained RBF interpolation network. Comput. Model. Eng. Sci. **122**(3), 757–777 (2020). http://dx.doi.org.ezproxy.libproxy.db.erau.edu/10.32604/cmes.2020.08164
8. Rogers, C.A., Kassab, A.J., Divo, E.A., Ostrowski, Z., Bialecki, R.A.: An inverse POD-RBF network approach to parameter estimation in mechanics. Inverse Probl. Sci. Eng. **20**(5), 749–767 (2012). https://doi.org/10.1080/17415977.2012.693080

Simulating the Dynamics of a Human-Exoskeleton System Using Kinematic Data with Misalignment Between the Human and Exoskeleton Joints

Divyaksh Subhash Chander[1(✉)], Max Böhme[2], Michael Skipper Andersen[3], John Rasmussen[3], and Maria Pia Cavatorta[1]

[1] Department of Mechanical and Aerospace Engineering, Politecnico Di Torino, Turin, Italy
chander.divyaksh@polito.it

[2] Faculty of Engineering, Leipzig University of Applied Sciences, Leipzig, Germany

[3] Department of Materials and Production, Aalborg University, Aalborg, Denmark

Abstract. Musculoskeletal model-based simulation can be a powerful tool in the evaluation of exoskeletons. An ideal exoskeleton model, perfectly aligned with the human joint axes, can be used to co-simulate the human and exoskeleton dynamics. However, human-exoskeleton joint misalignment is commonly observed during the use of an exoskeleton. Using misaligned motion data in the combined human-exoskeleton model can lead to unrealistic results. In this work, we present a new method to align human-exoskeleton models. This was achieved by introducing dummy segments that ensured kinetic alignment between the human and exoskeleton joints without altering the misaligned kinematics. The method was applied on an active lower-limb exoskeleton that assists the elderly in stair negotiation. In a pilot study, motion data of a single subject testing the exoskeleton in stair ascent were recorded using an optical marker-based system. Measured ground reaction force and exoskeleton assistive force were used as inputs in the human-exoskeleton model. The outputs from the model with the dummy segments were compared to those from a model with kinematic constraints and a reference model where the external forces were applied directly to the human model. The results of the knee compression force, knee flexion moment, and activation of vastus lateralis showed good agreement between the dummy segments and reference models. The use of the dummy segments allows the study of aligned kinetics and misaligned kinematics from the same model. The method will be used in a future study to evaluate the exoskeleton with more subjects.

Keywords: Musculoskeletal model · Human-exoskeleton interface · Contact model

1 Introduction

Exoskeletons are wearable mechanical devices that assist the users in performing tasks. Due to the wearable nature of exoskeletons, they interact closely with the users. They

J. M. R. S. Tavares et al. (Eds.): CMBBE 2021, LNCVB 38, pp. 65–73, 2023.
https://doi.org/10.1007/978-3-031-10015-4_6

are being increasingly used in industrial and medical contexts and there is an increasing focus on how to optimize an exoskeleton to ensure user acceptance. Musculoskeletal model-based simulation can be a very useful tool in the design and evaluation of exoskeletons. It allows estimating the loads in the human musculoskeletal structure and how an exoskeleton affects these loads. These models can be used at different stages of the design cycle to support the user-centered design process [1]. They have been used to analyze a specific task and identify the criticalities in the activity [2]. Abstract assistive moments have been applied directly at different joints of the human model in a parametric study to identify the optimal assistance for the task in question [3, 4]. Digital mock-ups of exoskeletons have been integrated into musculoskeletal models to make more detailed investigations such as kinematic compatibility between the human and exoskeleton [5, 6] or subject-specific optimization of the exoskeleton assistance [7, 8].

While musculoskeletal models have been used frequently in the virtual design phase, their use in the evaluation of real exoskeletons is less common. Musculoskeletal models are often driven by motion capture (mocap) data. In the virtual design phase, a model of the exoskeleton is added to the kinematics of the human without the exoskeleton. An ideal case is simulated by aligning the exoskeleton joint axes with the human joint axes. However, some misalignment between the human and exoskeleton joints is often present during the actual use of exoskeletons. This is easily verified through the mocap data collected with the exoskeleton. On the modelling side, the human-exoskeleton interface is typically modelled using kinematic constraints that restrain the relative movement and the associated reaction forces are set up between the human and exoskeleton. The numerical solvers, thus, require some assumptions when using mocap data consisting of both the human and exoskeleton. One way can be to assume the alignment at the joints and study just the idealized case [9]. Another approach can be to assume the kinematic alignment at the joints while considering the effects of the misalignment by modifying the exoskeleton force based on the observed misalignment [10]. The force compensation in [10] was based on the offset observed in a static pose.

In this work, we present an alternative method for modelling the human-exoskeleton system. The aim of the work was to develop a method allowing the use of mocap data with human-exoskeleton joint misalignment directly in a single model setup for the inverse dynamics analysis of the human-exoskeleton system. The dynamic effects of the misalignment were neutralized in the model by creating dummy exoskeleton joints that were aligned with the human joints. This allowed the human and exoskeleton joints in the model to be kinetically aligned while kinematically misaligned in the same model definition. The method did not use kinematic constraints between the human and exoskeleton, and the interaction forces were simulated using a rigid-body contact model [11–13]. The biomechanical outputs from the model were compared to those from a conventional model using kinematic constraints and a reference model, where the external forces were applied directly to the human model.

2 Methods

An active exoskeleton for the lower limbs was used in this study. The exoskeleton aims to assist the elderly in stair negotiation by assisting in knee extension. The unilateral exoskeleton is connected to the user through three interfaces: thigh, shank, and foot (Fig. 1). The thigh and shank interfaces are connected by a hinge joint that should be aligned with the human knee. Likewise, the shank and foot interfaces are connected by a hinge joint that should be coaxial with the human ankle. The exoskeleton ankle is a single degree of freedom (DOF) joint, allowing only plantarflexion-dorsiflexion. The eversion-inversion motion is accounted for via a compliant material at the foot interface. The exoskeleton can be adjusted to the user's anthropometry through multiple sliding rails that allow the adjustment of the distance between exoskeleton ankle and knee as well as the relative position of the three interfaces from the exoskeleton joints. Further, the interface attachments are also designed to accommodate users of different anthropometry. The assistive force of the exoskeleton is implemented through an active motor on the thigh interface. The motor pulls a cable connected to the shank interface, passing anteriorly to the human knee, thereby, assisting in the extension of the knee. The assistance is active only during the stance phase of the corresponding leg.

2.1 Experiments

One subject (27 years, male, 1.83 m, and 77.4 kg) from the exoskeleton development team participated voluntarily in this pilot study and signed an informed consent prior to participation. An ethical approval for this pilot study was not required. All procedures performed were in accordance with the 1964 Helsinki declaration and its later amendments or comparable ethical standards. The exoskeleton was tested in stair ascent on a purpose-built staircase, comprising of a force plate (Kistler MiniDyn type 9119AA2, Switzerland) in the second step to record ground reaction forces and moments (GRF) (Fig. 1). The assistive force of the exoskeleton was recorded by a strain gauge in the shank interface. A marker-based mocap system (Qualisys AB, Sweden) was used to record the kinematics at 100 Hz with markers fixed on both human and exoskeleton.

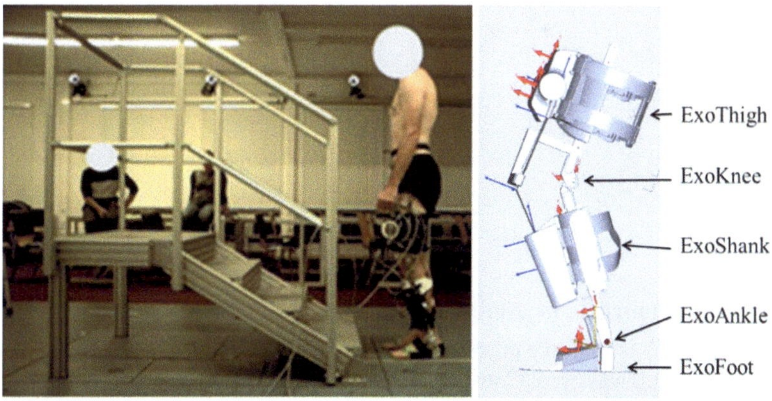

Fig. 1. Left: experimental setup with the custom-built stairs. Right: exoskeleton for the right leg showing the interface components and the hinge joints.

Relevant body parameters such as segment lengths were measured, and the exoskeleton was adjusted to the anthropometry of the user. The subject wore the exoskeleton on both legs and climbed the stairs several times before the actual trials to familiarize with the device. Then, the subject performed 13 ascents at a self-selected pace, starting with the left leg and climbing step-over-step without using handrails. However, only eight ascents were analysable due to the exclusion of incomplete data series. Further, two static recordings were taken of the subject and the exoskeleton in the standing neutral posture.

2.2 Musculoskeletal Modelling

The musculoskeletal model available in the AnyBody Managed Model Repository (AMMR) version 2.3.0 was used in this work [14]. The "Plug-in-gait_Multi-Trial_StandingRef" example in the repository was adapted and analysed in version 7.3.0 of the AnyBody Modeling Systems (AMS) [15]. The arms were disabled in the model and the standing trial was used to optimize the marker locations and segment lengths [16].

First, a baseline reference model was defined to observe the intended effect of the exoskeleton assistance by applying the external forces (GRF and exoskeleton assistive force) directly to the human model. This model did not account for the mass and inertia effects of the exoskeleton and served as a reference to check for significant discrepancies between the outputs of the reference model and the model with the kinetic alignment of the human and exoskeleton joints. The internal body loads of the statically indeterminate system due to the presence of more unknown forces than equilibrium equations in the model were solved through inverse dynamics analysis together with the muscle recruitment formulation [17]. The optimization problem to recruit the muscles was set up to minimize a polynomial function of the effort required:

$$\min_{\mathbf{f}} H\left(\mathbf{f}^{(\mathrm{M})}\right) = \sum_{i=1}^{n^{(\mathrm{M})}} \left(\frac{f_i^{(\mathrm{M})}}{N_i^{(\mathrm{M})}}\right)^p \tag{1}$$

subject to:

$$\mathbf{Cf} = \mathbf{d} \tag{2}$$

and

$$f_i^{(\mathrm{M})} \geq 0 \text{ for } i = 1 \ldots n^{(\mathrm{M})} \tag{3}$$

The effort required is measured by the ratio $f_i^{(\mathrm{M})}/N_i^{(\mathrm{M})}$, also known as muscle activity, where $f_i^{(\mathrm{M})}$ is the force produced by the ith muscle and $N_i^{(\mathrm{M})}$ is its strength. In this work, power $p = 3.0$ was used. Constraint (2) is the dynamic equilibrium equation where $\mathbf{C}$ is the coefficient matrix of the unknown internal forces (muscle and joint reaction), $\mathbf{f}$. $\mathbf{d}$ is the vector of the known external and inertial forces acting on the different segments. Constraint (3) restricts the muscles to pull only.

In the second model, the exoskeleton model was added and linked to the human model through kinematic constraints at the human-exoskeleton interfaces. A computer-aided design (CAD) model of the exoskeleton was translated into a script file using the version

1.2.0 of the AnyExp4SOLIDWORKS plugin for SolidWorks. The exoskeleton model required eight constraints as it consisted of three interfaces linked through two hinge joints and there are multiple ways to add these constraints. We added three constraints at the thigh interface (translation and rotation about anterior-posterior axis; translation in medial-lateral axis) and four constraints at the shank interface (translation in anterior-posterior and proximal-distal axes; translation and rotation about medial-lateral axis). Lastly, there was one constraint at the foot interface (translation in proximal-distal axis). These constraints linked the kinematics of the human-exoskeleton system and their associated reaction forces were included in Eq. (2) in the inverse dynamics analysis. Despite the use of kinematic constraints, the human-exoskeleton joint axes were not aligned in the model as we used the exoskeleton knee-ankle joint distance from the trials and it did not match the human knee-ankle distance of the model.

Finally, there was the third model with a rigid-body contact model at the interface and the method for kinetic alignment of the joint axes. Before the translation of the CAD model to the script file, reference frames were created at the thigh, shank, and foot interfaces for simulating the contact forces at the human-exoskeleton interface. The contact forces at the interface were simulated by adapting a method to predict GRF [18]. The reference frames at the interface were used to distribute several contact nodes and the corresponding contact detection zones. At each contact node, five contact elements were created to simulate the normal and shear contact forces. Each contact element, characterized by a maximum force-generating capacity (or strength), generated unidirectional force. Four contact elements were arranged in the positive and negative shear directions and the fifth contact element simulated normal force. The force of the contact elements was computed by extending Eq. (1) and constraint (3) to include the corresponding terms for the contact elements. Friction force was implemented through linear constraints between four of the five contact elements and included in constraint (2). The human and exoskeleton models were driven from their respective mocap data. This resulted in misalignments between the human and exoskeleton joint axes. The joint misalignment allowed the contact elements distributed at the different interfaces to contribute towards knee extension and flexion, resulting in an unrealistic unloading of the physiological muscles.

Method for Kinetic Alignment

The joint misalignments were compensated by introducing "dummy segments" in the contact model. These were mass-less and inertia-less segments that altered the kinetic path of the internal forces of the exoskeleton without altering the kinematics of the system. Three dummy segments were added in the model corresponding to the thigh, shank, and foot. They were kinematically constrained to the corresponding human segments to create a dummy leg with its dummy knee and ankle joints that were aligned with the human knee and ankle joints. Then, kinetic constraints were added between the corresponding exoskeleton and dummy segments in all six DOF. Finally, the reaction forces at the exoskeleton knee and ankle joints were disabled and substituted by the reaction forces at the dummy knee and ankle joints. Thus, a human-exoskeleton system was defined with the exoskeleton knee and ankle joints substituted kinetically by the dummy knee and ankle joints, which were kinematically aligned with the corresponding human joints. The concept is summarized in Fig. 2.

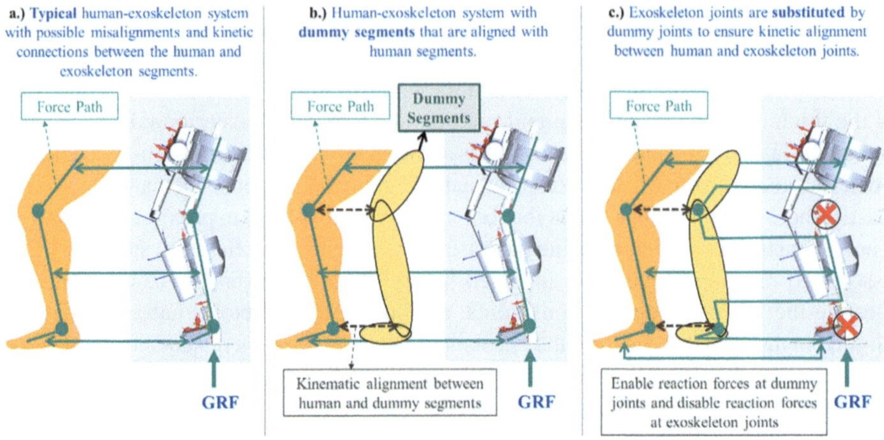

Fig. 2. Application of dummy segments to kinetically align human-exoskeleton joint axes.

2.3 Analyses

The stance phase of the right leg was analysed from the eight trials of stair ascent due to the availability of the recorded GRF at the second step. As the knee is the target joint for this exoskeleton, only outputs that concerned the knee (activation of vastus lateralis, knee flexion moment, and knee compression force) were studied. Point-based resampling technique (PBRT) was used to create 95% confidence bands about the outputs from the baseline model (denoted as IA for ideal assistance model) using bootstrapping [19]. The mean values for the eight trials from the second model with the kinematic constraints (KC) and the third model with the contact model and dummy segments (CMDS), were plotted on the same graphs with the IA model outputs.

3 Results and Discussion

Figure 3 shows the activation of the vastus lateralis, knee flexion moment and knee compression force from the baseline IA model and the KC and CMDS interface models. The CMDS model showed no significant difference with the IA model for the three outputs. The results show that the addition of dummy segments to kinetically align the human-exoskeleton joint axes resulted in an idealized system that was close to the simulation of the ideal assistance. The dummy segments allowed accounting for the mass and inertia effects of the exoskeleton that were not considered in the IA model. The KC model resulted in lower values of vastus lateralis activation and knee compression force when compared to the IA model. This difference in values can be attributed to a difference in the human-exoskeleton kinematics between the two models. The addition of kinematic constraints altered the naturally occurring motion of the human-exoskeleton system as recorded by the mocap system. However, there was no significant difference in knee flexion moment between the KC, CMDS and IA models.

Human-exoskeleton joint misalignment is a practical issue that affects the interface forces [20] and the comfort of the user [21, 22]. Moreover, it also results in a loss of

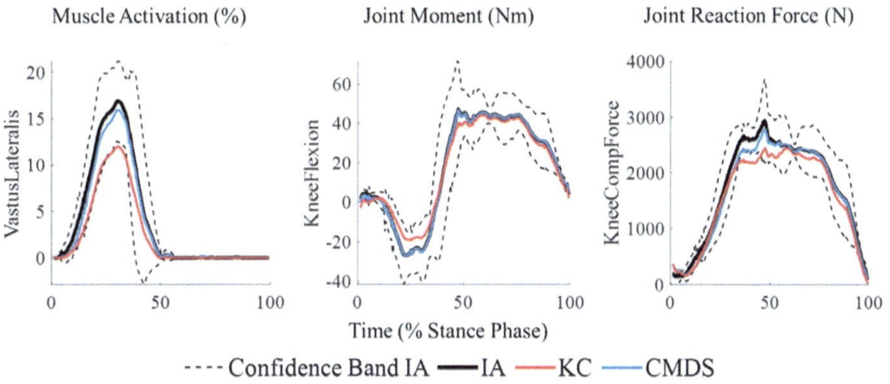

Fig. 3. Biomechanical outputs from the ideal assistance (IA) model, kinematic constraints (KC) model and the contact model with dummy segments (CMDS).

exoskeleton efficiency as a part of the exoskeleton force/moment is dissipated and does not contribute to assisting the user. In the past, joint misalignment has been accounted in the model by decomposing the force/moment from the exoskeleton into an assistive and a dissipative component based on the joint offset observed in a static pose [10]. The advantage of using dummy segments over kinematic constraints for joint alignment is that dummy segments allow using kinematic data containing joint misalignment in the same model. The misaligned kinematic data can be used to modulate the exoskeleton force in the model based on the misalignment at each time-step of the analysis. Modulation of the exoskeleton force based on the current misalignment can also be done using kinematic constraints in a two-step process. First, misaligned kinematics would need to be analysed to store the misalignment at each time step. Then, the stored misalignments can be used to modulate the exoskeleton assistance in the second step with the joint axes kinematically aligned.

4 Conclusions

Musculoskeletal models provide useful insights in the design and evaluation of exoskeletons. They are easier to use in the virtual prototyping phase where the exoskeleton is generally assumed to fit perfectly to the human model, resulting in human-exoskeleton joint axes alignment. However, during the evaluation of a prototype or commercial exoskeleton, joint misalignment can often be observed. Unrealistic solutions can be obtained from the model due to the joint misalignment and the closed-loop nature of human-exoskeleton systems. Joint misalignment can be corrected by adding kinematic constraints between the human and exoskeleton, but this will alter the actual kinematics of the human-exoskeleton system and can significantly affect the biomechanical outputs. In this work, a method was presented to align the joint axes kinetically. The biomechanical outputs from this method showed good agreement with an ideal assistance model where the external forces were applied directly to the human model. The method for kinetic alignment allows using the same model definition to study the misalignment in the kinematics and the dynamic effect of an idealized exoskeleton on the user. This is an advantage over aligning the exoskeleton with kinematic constraints as two separate models would be required to study the kinematic misalignment and dynamic effects of the exoskeleton. The results presented in this study were from a pilot study with a single subject. Further experiments with a greater number of subjects will be performed to study the effects of the exoskeleton using the modelling methods described in this work.

Acknowledgements. We would like to thank Johannes Zentner from Leipzig University of Applied Sciences for his support to this project.

References

1. Miehling, J., Wolf, A., Wartzack, S.: Musculoskeletal simulation and evaluation of support system designs. In: Karafillidis, A., Weidner, R. (eds.) Developing Support Technologies. BB, vol. 23, pp. 219–227. Springer, Cham (2018). https://doi.org/10.1007/978-3-030-01836-8_21
2. Tröster, M., Wagner, D., Müller-Graf, F., Maufroy, C., Schneider, U., Bauernhansl, T.: Biomechanical model-based development of an active occupational upper-limb exoskeleton to support healthcare workers in the surgery waiting room. Int. J. Environ. Res. Public Health. **17**, 1–16 (2020). https://doi.org/10.3390/ijerph17145140
3. Dembia, C.L., Silder, A., Uchida, T.K., Hicks, J.L., Delp, S.L.: Simulating ideal assistive devices to reduce the metabolic cost of walking with heavy loads. PLoS ONE **12**, 1–25 (2017). https://doi.org/10.1371/journal.pone.0180320
4. Shourijeh, M.S., Jung, M., Damsgaard, M.: Metabolic energy consumption in a box-lifting task: a parametric study on the assistive torque. In: Gonzalez-Vargas, J., Ibáñez, J., Contreras-Vidal, J.L., van der Kooij, H., Pons, J.L. (eds.) Wearable Robotics: Challenges and Trends. BB, vol. 16, pp. 143–148. Springer, Cham (2017). https://doi.org/10.1007/978-3-319-46532-6_24
5. Hansen, C., Gosselin, F., Mansour, K.B., Devos, P., Marin, F.: Design-validation of a hand exoskeleton using musculoskeletal modeling. Appl. Ergon. **68**, 283–288 (2018). https://doi.org/10.1016/j.apergo.2017.11.015
6. Ferrati, F., Bortoletto, R., Pagello, E.: Virtual modelling of a real exoskeleton constrained to a human musculoskeletal model. In: Lepora, N.F., Mura, A., Krapp, H.G., Verschure, P.F.M.J., Prescott, T.J. (eds.) Biomimetic and Biohybrid Systems. Living Machines 2013. Lecture Notes in Computer Science, vol. 8064, pp 96–107, Springer, Berlin (2013). https://doi.org/10.1007/978-3-642-39802-5_9
7. Jensen, E.F., Raunsbæk, J., Lund, J.N., Rahman, T., Rasmussen, J., Castro, M.N.: Development and simulation of a passive upper extremity orthosis for amyoplasia. J. Rehabil. Assist. Technol. Eng. **5**, 1 (2018). https://doi.org/10.1177/2055668318761525
8. Zhou, L., Li, Y., Bai, S.: A human-centered design optimization approach for robotic exoskeletons through biomechanical simulation. Rob. Auton. Syst. **91**, 337–347 (2017). https://doi.org/10.1016/j.robot.2016.12.012
9. Fournier, B.N., Lemaire, E.D., Smith, A.J.J., Doumit, M.: Modeling and simulation of a lower extremity powered exoskeleton. IEEE Trans. Neural Syst. Rehabil. Eng. **26**, 1596–1603 (2018). https://doi.org/10.1109/TNSRE.2018.2854605
10. Gordon, D.F.N., Henderson, G., Vijayakumar, S.: Effectively quantifying the performance of lower-limb exoskeletons over a range of walking conditions. Front. Robot. AI. **5**, 1–16 (2018). https://doi.org/10.3389/frobt.2018.00061

11. Cho, K., Kim, Y., Yi, D., Jung, M., Lee, K.: Analysis and evaluation of a combined human-exoskeleton model under two different constraints condition. Int. Summit Hum. Simul. (ISHS) **5**, 23–25 (2012)
12. Jung, M., et al.: Musculoskeletal simulation of SOLEUS ankle exoskeleton for countermeasure exercise in space. In: Gonzalez-Vargas, J., Ibáñez, J., Contreras-Vidal, J.L., van der Kooij, H., Pons, J.L. (eds.) Wearable Robotics: Challenges and Trends. BB, vol. 16, pp. 391–396. Springer, Cham (2017). https://doi.org/10.1007/978-3-319-46532-6_64
13. Chander, D.S., Cavatorta, M.P.: Modelling friction at the mechanical interface between the human and the exoskeleton. Int. J. Hum. Factors Model. Simul. **7**, 119 (2019). https://doi.org/10.1504/ijhfms.2019.105434
14. Lund, M.E., et al.: The AnyBody Managed Model Repository (AMMR) (2020). https://doi.org/10.5281/ZENODO.3932764
15. Damsgaard, M., Rasmussen, J., Christensen, S.T., Surma, E., de Zee, M.: Analysis of musculoskeletal systems in the AnyBody modeling system. Simul. Model. Pract. Theory. **14**, 1100–1111 (2006). https://doi.org/10.1016/j.simpat.2006.09.001
16. Andersen, M.S., Damsgaard, M., MacWilliams, B., Rasmussen, J.: A computationally efficient optimisation-based method for parameter identification of kinematically determinate and over-determinate biomechanical systems. Comput. Methods Biomech. Biomed. Engin. **13**, 171–183 (2010). https://doi.org/10.1080/10255840903067080
17. Rasmussen, J., Damsgaard, M., Voigt, M.: Muscle recruitment by the min/max criterion - a comparative numerical study. J. Biomech. **34**, 409–415 (2001). https://doi.org/10.1016/S0021-9290(00)00191-3
18. Skals, S., Jung, M.K., Damsgaard, M., Andersen, M.S.: Prediction of ground reaction forces and moments during sports-related movements. Multibody Sys. Dyn. **39**(3), 175–195 (2016). https://doi.org/10.1007/s11044-016-9537-4
19. Joch, M., Döhring, F.R., Maurer, L.K., Müller, H.: Inference statistical analysis of continuous data based on confidence bands—traditional and new approaches. Behav. Res. Methods **51**(3), 1244–1257 (2018). https://doi.org/10.3758/s13428-018-1060-5
20. Zanotto, D., Akiyama, Y., Stegall, P., Agrawal, S.K.: Knee joint misalignment in exoskeletons for the lower extremities: effects on user's gait. IEEE Trans. Robot. **31**, 978–987 (2015). https://doi.org/10.1109/TRO.2015.2450414
21. Naf, M.B., Junius, K., Rossini, M., Rodriguez-Guerrero, C., Vanderborght, B., Lefeber, D.: Misalignment compensation for full human-exoskeleton kinematic compatibility: state of the art and evaluation. Appl. Mech. Rev. **70**, 1–19 (2018). https://doi.org/10.1115/1.4042523
22. Mallat, R., Khalil, M., Venture, G., Bonnet, V., Mohammed, S.: Human-exoskeleton joint misalignment: a systematic review. In: International Conference on Advances in Biomedical Engineering, ICABME, October 2019, pp. 1–4 (2019). https://doi.org/10.1109/ICABME47164.2019.8940321

A 2D-FEM Model of Nonlinear Ultrasound Propagation in Trans-cranial MRgFUS Technique

Fabiano Bini[1(✉)], Andrada Pica[1], Maurizio Marrale[2], Cesare Gagliardo[3], and Franco Marinozzi[1]

[1] Department of Mechanical and Aerospace Engineering, "Sapienza" University of Rome, Rome, Italy
{fabiano.bini,andrada.pica,franco.marinozzi}@uniroma1.it
[2] Department of Physics and Chemistry, University of Palermo, 90133 Palermo, Italy
maurizio.marrale@unipa.it
[3] Section of Radiological Sciences, Department of Biomedicine, Neuroscience and Advanced Diagnostics, University of Palermo, 90127 Palermo, Italy
cesare.gagliardo@unipa.it

Abstract. Magnetic Resonance guided Focused Ultrasound (MRgFUS) is a non-invasive technique based on the thermal ablation of a target using high intensity focused ultrasound. MRgFUS treatment applied to brain is challenging due to the skull presence that attenuates ultrasound, leading to heating effects in bone region. In this study, we simulate trans-cranial nonlinear ultrasound propagation considering the detailed structure of bone tissue. We developed a 2D Finite Element (FE) model that mimics the propagation of focused ultrasound through skin, skull and brain tissue. The skull is represented as a three-layered system with two cortical tables packing a layer of trabecular bone. We assume that the space between the concave transducer and tissue is filled by water. Nonlinear ultrasound propagation is determined through Westervelt equation. To control reflection, absorbing layers have been implemented on the boundaries of the domains. The solution of the pressure equation is subsequently coupled with Pennes bioheat equation to determine the temperature distribution in the tissue region. The acoustic pressure, acoustic intensity and temperature distribution are achieved from FE simulation. Highest values of acoustic pressure occur in the focal area and in the bone tissue region. Ablative temperatures, i.e. superior to 55 °C, are achieved in the target zone and at the cortical-trabecular interface. The thermal response in the focal region is in agreement with available literature and allows to validate the model effectiveness. The FE model offers new insights to predict secondary heating effects of ultrasound propagation in the skull region and to improve treatment planning.

Keywords: Nonlinear acoustics · High intensity focused ultrasound · Brain biomechanics · Bone tissue · FEM

J. M. R. S. Tavares et al. (Eds.): CMBBE 2021, LNCVB 38, pp. 74–89, 2023.
https://doi.org/10.1007/978-3-031-10015-4_7

1 Introduction

Trans-cranial magnetic resonance guided focused ultrasound (MRgFUS) is an emerging, non-invasive stereotactic procedure based on the thermal ablation of the target using high intensity focused ultrasound beam delivered through the intact skull [1]. MRgFUS allows for target localization and for in vivo real time monitoring of temperature in the target [2]. The concentration of acoustic energy in the focal area induces a rapid increase of temperature with consequent ablation of the tissue, i.e. temperature superior to 55 °C [3]. Current experience of trans-cranial MRgFUS has been applied in functional neurosurgery for the treatment of movement disorders such as essential tremor, tremor dominant unilateral idiopathic Parkinson's disease and neuropathic pain [4].

Treatment based on high intensity focused ultrasound has been widely performed in other anatomical regions, e.g. thyroid [5], liver [6], but neurological applications of high intensity focused ultrasound have always been challenging due to the skull that represents a unique barrier to wave propagation. It is worth mentioning that preliminary preclinical and clinical applications required craniotomy [4]. In fact, the presence of skull compromises the high intensity ultrasound waves by affecting the focusing and the energy release into the planned target and, consequently, the treatment feasibility and duration. As observed from experimental studies [4, 7, 8], absorption of high intensity ultrasound beam in the skull region may result in harmful non-targeted heating and subsequently in local hyperthermia. Due to these disadvantages, trans-cranial application of high intensity ultrasound has been prevented in past decades. However, a return of interest in similar treatments has been observed recently due to the improvements of medical imaging that allow for targeting and feedback control of the procedure with quantitative temperature imaging [1, 9]. Moreover, support in this sense is given also by the increase of available computational power that has contributed to the development of physiological relevant predictive models. Computational models play an ever-increasing role in advancing the understanding of the complex behavior of biological tissues in physiological and pathological conditions [10–12]. Mathematical and computational modelling techniques are versatile and powerful tools that can contribute to the design optimization of devices [13], improve the safety and effectiveness of the treatment, the comfort of the patients and can decrease of the probability of side effects [9].

Models of trans-cranial focused ultrasound using skull models have been previously developed in literature [14–18]. Evaluations concerning linear [15] and nonlinear [16] wave propagation and low frequency ultrasound simulations [14] are also present in literature. It is worth mentioning that the studies available in literature adopted simplified geometrical models of the tissue region. In these studies, the skull was considered as a homogeneous medium and average values of the tissue properties, e.g. density, sound velocity, attenuation coefficient, have been assumed. Recently, the influence of the skull structural parameters, e.g. thickness, as well as acoustic parameters have been studied [18]. Analyzing the available literature, it emerges that the detailed skull structure and topology have not been deeply taken into account in the investigation of transcranial ultrasound propagation. In this study, as an advancement, the morphology of trabecular tissue is reconstructed from Computer Tomography (CT) images.

The propagation of acoustic waves through tissues for biomedical applications is usually linked to nonlinear propagation [19]. Most tissues show large acoustic attenuation coefficient with a power low dependence on the frequency [19] that leads to more pronounced heating effects. Acoustic nonlinear effects in soft biological tissues, e.g. liver [6, 20, 21] have been widely investigated in literature. Conversely nonlinear acoustic propagation through hard tissue, e.g. bone tissue, has not been extensively analyzed. Thus, in this work, we simulate trans-cranial nonlinear ultrasound propagation considering the detailed structure of bone tissue in order to investigate the induced heating in the multilayer system composed of skin, skull and brain tissue. The present study applies the nonlinear Westervelt equation to predict the acoustic pressure distribution in the heterogeneous environment of the trans-cranial tissues. The outcomes of the acoustic model are coupled with Pennes bioheat equation to predict the temperature distribution in the tissue medium. A comparison between the results of linear and nonlinear models of acoustic ultrasound propagation is also performed.

2 Methods

2.1 Geometrical Model

The simulation framework follows the trans-cranial MRgFUS procedure adopted for clinical treatment at "Paolo Giaccone" University Hospital – Palermo (Italy). The proposed finite element (FE) model is based on the technical features of a MRgFUS commercial transducer (ExAblate, 4000, InSightec, Haifa, Israel). The numerical simulations concerned a 2D axisymmetric domain around z-axis, assuming that the transducer was hemispherical and symmetric (Fig. 1). We modelled the transducer as a curved element with a radius of curvature and an aperture width [22, 23] such that to achieve a focalization depth equal to that used in the clinical treatment, i.e. 135 mm. The acoustic waves originate from the transducer and propagate along the z-axis direction. Ultrasound propagation was simulated for a working frequency $f_0 = 650$ kHz.

The developed model mimics the propagation of focused ultrasound through skin, skull and brain tissue. The skull is organized in a three-layered structure. The outer and inner tables are made of cortical bone and the middle layer, known also as diploë, is trabecular bone having a blood and fat-filled porous structure [24]. The viscous medium of blood and fat-filled inclusions within trabecular bone are indicated in the rest of the paper as diploë–soft tissue. Information concerning the morphology of the trabecular region has been reconstructed from CT images of the parietal region of a 64-years-old male patient. Specifically, the topology of the trabecular tissue is achieved as a 2D section (Fig. 1d) from a 3D reconstruction of the human skull from CT images (Fig. 1b-c). We also made the hypothesis that brain tissue is represented by the grey matter component. The regions mimicking skin, cortical tables and grey matter were considered homogenous in this study. We also assumed that the space between the multi-layered system and the transducer was filled by water.

For the evaluation of the acoustic pressure, the mesh is influenced by the minimum wavelength (λ) of the system. In this study, the nonlinear FE model adopts a quartic discretization with a maximum element size of $\lambda/4$ that leads to $1.8 \cdot 10^6$ degrees of freedom for the acoustic field. For the assessment of the temperature distribution, the

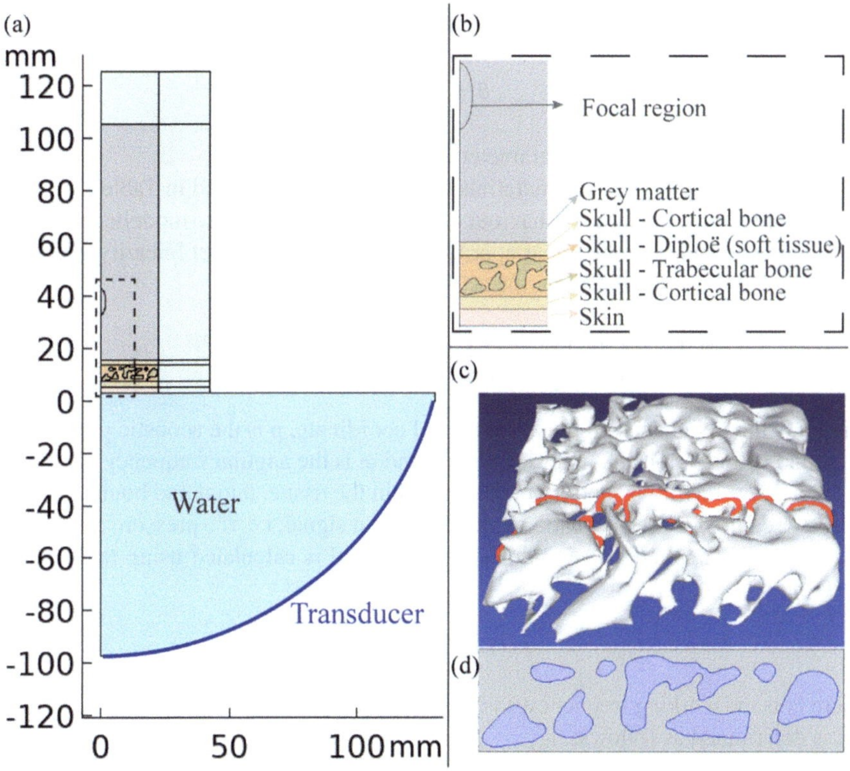

Fig. 1. Simulation setup: (a) schematic representation of the finite element model configuration; (b) enlarged view of the multi-layered structure mimicking trans-cranial region; (c) 3D reconstruction from CT scans of the skull trabecular region patient specific. Red line shows the site of the planar section (d) considered in the FEM model.

FE model is characterized by a quadratic discretization with a maximum element size of 0.2 mm that leads to $2.1 \cdot 10^5$ degrees of freedom.

2.2 Acoustic Model

We modelled the nonlinear ultrasound propagation by means of Westervelt equation [25] (Eq. 1):

$$\nabla^2 p - \frac{1}{c^2}\frac{\partial^2 p}{\partial t^2} + \frac{\delta}{c^4}\frac{\partial^3 p}{\partial t^3} = -\frac{\beta}{\rho c^4}\frac{\partial^2 p^2}{\partial t^2} \tag{1}$$

where p is the acoustic pressure, t is the time, c is the sound velocity, ρ is the material density and δ is the sound diffusivity which is expressed as:

$$\delta = \frac{2c^3\alpha}{\omega^2} \tag{2}$$

where α is the acoustic absorption coefficient and ω is the angular frequency.

The coefficient of nonlinearity (β) is achieved as:

$$\beta = 1 + \frac{B}{2A} \tag{3}$$

where B/A is the nonlinearity parameter of the medium.

Acoustic properties for the materials of the model are reported in Table 1.

To highlight the nonlinear behaviour of the skull region, we also modelled ultrasound propagation considering the linear approximation. In conditions of linearity, Helmholtz equation [26] is used to calculate the acoustic field:

$$\frac{\partial}{\partial r}\left[-\frac{r}{\rho}\left(\frac{\partial p}{\partial r}\right)\right] + r\frac{\partial}{\partial z}\left[-\frac{1}{\rho}\left(\frac{\partial p}{\partial z}\right)\right] - \left[\left(\frac{\omega}{c}\right)^2\right]\frac{rp}{\rho} = 0 \tag{4}$$

where r is the radial coordinate, z is the axial coordinate, p is the acoustic pressure, c is the sound velocity, ρ is the material density and ω is the angular frequency.

To achieve the ultrasound propagation within the tissue, initial and boundary conditions are required for the acoustic model. The input signal, i.e. the pressure at the acoustic interface, is applied as a Dirichlet condition and is calculated using the following equation:

$$p(t) = p_0 sin(2\pi f_0 t) \tag{5}$$

where f_0 is the working frequency, t is the time and p_0 is the source pressure amplitude, that is determined as follows:

$$p_0 = \sqrt{2\rho_{water} \cdot c_{water} \cdot \frac{Power}{Area}} \tag{6}$$

where ρ_{water} and c_{water} are the density and sound velocity of the water. The sonication power in this study is set to 300 W, in agreement with the clinical procedure. The area of the transducer is achieved from the ExAblate technical datasheet.

An interior boundary condition is considered to handle significant differences in the properties of adjacent materials of the multi-layered model. The condition implements continuity in pressure and normal velocity over the interior boundary, as follows:

$$\begin{cases} p_{inf} = p_{sup} \\ n \cdot u_{inf} = n \cdot u_{sup} \end{cases} \tag{7}$$

where subscripts inf and sup refer to the inferior and superior sides of the interior boundary and n is the normal.

Initially, the pressure and its derivative with respect to time was set to zero within the computational domain. In the FE model, the axial symmetry boundary condition was applied at the symmetry axis r = 0:

$$n \cdot u = 0 \tag{8}$$

where u is the velocity and n the normal.

To control the reflection of acoustic waves, an absorbing layer has been added to the tissue domains. The absorbing layer is an artificial domain introduced at the extremities of the tissue regions that combines a scaling system, filtering, and a simple low-reflecting impedance condition to achieve a decaying wave avoiding reflections [27]. At the outer boundary of the absorbing layer, an acoustic impedance condition (Eq. 9) is set representing non-reflecting boundary condition:

$$n \cdot u = \frac{p}{Z} \tag{9}$$

where u is the velocity, n the normal, p is the pressure and Z is the acoustic impedance.

In the linear model, the same boundary conditions are implemented, except for the condition concerning the discontinuity of the material which is not required by the algorithm.

2.3 Bioheat Model

In order to determine the temperature distribution in the tissue, the solution of the acoustic model needs to be coupled with a bioheat model. Pennes bioheat equation [28] has been extensively used to determine heat transfer in biological tissues and is expressed as follows (Eq. 10):

$$\rho C_p \frac{\partial T}{\partial t} = \nabla \cdot (k \nabla T) + Q + Q_{Bio} \tag{10}$$

where T is the temperature, ρ, C_p, k are the density, specific heat and thermal conductivity of the tissue, respectively. Q is the generated heat due to the exposure to the acoustic field and Q_{bio} is the source term that represent blood perfusion and metabolism contribution to model heat transfer in biological tissue.

For the calculation method applied to Q it is necessary to highlight the different time scales of the simulation. Sonication time during the clinical treatment is 10 s. This is roughly 10^5 times the period of the acoustic waves, that have a frequency $f_0 = 650$ kHz. Due to the considerable computational time required, acoustic field computations are performed for 100 cycles, the necessary time to reach the superior extremity of the brain region. Conversely, the heat transfer model addresses the longer times of the sonication procedure, i.e. 10 s. The heat source Q is achieved by means of time averaged values computed from the solution of the acoustic field as indicated by the following expression (Eq. 11) [29]:

$$Q = 2\alpha I = 2 \cdot \alpha \cdot \frac{\langle P^2 \rangle}{2 \cdot \rho \cdot c} \tag{11}$$

where α is the absorption coefficient, I is the intensity calculated by the acoustic model, the brackets denote time average and P is the pressure. The time average is computed considering the cycles determined after the ultrasound wave passed the focal region [29].

Q_{bio} is calculated from Eq. 12:

$$Q_{BIO} = \rho_b \cdot \omega_b \cdot C_b \cdot (T_b - T) + Q_{met} \tag{12}$$

Table 1. Acoustic and thermo-physical properties.

Property	Water	Skin	Skull -cortical bone	Skull - trabecular bone	Skull – diploë soft tissue	Grey matter
Density (ρ) (kg/m^3)	999.1[a]	1114[e]	1700 [e]	1000[e]	1057	1045[e]
Sound velocity (c) (m/s)	1465.5[b]	1615[f]	2393[i]	1500[i]	1500[p]	1450[f]
Nonlinear coefficient (B/A)	5[c]	7.87[d]	12[m]	286[m]	6.1[d]	6.55[d]
Absorption coefficient (α) - (Np/cm)	0.00021[c]	0.21[d]	0.54[h, m]	0.46[n]	0.014[p]	0.063[p]
Specific heat (C_p) (J/kg·K)	4186[d]	3250 [e]	1244[e]	2238[e]	3651[e]	3718[e]
Thermal conductivity (k) (W/m·K)	0.6[d]	0.342[g]	0.3[e]	0.29[e]	0.53[e]	0.55[e]
Blood perfusion rate (1/s)	–	0.00197[h]	0.00028[h]	0.0005 [h]	0.002 [h]	0.0133 [h]
Metabolic heat source (W/m^3)	–	1838 [h]	255[h]	460 [h]	1638 [h]	16239 [h]

[a]Ref. [31]; [b]Ref. [32]; [c]Ref. [33]; [d]Ref. [19]; [e]Ref. [34]; [f]Ref. [35]; [g]Ref. [36]; [h]Ref. [37]; [i]Ref. [38]; [m]Ref. [39]; [n]Ref. [40], [p]Ref. [41]

where $\rho_b = 1050$ kg/m^3 [19] is the density, $C_b = 3800$ J/kg·K [19] is the specific heat and ω_b (Table 1) is the perfusion rate of blood respectively. Q_{met} is the metabolic heat source. T and T_b are tissue and blood temperatures.

The initial temperature of the tissue regions is set equal to T = 37 °C while the water region between the transducer and skin has a temperature equal to T = 15 °C as in the ExAblate system [30]. Moreover, a Dirichlet boundary condition is set for the temperature on the tissue outer boundaries which is assumed to be constant at 37 °C.

Thermal properties for the materials involved in the model are listed in Table 1.

A time dependent simulation was used to calculate the nonlinear propagation of the acoustic field and the temperature distribution within the tissues. The solution of the Westervelt equation is achieved applying the discontinuous Galerkin method that uses a time-explicit solver based on Runge-Kutta fourth order algorithm. The linear model of the acoustic field is solved with MUMPS solver. The bioheat problem is discretized in time by means of Backward Difference Formulas (BDF). Subsequently, the system of equations achieved from BDF method is solved using PARDISO solver.

3 Results and Discussion

Trans-cranial MRgFUS is an appealing technique for its non-invasiveness and ability to selectively target a specific brain region [7]. In this paper, we developed a model for trans-cranial MRgFUS using a detailed skull model derived from medical images.

The study performs a comparison between the outcomes of linear and nonlinear models of ultrasound propagation providing an assessment of the heating effect in the skull region. Acoustic pressure, acoustic intensity and temperature distribution within the multi-layered model are presented.

Figure 2 shows the results of acoustic pressure in the direction of wave propagation. Due to the nonlinear nature of ultrasound propagation through tissues, the pressure predicted by means of the Westervelt based model shows higher values than the corresponding pressure achieved from the linear model. Effects of nonlinearity are noticeable in the skull region and in the focal area. Namely, in the outer cortical table, the peak pressure in nonlinear conditions reaches a value around 2.2 MPa which is roughly twice the value of the pressure determined by the linear model (Fig. 2b). In the focal region (Fig. 2c), the peaks of maximum acoustic pressure calculated in nonlinear conditions are higher with approximately 15 percent than the peak values estimated with the linear model. However, in the focal zone, significant differences are found when comparing the values of pressure predicted by the two models at the same distance from the transducer. For instance, at a distance $z = 135$ mm from the transducer, the nonlinear pressure is approximately five times higher than the pressure estimated from the linear model.

In Fig. 3 we reported the acoustic pressure in the r direction, at distances from the transducer where maximum peak values of pressure are achieved. Along the r direction, nonlinear ultrasound propagation leads to higher values of acoustic pressure in the focal region, and in the skull components, i.e. outer cortical table, skull trabecular region and inner cortical table. In the focal region (Fig. 3a) the peak pressure determined in nonlinear conditions is roughly 2.4 MPa which is approximately four times the pressure predicted by the linear model. Similarly, the peak pressure estimated from nonlinear conditions in the outer cortical table (Fig. 3b) corresponds to roughly twice the value of the pressure achieved from the linear model. Conversely, in the trabecular layer and inner cortical table, the differences among the findings of the two models are less marked. In fact, in the trabecular region (Fig. 3c), the maximum pressure determined from the nonlinear field is roughly 1.4 MPa which represents an increase of 27 percent with respect to the linear outcomes. In the inner cortical table (Fig. 3d), the maximum pressure achieved by the nonlinear model is higher with roughly 42 percent than the peak value determined by linear predictions. Overall, also from Fig. 3 emerges that the linear model underestimates the acoustic field.

In addition, also the comparison of the acoustic intensity distributions achieved from both models highlights the differences between linear and nonlinear ultrasound propagation (Fig. 4). In line with the outcomes of the pressure field, the nonlinear model predicts higher intensity values with respect to the outcomes of the linear model. In

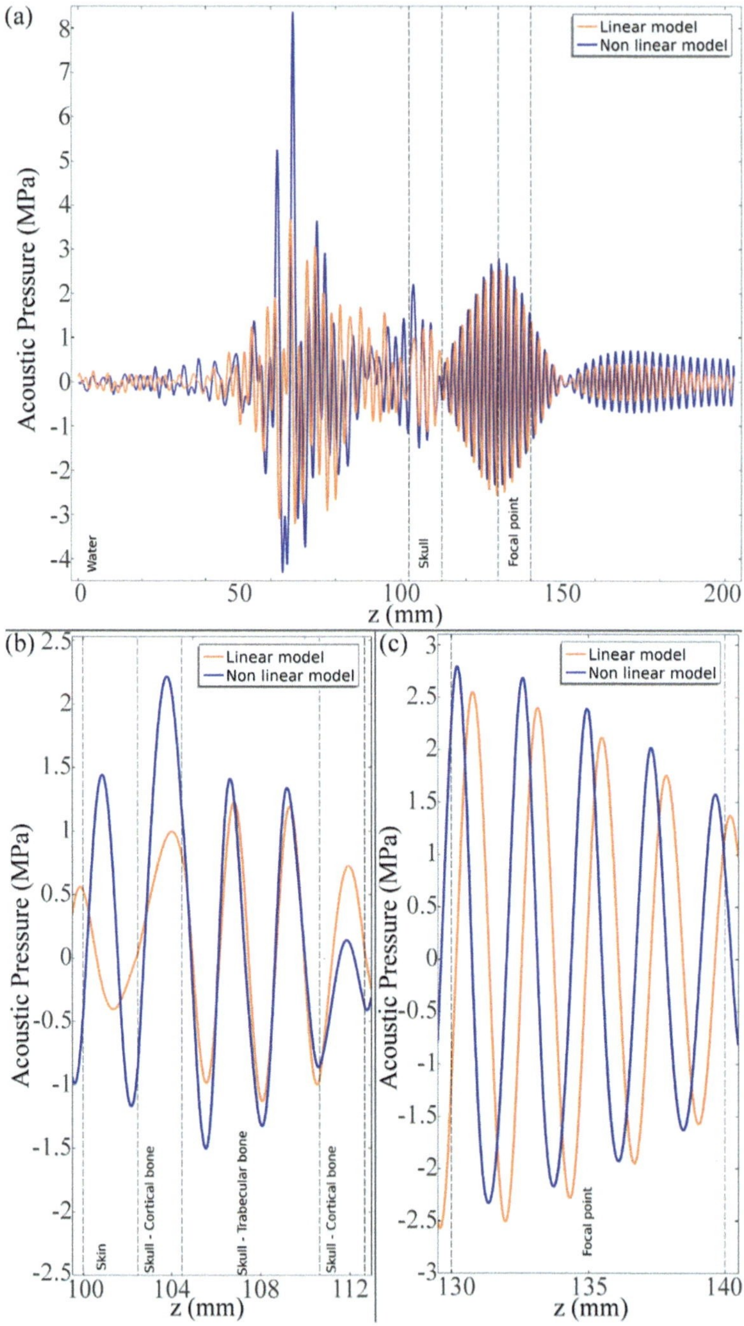

Fig. 2. (a) Pressure line plots along the z direction of propagation of ultrasound waves. (b) Enlarged view of the pressure variation in the tissues domains. (c) Enlarged view of the pressure variation in the focal region. Comparison between the results of linear model (orange line) and nonlinear model (blue line) is reported.

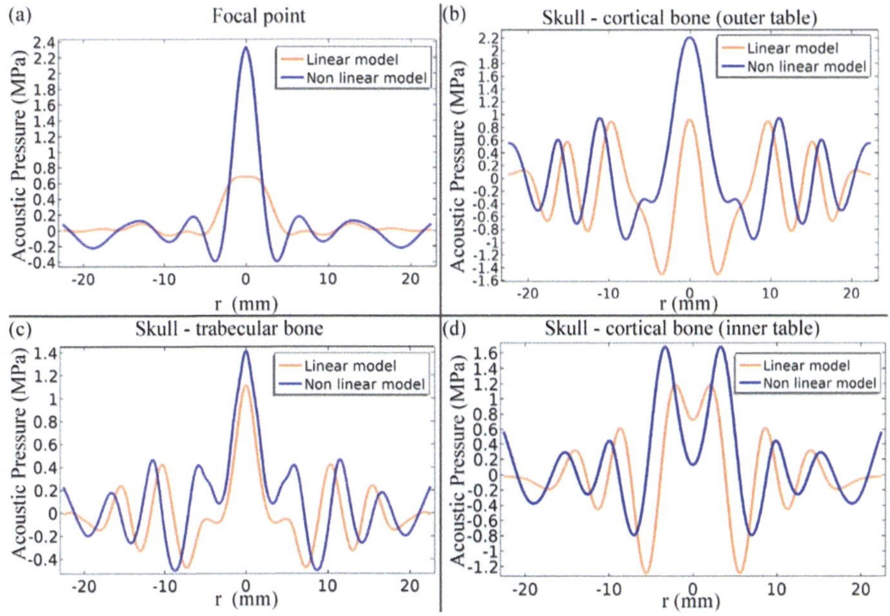

Fig. 3. Acoustic pressure along the r direction (a) in the focal point, at a distance z = 135 mm from the transducer, (b) outer cortical table, at a distance z = 103.75 mm from the transducer (c) trabecular bone region, at a distance z = 106.7 mm from the transducer, and (d) inner cortical table, at a distance z = 112 mm from the transducer, calculated with linear (orange line) and nonlinear (blue line) models of the acoustic field.

the focal area, the intensity achieved from nonlinear conditions has a maximum value around 485 W/cm^2, which is approximately twice the maximum intensity calculated from the linear model. Increased intensity values occur also in the skull when considering nonlinear ultrasound propagation. In the cortical regions of the skull, the intensity reaches values up to 40 W/cm^2 in the nonlinear conditions, that corresponds to twice the value achieved from the linear model. In the trabecular layer, the nonlinear model leads to intensity values up to 80 W/cm^2, that are almost five times higher than the intensity predicted in linear conditions.

Overall good agreement in terms of focal pressure and acoustic intensity can be observed among the present model and similar numerical studies [14, 17]. Kyriakou et al. [14] simulated the acoustic field at a working frequency of 500 kHz and achieved, in the focal area, peak pressure values in the range 1.5–5 MPa. Voie et al. [42] investigated the effects of the skull on the acoustic field, estimating a focal pressure in the interval 0.3–1.5 MPa. Nonetheless, it should be noticed that previous publications analysed systems with lower working frequency, e.g. 500 kHz [14], 230 kHz [42], or considered less detailed skull models [15–18].

Figure 5 shows the estimated temperature distribution within the model after 10 s of sonication. Significative rise of temperature is observed in the target zone and at the cortical-trabecular interface. In these regions ablative temperatures, i.e. temperature superior to 55 °C, are achieved. Moreover, the nonlinear model leads to an increase of the temperature especially in the skull region and in the focal region. In the outer cortical table, the maximum temperature predicted applying the nonlinear model is around 50 percent higher than the outcomes achieved with the linear method. Zones of the trabecular layer are characterized by temperature values that are higher of approximately 42 percent in the nonlinear model with respect to the linear conditions.

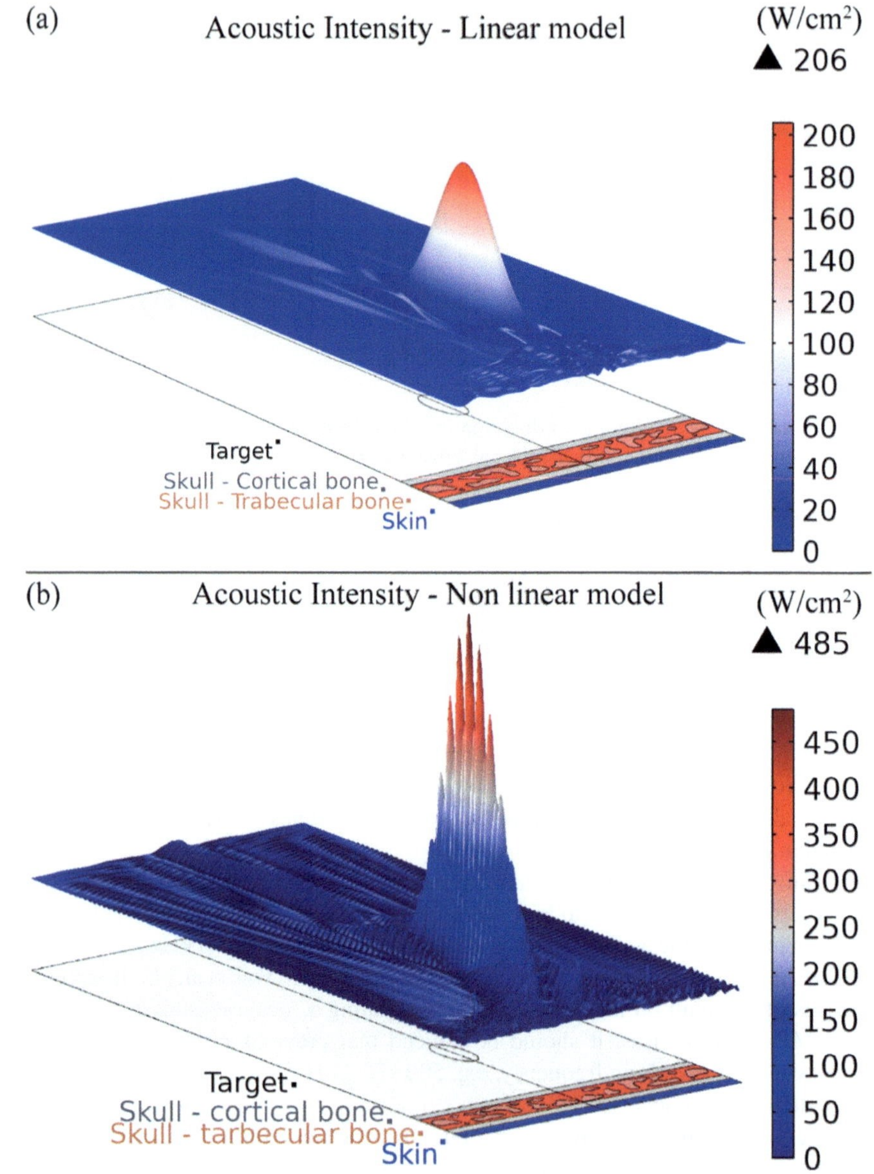

Fig. 4. Acoustic intensity distribution within the trans-cranial model achieved with the linear (a) and nonlinear (b) model of the acoustic field.

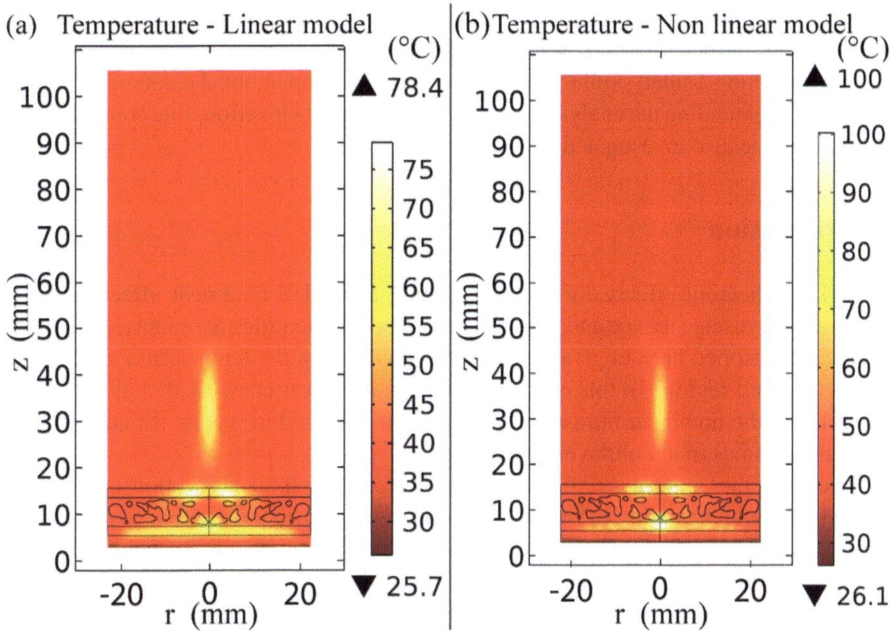

Fig. 5. Temperature distribution after the sonication time (t = 10 s) achieved considering (a) linear model and (b) nonlinear model of the acoustic field.

In the inner cortical table, the local maximum temperature achieved from the nonlinear case increases with roughly 28 percent with respect to the value determined from the linear model. In the focal region, the nonlinear effects increase the temperature with roughly 10 percent. In fact, the nonlinear model estimates a maximum temperature around 74 °C in the focal area, while in linear conditions the peak value is almost 67 °C. The outcomes of the temperature distribution are in agreement also with the available literature. For instance, McDannold et al. [43] predicted that after the sonication time, in the target zone the temperature may reach up to 80 °C. Increase of peak target temperature up to 100 °C are estimated also by [14].

Efficiency of clinical procedures based on high intensity focused ultrasound are highly dependent on the good understanding of the trans-cranial acoustic wave propagation. These findings, in agreement with those of previous literature, emphasize the crucial role of using detailed model of the skull and accurate representation of the ultrasound propagation. In low intensity acoustic wave treatments [15, 16], negligible variation is observed between analysis performed with linear and nonlinear models. Conversely, for treatments utilizing high intensity focused ultrasound procedures, the linear acoustic theory no longer represents the wave propagation in the biological system [7]. The findings reported in this study highlight the importance of analysing ultrasound effects in the trans-cranial region under the perspective of nonlinear acoustic theory.

Some limitations of the FE model should be highlighted. Firstly, we assumed homogeneous regions for skin, skull cortical tables and grey matter. In order to develop more detailed skull models, a future study will take into account the impact of the heterogeneity of all tissues involved. Secondly, the investigation is performed with a 2D FE model concerning a small skull region, due to the high computational resources required. Strategies of extending the analysis to a wider skull region with affordable computational costs should be also investigated.

4 Conclusion

Remarkable heating effects during trans-cranial MRgFUS treatment affect the skull region and represent an issue under investigation for the scientific community. A FE study has been developed here to give new insights concerning the temperature distribution within the skull region. To this end, a new FE simulation framework that allow to take into account the nonlinear ultrasound propagation was used to predict the acoustic and thermal outcomes in a multilayer tissue.

The outcomes contribute to evaluate the impact of skull absorption and to provide a better understanding of modelling temperature distribution within tissues during transcranial MRgFUS. As discussed above, it is crucial considering the nonlinear ultrasound propagation through bone tissue when analysing trans-cranial MRgFUS procedures, otherwise the treatment planning results underestimated. This numerical method could represent a promising approach and an alternative to purely analytic or time-consuming experimental procedure. It is necessary to develop more physiological relevant FE models to allow computational modelling techniques to be adopted for treatment planning.

Acknowledgements. . The authors wish to acknowledge eng. Andrea Bergomi and eng. Arianna Ruffo for their contribution during the development of the preliminary study.

References

1. Gagliardo, C., et al.: Intraoperative imaging findings in transcranial MR imaging-guided focused ultrasound treatment at 1.5T may accurately detect typical lesional findings correlated with sonication parameters. Eur. Radiol. **30**(9), 5059–5070 (2020). https://doi.org/10.1007/s00330-020-06712-0
2. Pozzi, S., Borrazzo, C., Carnì, M., Di Castro, E., Valentini, S., Caccia, B.: A computational tool for evaluating HIFU safety. Ann. Ist. Super Sanita. **52**(2), 256–260 (2016). https://doi.org/10.4415/ANN_16_02_18
3. Lopresto, V., Argentieri, A., Pinto, R., Cavagnaro, M.: Temperature dependence of thermal properties of ex vivo liver tissue up to ablative temperatures. Phys. Med. Biol. **64**(10), 105016 (2019). https://doi.org/10.1088/1361-6560/ab1663
4. Gagliardo, C., et al.: Transcranial magnetic resonance-guided focused ultrasound surgery at 1.5T: a technical note. Neuroradiol J. **32**(2), 132–138 (2019). https://doi.org/10.1177/1971400918818743
5. Trimboli, P., Bini, F., Marinozzi, F., Baek, J.H., Giovanella, L.: High-intensity focused ultrasound (HIFU) therapy for benign thyroid nodules without anesthesia or sedation. Endocrine **61**(2), 210–215 (2018). https://doi.org/10.1007/s12020-018-1560-1

6. Gupta, P., Srivastava, A.: Numerical analysis of thermal response of tissues subjected to high intensity focused ultrasound. Int. J. Hyperthermia. **35**(1), 419–434 (2018). https://doi.org/10.1080/02656736.2018.1506166
7. Pinton, G., Aubry, J.F., Fink, M., Tanter, M.: Effects of nonlinear ultrasound propagation on high intensity brain therapy. Med. Phys. **38**(3), 1207–1216 (2011). https://doi.org/10.1118/1.3531553
8. Leung, S.A., Webb, T.D., Bitton, R.R., Ghanouni, P., Butts Pauly, K.: A rapid beam simulation framework for transcranial focused ultrasound. Sci. Rep. **9**(1), 7965 (2019). https://doi.org/10.1038/s41598-019-43775-6
9. Iacopino, D.G., et al.: Preliminary experience with a transcranial magnetic resonance-guided focused ultrasound surgery system integrated with a 1.5-T MRI unit in a series of patients with essential tremor and Parkinson's disease. Neurosurg Focus. **44**(2), E7 (2018). https://doi.org/10.3171/2017.11.FOCUS17614
10. Bini, F., Pica, A., Novelli, S., Marinozzi, A., Marinozzi, F.: 3D-FEM Modeling of iso-concentration maps in single trabecula from human femur head. In: Tavares, J., Natal Jorge, R. (eds.) VipIMAGE 2019. VipIMAGE 2019, Lecture Notes in Computational Vision and Biomechanics, vol. 34, pp. 509–518. Springer, Cham (2019). https://doi.org/10.1007/978-3-030-32040-9_52
11. Bini, F., et al.: 3D FEM model to simulate Brownian motion inside trabecular tissue from human femoral head. Comput. Methods Biomech. Biomed. Eng. Imaging Vis. (2021). https://doi.org/10.1080/21681163.2021.1956370
12. Bini, F., Pica, A., Marinozzi, A., Marinozzi, F.: Prediction of stress and strain patterns from load rearrangement in human osteoarthritic femur head: finite element study with the integration of muscular forces and friction contact. In: Tavares, J., Fernandes, P. (eds.) New Developments on Computational Methods and Imaging in Biomechanics and Biomedical Engineering. Lecture Notes in Computational Vision and Biomechanics, vol. 33, pp 49–64. Springer, Cham (2019). https://doi.org/10.1007/978-3-030-23073-9_4
13. Araneo, R., Bini, F., Rinaldi, A., Notargiacomo, A., Pea, M., Celozzi, S.: Thermal-electric model for piezoelectric ZnO nanowires. Nanotechnology **26**(26), 265402 (2015). https://doi.org/10.1088/0957-4484/26/26/265402
14. Kyriakou, A., Neufeld, E., Werner, B., Székely, G., Kuster, N.: Full-wave acoustic and thermal modeling of transcranial ultrasound propagation and investigation of skull-induced aberration correction techniques: a feasibility study. J. Ther. Ultrasound. **3**, 11 (2015). https://doi.org/10.1186/s40349-015-0032-9
15. Mueller, J.K., Ai, L., Bansal, P., Legon, W.: Numerical evaluation of the skull for human neuromodulation with transcranial focused ultrasound. J. Neural Eng. **14**(6), 066012 (2017). https://doi.org/10.1088/1741-2552/aa843e
16. Rosnitskiy, P.B., Yuldashev, P.V., Sapozhnikov, O.A., Gavrilov, L.R., Khokhlova, V.A.: Simulation of nonlinear trans-skull focusing and formation of shocks in brain using a fully populated ultrasound array with aberration correction. J. Acoust. Soc. Am. **146**(3), 1786 (2019). https://doi.org/10.1121/1.5126685
17. Samoudi, M.A., Van Renterghem, T., Botteldooren, D.: Computational modeling of a single-element transcranial focused ultrasound transducer for subthalamic nucleus stimulation. J. Neural Eng. **16**(2), 026015 (2019). https://doi.org/10.1088/1741-2552/aafa38
18. Zhang, H., et al.: The effects of the structural and acoustic parameters of the skull model on transcranial focused ultrasound. Sensors (Basel). **21**(17), 5962 (2021). https://doi.org/10.3390/s21175962
19. Duck, F.A.: Physical Properties of Tissue. A Comprehensive Reference Book. Academic Press, London, UK (1990). https://doi.org/10.1016/C2009-0-02755-X

20. Huang, J., Holt, R.G., Cleveland, R.O., Roy, R.A.: Experimental validation of a tractable numerical model for focused ultrasound heating in flow-through tissue phantoms. J. Acoust. Soc. Am. **116**, 2451–2458 (2004). https://doi.org/10.1121/1.1787124
21. Haddadi, S., Ahmadian, M.T.: Numerical and experimental evaluation of high-intensity focused ultrasound-induced lesions in liver tissue ex vivo. J Ultrasound Med. **37**, 1481–1491 (2018). https://doi.org/10.1002/jum.14491
22. McDannold, N., et al.: Blood-brain barrier disruption and delivery of irinotecan in a rat model using a clinical transcranial MRI-guided focused ultrasound system. Sci. Rep. **10**, 8766 (2020). https://doi.org/10.1038/s41598-020-65617-6
23. Albin, N., Bruno, O.P., Cheung, T.Y., Cleveland, R.O.: Fourier continuation methods for high-fidelity simulation of nonlinear acoustic beams. J. Acoust. Soc. Am. **132**(4), 2371–2387 (2012). https://doi.org/10.1121/1.4742722
24. Fry, F.J., Barger, J.E.: Acoustical properties of the human skull. J. Acoust. Soc. Am. **63**(5), 1576–1590 (1978). https://doi.org/10.1121/1.381852
25. Westervelt, P.J.: Parametric acoustic array. J. Acoust. Soc. Am. **35**(4), 535–537 (1963). https://doi.org/10.1121/1.1918525
26. Pierce, A.D.: Acoustics: An Introduction to its Physical Principles and Applications. 3rd edn. Springer, Cham. Springer Nature Switzerland AG vol. 2019 (2019). https://doi.org/10.1007/978-3-030-11214-1
27. Jin, J.M.: Theory and Computation of Electromagnetic Fields. John Wiley, Hoboken (2010). https://doi.org/10.1002/9780470874257
28. Pennes, H.H.: Analysis of tissue and arterial blood temperature in the resting human forearm. J. Appl. Physiol. **1**, 93–122 (1948). https://doi.org/10.1152/jappl.1948.1.2.93
29. Gnanaskandan, A., Hsiao, C.T., Chahine, G.: Modeling of microbubble-enhanced high-intensity focused ultrasound. Ultrasound Med. Biol. **45**(7), 1743–1761 (2019). https://doi.org/10.1016/j.ultrasmedbio.2019.02.022
30. Quadri, S.A., et al.: High-intensity focused ultrasound: past, present, and future in neurosurgery. Neurosurg. Focus. **44**(2), E16 (2018). https://doi.org/10.3171/2017.11.FOCUS17610
31. Venard, J.K., Street R.L: Elementary Fluid Mechanics, 5th ed., Wiley, New York (1975)
32. Chavez, M., Sosa, V.: Speed of sound in saturated pure water. J. Acoust. Soc. Am. **77**, 420 (1985). https://doi.org/10.1121/1.391861
33. Lauterborn, W., Kurz, T., Akhatov, I.: Nonlinear Acoustics in Fluids. In: Rossing T. (eds.) Springer Handbook of Acoustics. Springer Handbooks. Springer, New York (2007). https://doi.org/10.1007/978-0-387-30425-0_8
34. Mcintosh, R.L., Anderson, V.: A comprehensive tissue properties database provided for the thermal assessment of a human at rest. Biophys. Rev. Lett. **5**, 129–151 (2010). https://doi.org/10.1142/S1793048010001184
35. Mast, T.D.: Empirical relationships between acoustic parameters in human soft tissues. Acoust. Res. Lett. Online **1**, 37–42 (2000). https://doi.org/10.1121/1.1336896
36. Van Leeuwen, G.M., Lagendijk, J.J., Van Leersum, B.J., Zwamborn, A.P., Hornsleth, S.N., Kotte, A.N.: Calculation of change in brain temperatures due to exposure to a mobile phone. Phys. Med. Biol. **44**(10), 2367–2379 (1999). https://doi.org/10.1088/0031-9155/44/10/301
37. Hasgall, P.A. et al.: IT'IS Database for Thermal and Electromagnetic Parameters of Biological Tissues, Version 4.0, 15 May 2018. https://doi.org/10.13099/VIP21000-04-0
38. Connor, C.W., Clement, G.T., Hynynen, K.: A unified model for the speed of sound in cranial bone based on genetic algorithm optimization. Phys. Med. Biol. **47**(22), 3925–3944 (2002). https://doi.org/10.1088/0031-9155/47/22/302
39. Renaud, G., Calle, S., Remenieras, J.P., Defontaine, M.: Exploration of trabecular bone non-linear elasticity using time-of-flight modulation. IEEE Trans. Ultrason. Ferroelectr. Freq. Control. **55**(7), 1497–1507 (2008). https://doi.org/10.1109/TUFFC.2008.825

40. Clarke, A.J., Evans, J.A., Truscott, J.G., Milner, R., Smith, M.A.: A phantom for quantitative ultrasound of trabecular bone. Phys. Med. Biol. **39**(10), 1677–1687 (1994). https://doi.org/10.1088/0031-9155/39/10/011
41. Petkus, V., Ragauskas, A., Jurkonis, R.: Investigation of intracranial media ultrasonic monitoring model. Ultrasonics **40**(1–8), 829–833 (2002). https://doi.org/10.1016/s0041-624x(02)00216-0
42. Voie, A., Fisher, D., Ahadi, G., Holscher, T.: Transcranial measurements of the acoustic field produced by a low frequency focused ultrasound system. AIP Conf. Proc. **1503**, 256 (2012). https://doi.org/10.1063/1.4769954
43. McDannold, N., White, P.J., Cosgrove, R.: Elementwise approach for simulating transcranial MRI-guided focused ultrasound thermal ablation. Phys Rev Res. **1**(3), 033205 (2019). https://doi.org/10.1103/physrevresearch.1.033205

Predicting Neurological Effects Associated with Traumatic Brain Injuries Using Video Analysis and Finite Element Modeling

Bianca Acot[1(✉)], Branko Glisic[1], Annegret Dettwiler[1], and Michael D. Gilchrist[2]

[1] Princeton University, Princeton, NJ 08544, USA
bacot@alumni.princeton.edu
[2] University College Dublin, Dublin D4, Ireland

Abstract. The topic of concussion has become controversial due to the recent discoveries of long-term neurodegenerative diseases in former football players related to concussions they sustained earlier in life. In order to prevent concussion, the mechanism of concussive head impacts in football must be understood. However, studying concussion is difficult because of the ethical issues related to studies involving living subjects. Finite element simulations enable researchers to study the relation between predicted injury in the brain and head impact mechanics as surrogates to live subjects. To improve the accuracy of these simulations, this paper developed a methodology to cyclically analyze and improve the process of finite element modelling of head impacts. Case studies of two college football players were analyzed through the following steps: video analysis of representative concussive hits to elucidate head impact mechanics, finite element simulation of head impact to garner biomechanical metrics in the brain, and correlation of biomechanical metrics with neuroimaging metrics. A relationship was found between these metrics while areas of improvement within the accident reconstruction and finite element simulation process were also found. Through acknowledging areas of further work, researchers can continue to develop the process of head impact simulations to eventually use this tool for diagnostic purposes.

Keywords: Concussion · Finite element modeling · Accident Reconstruction · Diffusion Tensor Imaging

1 Introduction

Concussion is an important form of injury in sports medicine that can be studied through the lenses of neuroscience, neurology, and engineering. Head injuries remain prevalent in many popular sports, but are most prevalent in contact sports like football, soccer, and hockey [1]. Due to the increase in concussions' rising societal exposure, explaining how concussions occur and the short and long-term effects of concussion are current research topics.

Studies of concussion span various areas of research using different technologies such as neuroimaging [2], video analysis of head impact [3], brain modeling [4], clinical

J. M. R. S. Tavares et al. (Eds.): CMBBE 2021, LNCVB 38, pp. 90–100, 2023.
https://doi.org/10.1007/978-3-031-10015-4_8

and neurocognitive assessment [5], and blood serum-based biomarkers [6]. Concussion research has become both interdisciplinary and multi-faceted. However, mainly due to the challenges of data collection, concussion research in live subjects is difficult to design and execute. Very few studies have been able to connect the mechanics of head impact to structural, neurocognitive, blood serum based, behavioral, and clinical/neurocognitive outcomes. Thus, there is a need for a different method of studying how concussions occur. Numerical methods, like finite element modeling (FEM) [7–9], have recently been proposed as a method of studying concussive hits without the necessity of having live subjects [4].

FEM is a computational method used by engineers to model structures and garner the stress and strain concentrations in the structure due to certain restraints and loading. FEM has been used to model the head, including the skull and brain. By simulating head impact conditions on a finite element model of the head, researchers aim to find correlation between the areas of high stress and strain with the occurrence of concussion and to identify the locations and magnitudes of stress and strain caused by head impacts without live testing.

FEM reconstructions of concussive head impacts can help to establish concussive thresholds for stress and strain in the brain [10] and be used as a future diagnostic tool. This could eventually help physicians understand the severity of traumatic brain injuries and make clinical decisions on when concussed athletes may return to play. Additionally, the results of FEM research can improve the design of sports helmets, to protect the brain during head impacts, or the rules of football, to decrease and prevent head impacts. The prevalence of head injuries and their severe neuropsychological and clinical effects make the understanding of traumatic brain injuries relevant not just to athletes, but to sports fans and the general public as well.

2 Objectives

The overall aim of this research was to create a methodology for correlating head impact reconstruction using FEM with clinical outcomes on an individual scale. The primary objectives were:

1. Use video analysis of game footage to infer (quantify) the mechanics of head impacts in subjects.
2. Create FEM simulations of the head impacts of subjects by inputting the found head impact mechanics.
3. Identify areas of high stress and strain in the FEM of the white matter as regions of interest.
4. Using diffusion tensor imaging (DTI) results of subjects, identify changes in diffusivity values in the found regions of interest to correlate clinical and FEM results. Additionally, identify a new set of regions of interest from the areas of high diffusivity change in the DTI scans to mark areas that the FEM simulation missed.

By developing this methodology, areas of improvement may be identified to refine the FEM simulation process.

3 Methodology

3.1 Accident Reconstruction

In a study titled 'Longitudinal assessment of brain structure and function in sports-related concussion' by Cubon et al., 13 participants from the Princeton University ice hockey and football varsity teams who experienced concussions were identified for whom diffusion tensor imaging (DTI) pre- and post-concussion was performed [2]. Two subjects were identified for whom game footage of the games in which subjects incurred concussions was available. Due to the lack of an obvious concussive hit, an assumption was made that the concussion was caused by an accumulation of subconcussive hits. Thus, the test case for each subject was video of their 'hardest hit' by subjective visual appearance.

The test cases were isolated and a frame-by-frame analysis was performed on the test case video clip. Kinovea version 0.8.15 was used to perform the video analysis in this study [11]. Kinovea and MATLAB code were used to measure and calculate head impact linear and rotational velocities, orientation, and location of impact in these cases. Statistical analysis was performed to measure the accuracy of velocities found. These velocities were then used as impact velocity inputs for FEM simulation reconstructions.

3.2 Finite Element Model Simulation

The University College Dublin Brain Trauma Model (UCDBTM) is a finite element model of the head in Abaqus [12] and was developed by Horgan et al. [4, 13, 14]. It is comprised of 28,286 hexahedral elements. The geometry of the UCDBTM is based on CT scans of a male cadaver head. This model has been verified by both cadaver experiments [4] and real life simulations [15]. The orientation of the head during simulation was chosen to match the head impact location found in Accident Reconstruction. An image of the orientation of head impact using the finite element model of the head and helmet is illustrated in Fig. 1.

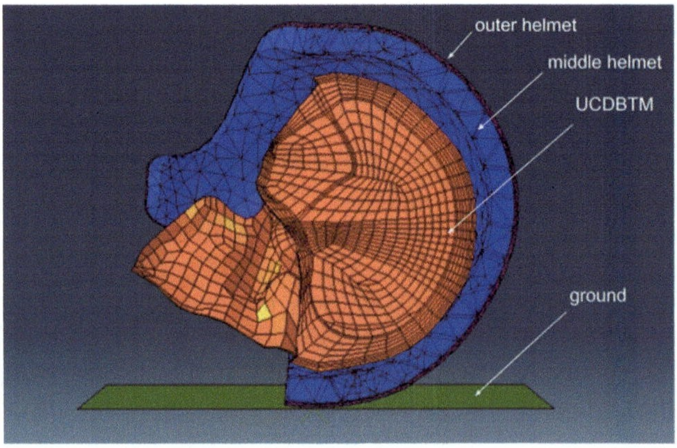

Fig. 1. Section cut of the UCDBTM and the football helmet model.

A basic helmet model was created using Rhino [16] and material and interaction properties were assigned in Abaqus [12]. The helmet is comprised of 28,305 tetrahedral elements. The goals of the helmet model used in this study were to have similar material properties to the various parts of the helmet and to have a similar geometry and mass. The head was positioned close to the location of impact within the helmet in order to simulate a close-fitting helmet. The geometry and material properties of the helmet were similar to those of a football helmet model developed by Darling et al., [17].

The chosen output variables for the simulation were maximum principal strain and von Mises stress, which have been used in previous accident reconstructions with the UCDBTM [8, 18] and used in verification of the UCDBTM [4, 13, 14]. Using Abaqus, the output variables were calculated and mapped onto the brain for each element with the color gradient correlating with the magnitude of output variable, as shown in Fig. 2.

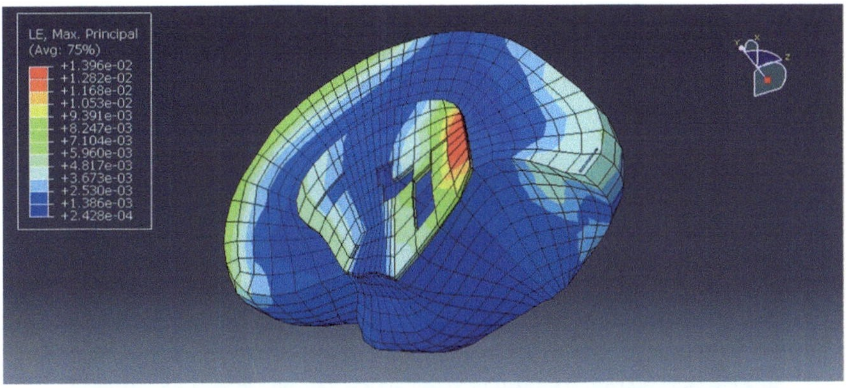

Fig. 2. Example of maximum principal strain gradation mapped on to the model of the brain.

Areas with high von Mises stress and maximum principal strain, indicating potential damage, represented regions of interest. Regions of interest were chosen through a subjective visual analysis of the white matter part of the brain. The elements chosen for the region were based on an atlas of a different finite element model of the white matter tract, provided by Post of the University of Ottawa [19].

After regions of interest were identified, the von Mises stress and maximum principal strain for each element within the region were exported. Elements with low values were removed from analysis using a threshold that was arbitrarily assigned as 25% of the maximum of the gradation scale, as only parts of some regions contained high output values and low value elements skewed the data. The average, standard deviation, and maximum were calculated for each region of interest and for the rest of the brain that was not included in regions of interest.

3.3 Comparison of Engineering and Neuroimaging Metrics of Injury

DTI measurements of subjects were acquired pre-season and within two days post-concussion [2]. A Tract Based Spatial Statistic (TBSS) analysis revealed that radial diffusivity (RD) and mean diffusivity (MD) were significantly increased at 2-days post-concussion compared to the same subjects baseline (corrected $p < 0.02$) in the study from which this data was extracted [2]. These DTI measurements were visually compared to the FEM results to establish a relationship between the two. The strength of this relationship was used to evaluate and identify areas of improvement within the FEM simulation process. To that end, the DTI measurements was visualized. FMRIB Software Library (FSL) [20] was used to create a whole brain white matter skeleton [2]. The white matter skeleton consists of 3D pixel-like boxes called voxels as shown in Fig. 3. For each voxel, the percentage change in radial diffusivity was calculated and then mapped using color gradation from red (low) to yellow (high) [2]. Within each of the regions of interest as indicated by the FEM simulation, slices of DTI images were analyzed to find clusters of high percentage change of radial diffusivity. Finding these clusters is indicative of a relationship between engineering and neuroimaging metrics.

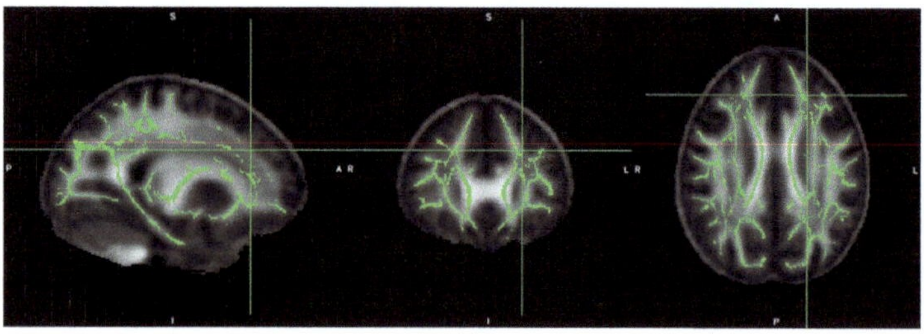

Fig. 3. White matter skeleton shown in green [2].

4 Results

4.1 Accident Reconstruction

Video of the test case from two orthogonal viewpoints were provided for each subject; the orthogonal views of the paths tracked are shown in Figs. 4 and 5. Kinematics of head impact are shown in Table 1.

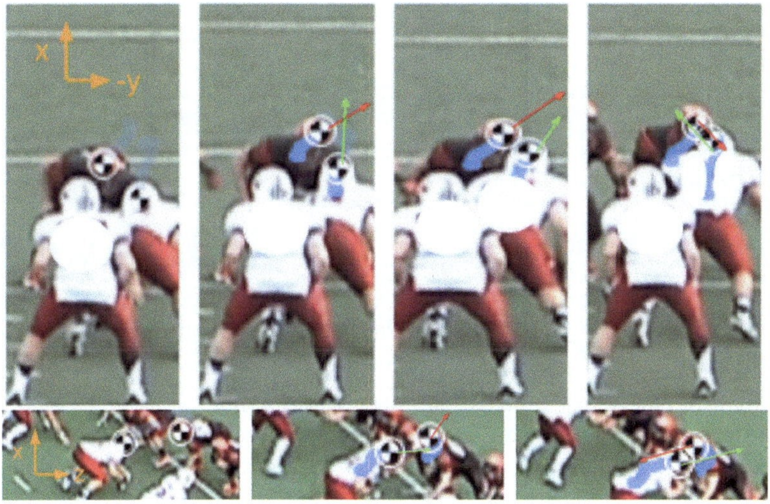

Fig. 4. Images of the test case path for Subject A from orthogonal views. The axes show the directionality. The red arrow represents the direction of movement for the subject and the green arrow represents the direction of movement for the opposing player of the helmet hit.

Fig. 5. Images of the test case path for Subject B from orthogonal views. The axes show the directionality. The red arrow represents the direction of movement for the subject and the green arrow represents the direction of movement for the opposing player of the helmet hit.

Table 1. Kinematic Inputs (linear and angular velocity components) for FEM Simulation.

	V_x (m/s)	V_y (m/s)	V_z (m/s)	w_x (rads/s)	w_y (rads/s)	w_z (rads/s)
Subject A	0.23	−1.73	−4.20	−0.81	1.13	−0.35
Subject B	1.23	3.47	−3.96	0.59	0.98	0.74

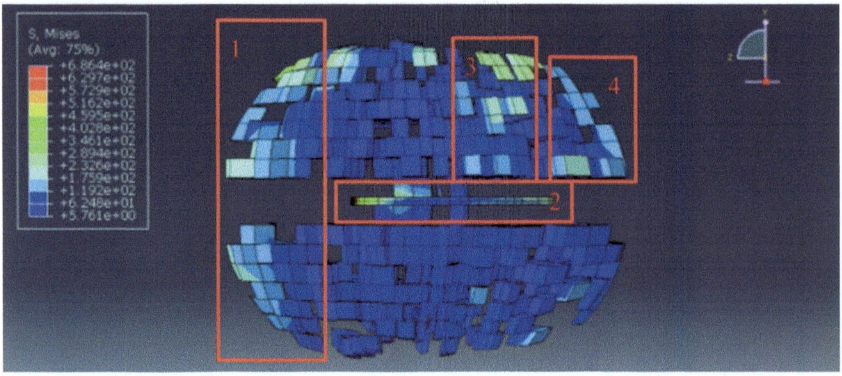

Fig. 6. Von Mises stresses and regions of interest for subject A. Box 1 is the occipital lobe region, Box 2 is the corpus callosum, Box 3 is the left motor association area, and Box 4 is the left prefrontal cortex.

4.2 Finite Element Simulation

Images of the gradated white matter tract FEMs for Subjects A and B are shown in Figs. 6 and 7. The regions of interest for Subject A were the occipital lobes, corpus callosum, left motor association area, and left prefrontal cortex. The regions of interest for Subject B were the right occipital lobe, corpus callosum, and left motor association area.

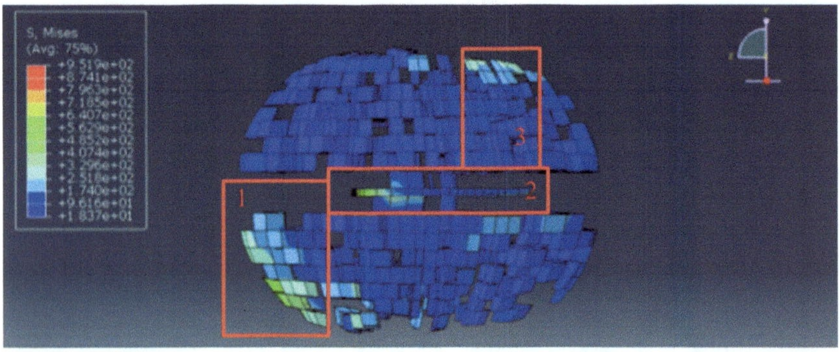

Fig. 7. Von Mises stresses and regions of interest for Subject B. Box 1 is the right occipital lobe region, Box 2 is the corpus callosum, and Box 3 is the left motor association area.

The average, standard deviation, maximum, and threshold of von Mises stress and maximum principal strain were calculated for the elements in each region and for all elements not included in the regions of interest, as shown in Tables 2 and 3.

Table 2. Maximum principal strain and von Mises stress results for subject A.

Region	Engineering metric	Average	Standard Deviation	Maximum	Threshold
Corpus Callosum	Maximum principal strain (m/m)	0.007	0.003	0.014	0.004
	von Mises stress (Pa)	366.960	155.613	686.918	191.900
Left motor association cortex	Maximum principal strain (m/m)	0.003	0.001	0.005	0.002
	von Mises stress (Pa)	175.571	59.766	255.041	77.150
Occipital lobe	Maximum principal strain (m/m)	0.002	0.001	0.006	0.002
	von Mises stress (Pa)	131.704	60.747	353.729	118.600
Left prefrontal cortex	Maximum principal strain (m/m)	0.002	0.001	0.003	0.001
	von Mises stress (Pa)	122.042	33.559	181.629	64.760
Regions of Non-Interest	Maximum principal strain (m/m)	0.001	0.001	0.006	
	von Mises stress (Pa)	60.280	43.105	312.971	

Table 3. Maximum principal strain and von Mises stress results for subject B.

Region	Engineering metric	Average	Standard deviation	Maximum	Threshold
Corpus Callosum	Maximum principal strain (m/m)	0.011	0.004	0.019	0.005
	von Mises stress (Pa)	537.912	214.714	964.678	263.100
Left motor association cortex	Maximum principal strain (m/m)	0.002	0.001	0.004	0.001
	von Mises stress (Pa)	135.833	59.017	260.559	76.430
Right occipital	Maximum principal strain (m/m)	0.006	0.002	0.012	0.003
	von Mises stress (Pa)	341.601	129.168	721.551	195.300
Regions of Non-Interest	Maximum principal strain (m/m)	0.002	0.001	0.009	
	von Mises stress (Pa)	82.853	55.412	472.013	

4.3 Comparison of Engineering and Neuroimaging Metrics of Injury

Slices of the DTI scans showing percent difference of radial diffusivity and mean diffusivity pre- and post-concussion were thresholded to show clusters of significant radial diffusivity increase in a TBSS analysis [2]. The minimum threshold for Subject A was

25% increase in radial diffusivity and the minimum threshold for Subject B was 10% increase in radial diffusivity. Regions of interest were then examined to find these clusters, which would indicate a relationship between FEM simulation results and DTI results; an example of this DTI scan analysis is shown in Fig. 8. Areas of high stress and strain are indicative of areas of injury; increase of radial diffusivity have been correlated with concussion [2]. Thus, the regions of interest as indicated by the FEM simulation results are expected to coincide with areas of high percent difference of radial diffusivity as indicated by the DTI results. Corresponding clusters were found for six of seven regions of interest overall, except for the left motor association areas in Subject A. The DTI results were also analyzed to identify regions of high diffusivity increase and the corticospinal was a region of interest found that was not identified by the FEM results.

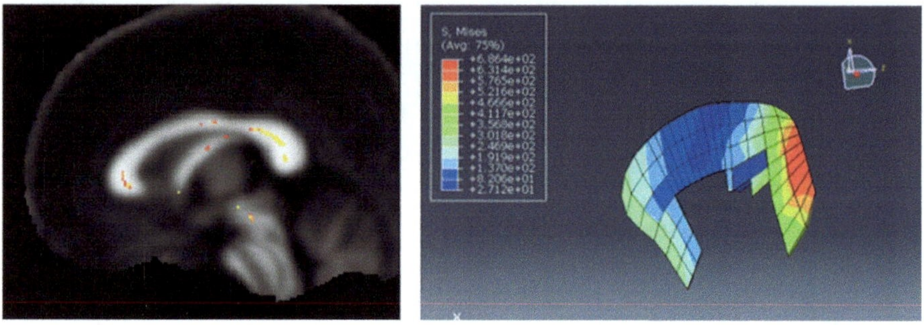

Fig. 8. Comparison of subject A DTI scan and FEM of corpus callosum

5 Conclusions and Future Work

In order to understand the mechanism by which concussion occurs, the kinematics of concussive head impact were first elucidated. The accuracy of these estimated velocities can be improved by using multiple methods to find the kinematics of head impacts, like physical experiments and multibody dynamics studies, and clarifying which hit caused the concussion. The velocities of head impact for the two test cases were input into the UCDBTM in order to estimate the von Mises stress and maximum principal strain of the brain immediately post-injury. The FEM simulations of the test cases were used to identify regions of interest where engineering metrics indicated damage. Advances in finite element modeling of the head, including developing constitutive equations describing the stress-strain relationship of the various parts of the brain, could improve the accuracy of this step. These regions of interest were investigated within the DTI scans to find indications of structural changes in the white matter microstructure in the form of clusters of large increases in radial diffusivity. Six out of seven of these regions of interest also had clusters within the regions in the DTI scans, indicating a potential relationship between biomechanical and neuroimaging metrics. Accuracy can be improved by gathering more data to provide relationships found with statistical significance.

From these case studies, a workflow was created in order to improve replicability of this methodology. This workflow is shown in Fig. 9, with the squares representing the action and the diamonds representing the inputs/outputs of the actions. The methodology consists of two general phases: Data Collection and Analysis. Data Collection describes the process to produce a FEM simulation of head impact. The Analysis sections consists of an evaluation methodology for the Data Collection phase. Through this analysis, issues in the Data Collection phase can be identified and fixed. Eventually, when the Data Collection methodology becomes acceptably accurate, the Data Collection methodology could eventually be used to reconstruct head impacts of potentially concussed individuals and help diagnose them for concussion.

Currently, little is known about the connection between the kinematics of a concussive head impact and structural changes within the brain due to concussion. Thus, this project aimed to serve as a study on the relationship between FEM simulation results of head impact and DTI results on an individual basis. The use of this methodology can improve FEM simulation methodology to the point of establishing a critical threshold for biomechanical metrics that will elicit structural changes in the white matter of the brain. Eventually, improvements in data collection and computational power can allow for head impact kinematics to be automatically input into finite element simulations of the head and identify concussive hits based on biomechanical metric thresholds. Thus, these biomechanical metrics, along with neuroimaging and clinical data, could be used by physicians as a sideline diagnostic tool. This research provides a workflow for cyclically evaluating FEM simulations and improving FEM reconstructions of head impacts to the point of concussion diagnostic use.

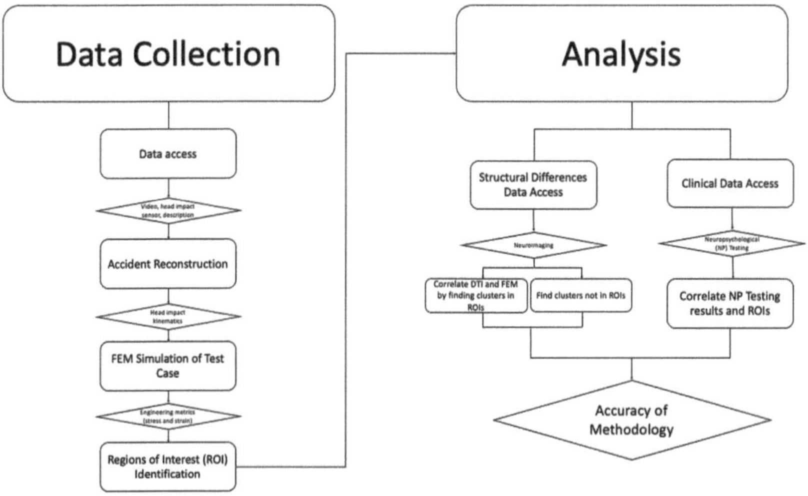

Fig. 9. Schematic of proposed workflow for analysis of FEM simulation accuracy

References

1. Delaney, J.S., Puni, V., Rouah, F.: Mechanisms of injury for concussions in university football, ice hockey, and soccer: a pilot study. Clin. J. Sport Med. **16**(2), 162–165 (2006)
2. Cubon, V.A., Murugavel, M., Holmes, K.W., Dettwiler, A.: Preliminary evidence from a prospective DTI study suggests a posterior-to-anterior pattern of recovery in college athletes with sports-related concussion. Brain Behav. **8**(12), e01165 (2018). https://doi.org/10.1002/brb3.1165
3. Post, A., et al.: Analysis of speed accuracy using video analysis software. Sports Eng. **21**(3), 235–241 (2018). https://doi.org/10.1007/s12283-018-0263-4
4. Horgan, T.J., Gilchrist, M.D.: The creation of three-dimensional finite element models for simulating head impact biomechanics. Int. J. Crashworthiness **8**(4), 353–366 (2003). https://doi.org/10.1533/ijcr.2003.0243
5. Dettwiler, A., et al.: Persistent differences in patterns of brain activation after sports-related concussion: a longitudinal functional magnetic resonance imaging study. J. Neurotrauma **31**(2), 180–188 (2013). https://doi.org/10.1089/neu.2013.2983
6. Meier, T.B., et al.: Prospective study of the effects of sport-related concussion on serum kynurenine pathway metabolites. Brain Behav. Immun. **87**, 715–724 (2020). https://doi.org/10.1016/j.bbi.2020.03.002
7. Post, A., et al.: The biomechanics of concussion for ice hockey head impact events. Comput. Methods Biomech. Biomed. Engin. **22**(6), 631–643 (2019). https://doi.org/10.1080/10255842.2019.1577827
8. Clark, J.: In-depth Analysis and Reconstruction of Equestrian Jockey Accidents and their Helmets, PhD thesis, University College Dublin (2019)
9. Acot, B.: Predicting neurological effects associated with traumatic brain injuries using video analysis and finite element modeling (2021). http://arks.princeton.edu/ark:/88435/dsp019g54xm76j
10. Clark, J., et al.: Proposed injury thresholds for concussion in equestrian sports. J. Sci. Med. Sport **23**(3), 222–236 (2020). https://doi.org/10.1016/j.sams.2019.10.006
11. Kinovea. https://www.kinovea.org/. Accessed 02 Oct 2020
12. Abaqus Unified FEA - SIMULIA^{TM} by Dassault Systèmes®. https://www.3ds.com/products-services/simulia/products/abaqus/. Accessed 05 Apr 2021
13. Horgan, T.J., Gilchrist, M.D.: Influence of FE model variability in predicting brain motion and intracranial pressure changes in head impact simulations. Int. J. Crashworthiness **9**(4), 401–418 (2004). https://doi.org/10.1533/ijcr.2004.0299
14. Trotta, A., et al.: Biofidelic finite element modelling of brain trauma: Importance of the scalp in simulating head impact. Int. J. Mech. Sci. **173**, 105448 (May 2020). https://doi.org/10.1016/j.ijmecsci.2020.105448
15. Doorly, M.C., Gilchrist, M.D.: The use of accident reconstruction for the analysis of traumatic brain injury due to head impacts arising from falls. Comput. Methods Biomech. Biomed. Eng. **9**(6), 371–377 (2006). https://doi.org/10.1080/10255840601003551
16. Rhino - Rhinoceros 3D. https://www.rhino3d.com/. Accessed 05 Apr 2021
17. Darling, T., Muthuswamy, J., Rajan, S.D.: Finite element modeling of human brain response to football helmet impacts. Comput. Methods Biomech. Biomed. Engin. **19**(13), 1432–1442 (2016). https://doi.org/10.1080/10255842.2016.1149574
18. Borland, R.: Computer Simulation of Protected and Unprotected Sports Related Head Impacts, ME in Biomedical Engineering, University College Dublin, Dublin, Ireland (2020)
19. Post, A., et al.: A preliminary examination of the relationship between biomechanical measures and structural changes in the brain. Trauma **23**(1), 24–32 (2021). https://doi.org/10.1177/1460408620916578
20. FSL - FslWiki. https://fsl.fmrib.ox.ac.uk/fsl/fslwiki. Accessed 07 Apr 2021

Dense-Discrete Phase Simulations of Blood Flow in a Stenotic Coronary

Violeta Carvalho[1,2(✉)], Nelson Rodrigues[1,2], José C. Teixeira[1], Rui Lima[1,3], and Senhorinha Teixeira[2]

[1] MEtRICs, Minho University, Guimarães, Portugal
violeta.carvalho@dem.uminho.pt
[2] Algoritmi, Minho University, Guimarães, Portugal
[3] CEFT, Faculty of Engineering of the University of Porto, Porto, Portugal

Abstract. Atherosclerotic cardiovascular disease is a silent and common pathology that affects millions of people around the world. Over time, lipids are deposited in the arterial walls reducing blood flow and as a result can lead to several dangerous and life-threatening cardiovascular issues including myocardial infarction. Due to the high mortality caused by this disease, over the years, great progress has been achieved in understanding the blood fow phenomena through extensive experimental and numerical research. Although in vivo and in vitro experiments have played an important role in the validity of new treatment techniques, due to the great progress that has been made in computational power, numerical methods have been frequently applied in atherosclerosis' research. Typically, blood flow modeling is performed considering the blood as a homogeneous Newtonian or non-Newtonian fluid. However, blood is a complex fluid that contains a mixture of plasma, red blood cells (RBCs), white blood cells, and platelets. Accordingly, in the present study, the dense-discrete phase model (DDPM), which consists of a hybrid Eulerian-Lagrangian method, is used to simulate the presence of RBCs in blood when flowing through a stenotic coronary arteries. Moreover, the same simulation was performed resorting to the mixture model in order to verify if the differences are considerable. In general, it was found that the generation of recirculation zones is intensified in the DDPM model when compared to the mixture model, but in terms of velocity no significant differences were found.

Keywords: Blood flow · Hemodynamics · Multiphase modelling · DDPM

1 Introduction

Atherosclerosis is a gradual and complex disease characterized by the accumulation of fibrofatty deposits in the intima of arteries, and it is the pathological basis of several cardiovascular diseases, the largest cause of mortality and morbidity worldwide [1–3]. Nevertheless, with the appearance of COVID-19 disease, the diagnosis and prevention of various diseases have been, unwittingly, left behind, and atherosclerosis is one of

R. Lima and S. Teixeira—Shared senior authorship.

J. M. R. S. Tavares et al. (Eds.): CMBBE 2021, LNCVB 38, pp. 101–111, 2023.
https://doi.org/10.1007/978-3-031-10015-4_9

them. Superimposed to this, recent studies have proven that patients with COVID-19 are at high risk for thrombotic occlusions and also that pre-existing atherosclerosis has an increased tendency for worse outcomes from COVID-19 [4–7]. For all these reasons, the study of atherosclerosis is of utmost importance either experimentally and numerically.

With the great advances in computational power, modeling of blood flow using numerical simulations has recently become more popular and wider available [8, 9]. Usually, hemodynamic studies adopt simplified idealistic blood rheology, namely assuming blood as a single-phase fluid [10–13]. Although this assumption has shown to be adequate and relevant results have been obtained allowing a reliable simulation, in reality, blood is a two-phase fluid. The first phase is the plasma, and the second phase is composed of the blood cells, red blood cells, white blood cells, and platelets [14, 15]. Taking this into consideration, some researchers have investigated and applied mainly Eulerian multiphase models in hemodynamic studies [8, 16–20], in which both phases are treated as continuous fluids, due to the inability of the discrete phase model (DPM) to simulate a volume of particles above 10% [20, 21]. In DPM model, it is assumed that the volume fraction of the discrete phase is sufficiently low and it is not taken into account when assembling the continuous phase equations. Nevertheless, recent improvements were made to the discrete phase model, and the dense discrete phase model (DDPM) emerged. This model, like the discrete phase model, is a combination of the Eulerian and Lagrange approaches in which the primary phase is treated as a continuum by solving the Navier-Stokes equations, while the secondary discrete dispersed-phase is solved by using Newton's second law of motion [22, 23]. However, the DDPM model allows simulating flows with a volume fraction of solids higher than 10% [24], which makes it interesting to simulate blood flow, due to the high volume fraction of RBCs (45%–47%) [25, 26]. Hence, by using this approach, the disadvantages associated with the DPM and the Eulerian models can be overcome.

Motivated by the progress made in the improvement and development of new multiphase models, in the present work, a hemodynamic study in a stenotic coronary artery is presented by using the DDPM multiphase model and a comparison to the mixture model is made. To the best of the authors' knowledge, the DDPM model has not been yet applied in hemodynamic studies. It should be noted that since DDPM simulations require great computational power, simulations were performed in a 2D model.

2 Materials and Methods

The geometry is the first step to be accomplished in numerical simulations, followed by mesh development. In the present work, a 2D idealized coronary artery was generated considering a diameter reduction of 50%. The respective dimensions and mesh are depicted in Fig. 1. Note that the selected dimensions are in accordance with previous studies [10, 27].

Regarding the mesh, to the present model, a hexahedral mesh was created with 57547 elements and 56492 nodes. The quality of the mesh was assured by evaluating the skewness and orthogonal quality, which corresponded to an average value of 5.37 $\times 10^{-2}$ and 0.99, proving that the quality of the mesh is adequate [21].

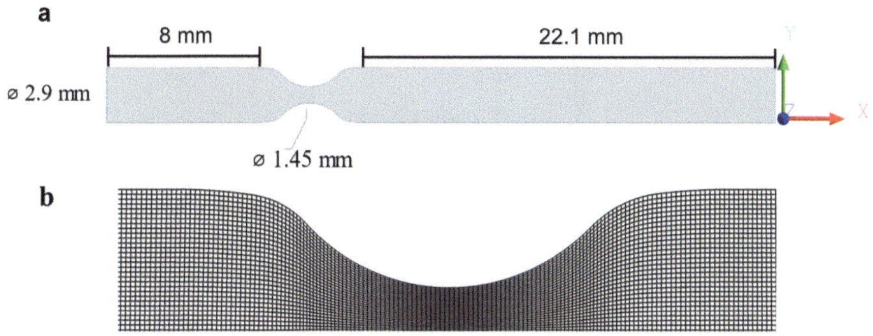

Fig. 1. a) Geometry of the artery model and b) the respective mesh in the stenotic zone.

2.1 Numerical Approach

Numerical simulations were performed in ANSYS® software and in the following are presented the models and boundary conditions assumed in the present study.

Dense Discrete Phase Model (DDPM)

When using the DDPM multiphase model, the mass and momentum conservation equations for the fluid phase q are given by Eqs. 1 and 2, respectively:

$$\frac{\partial}{\partial t}(\alpha_q \rho_q) + \nabla \cdot (\alpha_q \rho_q \vec{v}_q) = 0 \tag{1}$$

$$\frac{\partial}{\partial t}(\alpha_q \rho_q \vec{v}_q) + \nabla \cdot (\alpha_q \rho_q \vec{v}_q \vec{v}_q) = -\alpha_q \nabla \mathrm{p} + \nabla \cdot \overline{\overline{\tau}}_q + \alpha_q \rho_q + \vec{K}_{pq}(\vec{v}_p - \vec{v}_q) \tag{2}$$

where α_q is the volume fraction, $\vec{v}_q$ is the velocity vector, ρ_q is the fluid density, p is the static pressure, $\overline{\overline{\tau}}_q$ is the stress tensor, $\vec{K}_{pq}$ is the momentum exchange coefficient between fluid q and solid phase p, $\vec{v}_p$ is the solid phase velocity vector. $\vec{v}_{DPM}$ represents the particle averaged velocity of the considered discrete phase and K_{DPM} represents its particle averaged interphase momentum exchange coefficient [25].

As previously mentioned, Newton's second law of motion is used to track the path of each particle as follows in Eq. 3:

$$\frac{\partial v_p}{\partial t} = \mathrm{F}_D(\vec{v}_q - \vec{v}_p) + \frac{(\rho_q - \rho_p)}{\rho_p} + F_x \tag{3}$$

where F_x is the additional acceleration (force/unit particle mass) term, and $\mathrm{F}_D(\vec{v}_q - \vec{v}_p)$ is the acceleration due to the drag force. F_D is given by Eq. 4,

$$\mathrm{F}_D = \frac{3}{4} C_d Re_p \frac{\mu_q}{\rho_p d_p^2} \tag{4}$$

where μ_q is the fluid viscosity, d_p is the particle diameter, C_d is the drag coefficient, and Re_p is the particle Reynolds number, which can be expressed as follows in Eq. 5:

$$Re_p = \frac{\rho_p d_p (v_p - v_q)}{\mu_q} \tag{5}$$

The number of particles (n_p) in each parcel is calculated by Eq. 6,

$$n_p = \frac{\dot{m}_{parcel} \Delta t}{m_p} \tag{6}$$

where $\dot{m}_{parcel}$ is the mass flow rate and m_p is the mass of the particle.

Mixture Model

The mixture model is an Euler-Euler multiphase model that has been used by several authors for modelling blood flow [18, 28, 29]. Both phases are considered continuous, the interaction between different phases is considered, and the viscosity of each phase is a function of shear rate and volume fraction. The continuity and momentum equations are given by Eqs. 7 and 8,

$$\frac{\partial}{\partial t}(\rho_m) + \nabla \cdot \left(\rho_m \vec{v}_m\right) = 0 \tag{7}$$

$$\frac{\partial}{\partial t}(\rho_m) + \nabla \cdot (\rho_m \vec{v}_m \vec{v}_m) = -\nabla p + \nabla \cdot \left[a_k \rho_k \left(\nabla \vec{v}_m + \nabla \vec{v}_m^T\right)\right] + \vec{F} + \nabla \cdot \left(\sum_{k=1}^{n} a_k \rho_k \vec{v}_{dr,k} \vec{v}_{dr,k}\right) \tag{8}$$

In the continuity equation n, a_k, $\vec{v}_m$, ρ_m and ρ_k, are the number of phases, volume fraction of phase k, mass-averaged velocity, mixture density, and the density of phase k, respectively. In the momentum equation $\vec{F}$, μ_m and $\vec{v}_{dr,k}$ correspond to the body force, mixture viscosity, and the drift velocity for secondary phase k, respectively.

The drift velocity for secondary phase k ($\vec{v}_{dr,k}$) is calculated through Eq. 9:

$$\vec{v}_{dr,k} = \vec{v}_k - \vec{v}_m \tag{9}$$

The relative velocity is defined as the velocity of a secondary phase (p) relative to the velocity of the primary phase (q). Furthermore, in the present work, the Gidaspow model [30] was used to simulate the drag force between RBCs and plasma, since it has been commonly applied in hemodynamic studies [8, 19].

Boundary Conditions and Physical Properties of the Fluid and Particles

In the present study, transient simulations were performed considering a velocity inlet of 0.18 m/s, and assuming a laminar regime. This input value was selected since it represents the average velocity in coronary arteries [10, 20]. At the outlet, a zero gauge pressure condition was applied, and the artery walls were considered rigid with a no-slip boundary condition.

The physical properties set for RBCs for the multiphase model were based on published data [14, 31]. Accordingly, a density of 1080 kg/m^3 was considered and their viscosity was simulated resorting to a non-Newtonian model. In this case, the Carreau model was selected as used in previous works [19, 20]. In this model, the viscosity (μ) is given by Eq. 10:

$$\mu_p = \mu_\infty + (\mu_0 - \mu_\infty)\left[1 + \lambda\gamma^2\right]^{\frac{n-1}{2}} \quad (10)$$

where, $\mu_\infty = 0.00345$ Pa · s is the infinite shear viscosity,$\mu_0 = 0.0560$ Pa · s is the blood viscosity at zero shear rate, $\dot{\gamma}$ is the instantaneous shear rate, $\lambda = 3.313$ s is the time constant and n = 0.3568 is the power-law index.

For both DDPM and mixture simulations, an 8 μm diameter, and a volume fraction equal to 47% were considered for RBCs, and the plasma was assumed to be a Newtonian fluid with a density and viscosity equal to 1000 kg/m^3 and 0.0012 Pa · s, respectively [14, 31].

Solver

The commercial software ANSYS FLUENT was used to solve the governing equations for the conservation of mass and momentum. In this software, the domain is divided into discrete control volumes by using a computational mesh, previously presented, and the governing equations are integrated on each control volume to construct algebraic equations for the discrete dependent variables. Then, the discretized equations are linearized and the solution of the resulting linear system to return updated values of the dependent variables is obtained [21]. In this study, the Coupled algorithm which uses the pressure-based coupled solver was applied [21]. A constant time step of 1e-5 s and a residual target of 10^{-3} were considered, and a maximum of 200 iterations per time-step was set. These conditions were applied for both multiphase models, DDPM and mixture.

3 Results and Discussion

3.1 RBCs Velocity and Plasma Velocity Streamlines

Figure 2 shows the velocity of the RBCs (A), the velocity streamlines of the plasma (B) predicted by the DDPM model, and the velocity streamlines computed by using the mixture model, for different instants (0.02, 0.07, 0.1, and 0.2 s). Note that in the mixture model, it can only be evaluated the velocity of the mixture and not the separate phases.

Looking at the results it can be seen that the velocities predicted by both models are similar, with higher values in the stenosis throat and recirculation zones in the post-stenotic area. These observations are expected due to the sudden reduction of the vessel diameter, a phenomenon that has been observed and proven in several experimental and numerical studies [11, 32, 33]. Additionally, one can see that the plasma and mixture streamlines are similar for both multiphase models, but in the DDPM model, there are more small vortices while in the multiphase model the recirculation zones tend to be more regular.

From these results, it can be observed that the presence of particles significantly affects the blood flow, creating more disturbed and irregular recirculation zones, a situation that is more close to reality. Moreover, the consideration of two different phases showed to have a great impact on the hemodynamic behavior since previous steady-state studies in similar geometries do not present recirculation zones when the diameter reduces 50% [11].

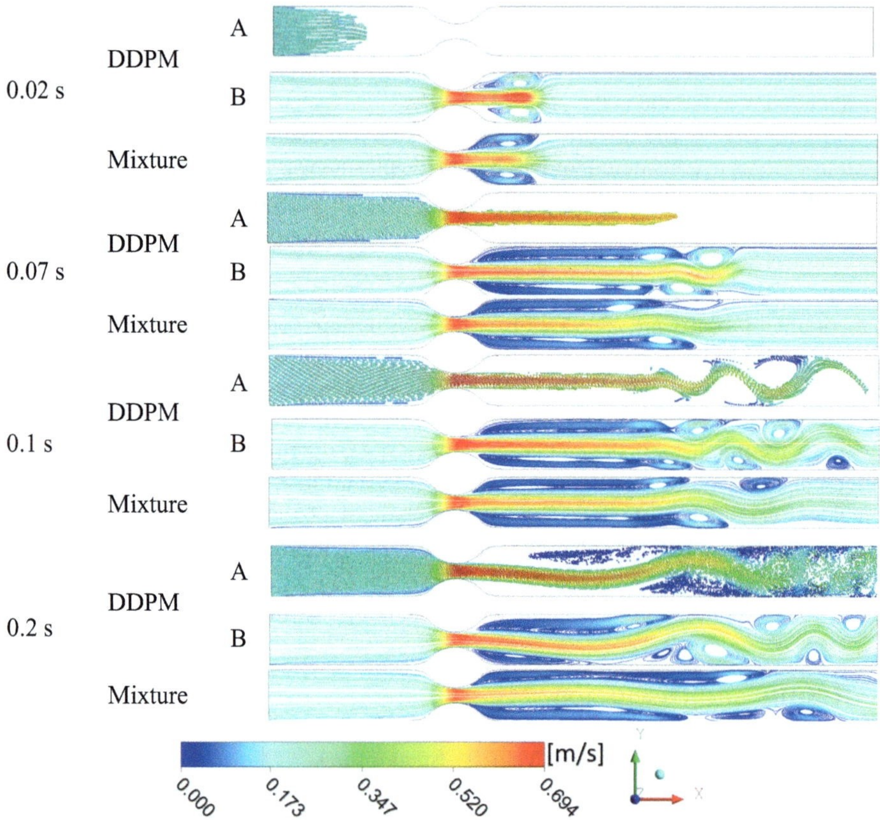

Fig. 2. Velocity of RBCs, plasma and mixture, for different instants in the two multiphase models.

3.2 Pressure Contours

In Fig. 3, the pressure contours observed in the different time instants for the two multiphase models are represented.

In general, a sudden pressure drop is observed in all cases in the post stenotic section. This phenomenon has already been observed in several studies [34, 35] and is due to the reduction of the artery diameter. The greater the reduction in diameter, the greater the pressure drop. Moreover, the results obtained for the mixture model are very similar over time, with only slight variations, but the observed values of pressure are slightly higher than in the DDPM model.

Regarding the DDPM model, when t = 0.02 s the pressure drop is similar to the mixture model because the particles have not yet reached stenosis. From t = 0.07 s, the results are quite distinct. The pressure variations observed are due to the presence of the particles. This can be understood by looking at the RBCs velocities previously represented in Fig. 2.

Overall, the pressure drop observed is in agreement with a recent study [35], however, the predicted values are considerably lower. This can be explained by the fact that in the present study a constant inlet velocity was considered, while in the previous study a pulsatile condition was considered.

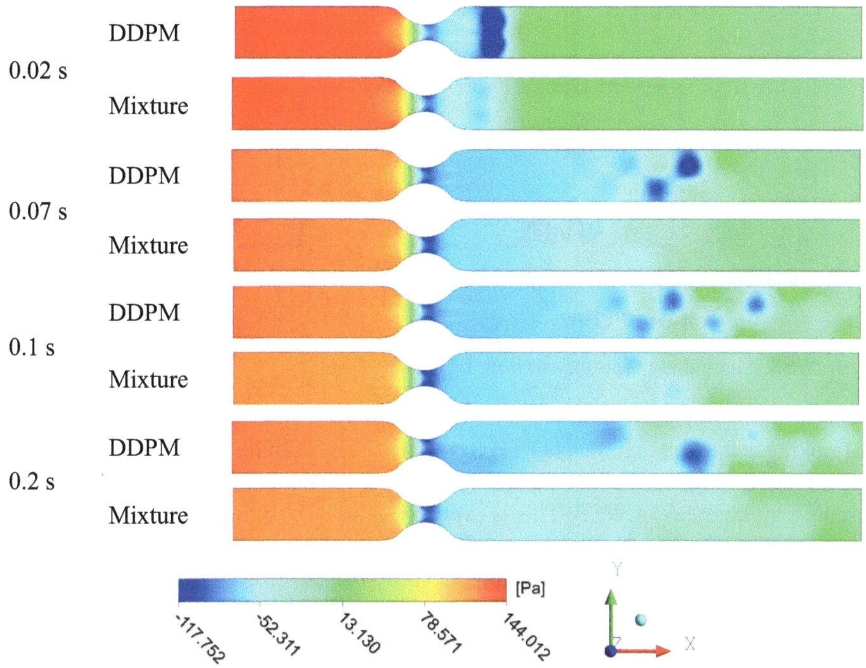

Fig. 3. Pressure contours for DDPM and mixture model at different time instants.

3.3 Wall Shear Stress Evaluation

The wall shear stress is an important biomechanical factor that is directly related to the formation of atherosclerosis, and as a result, is a common variable addressed in hemodynamic studies [36]. Low wall shear stresses are usually associated with early plaque growth, while high WSS values are observed in the stenotic areas [36]. Figure 4 represents the WSS measured along the artery model, for the mixture and DDPM model for different time instants.

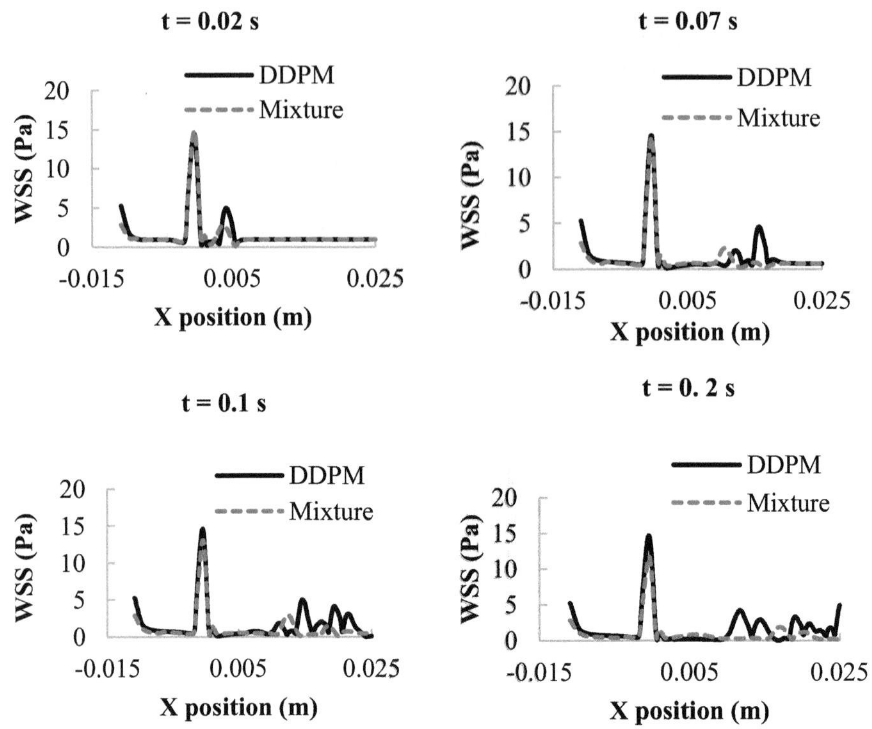

Fig. 4. WSS values along the artery wall for DDPM and mixture model at different time instants.

Looking at the first Figure, when t = 0.02 s, the plots are quite similar for both models. Only a small difference is observed after the stenotic section, in which the DDPM model presents a higher peak of WSS. This is explained by the slightly higher velocities in the recirculation zone presented in this model. When t = 0.07 s the differences are more notorious with an extra peak of WSS predicted by the DDPM model. This is due to the additional vortices observed in Fig. 2 at this instant. When t = 0.1 s and t = 0.02 s more peaks of WSS can be noticed, which appears with the recirculation zones that extend until the end of the artery, as can be observed in Fig. 2.

Despite the previous differences pointed out, in general, the previous results show that the WSS peak predicted by both models is quite similar, with a value near 15 Pa which appears in the stenotic section, as expected due to the diameter reduction, and in the post-stenotic section the WSS reduces immediately, and only small peaks of WSS are observed due to the vortices created [11].

4 Conclusions

In the present research, a hemodynamic study was performed with two different multiphase models where it was taken into account the presence of RBCs within the blood flow.

The obtained results show that in terms of recirculation zones, the DDPM model predicts more vortices than the mixture model, and significant differences were observed between the computed pressure distribution. The mixture model allows to obtain a simplification of what occurs when a discrete solid phase is considered, while the DDPM model allows to simulate the flow of RBCs closer to reality due to the ability to track the particles. However, this model requires high computational power and time, even considering a 2D geometry. This is due to the need of implementing a very small time-step and a high number of iterations per time-step to perform the simulation with high precision. Hence, the selection of the multiphase model should be done carefully considering the pros and cons of each model.

Acknowledgments. This work was supported by FCT—Fundação para a Ciência e Tecnologia within the R&D Units Project Scope: UIDB/00319/2020, UIDB/04077/2020. Violeta Carvalho acknowledges the PhD scholarship UI/BD/151028/2021 attributed by FCT.

References

1. Wang, T.: Pathogenesis of atherosclerosis. Diagn. Histopathol. **23**, 473–478 (2017). https://doi.org/10.1016/j.mpdhp.2017.11.009
2. World Health Organization (WHO). https://www.who.int/news-room/fact-sheets/detail/cardiovascular-diseases-(cvds)
3. Carvalho, V., et al.: In vitro stenotic arteries to perform blood analogues flow visualizations and measurements: a review. Open Biomed. Eng. J. **14**, 87–102 (2020). https://doi.org/10.2174/1874120702014010087
4. Parini, P., et al.: Taking action: European atherosclerosis society targets the United Nations sustainable development goals 2030 agenda to fight atherosclerotic cardiovascular disease in Europe. Atherosclerosis (2021). https://doi.org/10.1016/j.atherosclerosis.2021.02.007
5. Bae, S.A., Kim, S.R., Kim, M.N., Shim, W.J., Park, S.M.: Impact of cardiovascular disease and risk factors on fatal outcomes in patients with COVID-19 according to age: a systematic review and meta-analysis. Heart. 373–380 (2020). https://doi.org/10.1136/heartjnl-2020-317901
6. Zuo, Y., et al.: Prothrombotic autoantibodies in serum from patients hospitalized with COVID-19. Sci. Transl. Med. **3876**, 1–17 (2020). https://doi.org/10.1126/scitranslmed.abd3876
7. Carvalho, V., Pinho, D., Lima, R.A., Teixeira, J.C., Teixeira, S.: Blood flow modeling in coronary arteries : a review. Fluids **6**, 53 (2021). https://doi.org/10.3390/fluids6020053
8. Ostrowski, Z., Melka, B., Adamczyk, W., Rojczyk, M., Golda, A., Nowak, A.J.: CFD analysis of multiphase blood flow within aorta and its thoracic branches of patient with coarctation of aorta using multiphase Euler - Euler approach. J. Phys: Conf. Ser. **745**, 32112 (2016). https://doi.org/10.1088/1742-6596/745/3/032112
9. Pandey, R., Kumar, M., Majdoubi, J., Rahimi-Gorji, M., Srivastav, V.K.: A review study on blood in human coronary artery: numerical approach. Comput. Methods Programs Biomed. **187**, 105243 (2020). https://doi.org/10.1016/j.cmpb.2019.105243
10. Carvalho, V., Rodrigues, N., Lima, R.A., Teixeira, S.F.C.F.: Modeling blood pulsatile turbulent flow in stenotic coronary arteries. Int. J. Biol. Biomed. Eng. **14**, 1998–4510 (2020). https://doi.org/10.46300/91011.2020.14.22
11. Carvalho, V., Rodrigues, N., Ribeiro, R., Costa, P.F., Lima, R.A., Teixeira, S.F.C.F.: 3D Printed biomodels for flow visualization in stenotic vessels : an experimental and numerical study. Micromachines **11**, 549 (2020). https://doi.org/10.3390/mi11060549

12. Pandey, R., Kumar, M., Srivastav, V.K.: Numerical computation of blood hemodynamic through constricted human left coronary artery: Pulsatile simulations. Comput. Methods Programs Biomed. **197**, 105661 (2020). https://doi.org/10.1016/j.cmpb.2020.105661
13. Karimi, A., Razaghi, R., Koyama, M.: A patient-specific numerical modeling of the spontaneous coronary artery dissection in relation to atherosclerosis. Comput. Methods Program. Biomed. **182**, 105060 (2019). https://doi.org/10.1016/j.cmpb.2019.105060
14. Rubenstein, D.A., Yin, W., Frame, M.: Biofluid Mechanics. In: Vitro Biofluid Mechanics PublisherName, pp. 495–508, Elsevier, Amsterdam (2015)
15. Formaggia, L., Perktold, K., Quarteroni, A.: Cardiovascular mathematics- modeling and simulation of the circulatory system (2009). https://doi.org/10.1017/CBO9781107415324.004
16. Jung, J., Hassanein, A., Lyczkowski, R.W.: Hemodynamic computation using multiphase flow dynamics in a right coronary artery. Ann. Biomed. Eng. **34**, 393–407 (2006). https://doi.org/10.1007/s10439-005-9017-0
17. Wen, J., Liu, K., Khoshmanesh, K., Jiang, W., Zheng, T.: Numerical investigation of haemodynamics in a helical-type artery bypass graft using non- newtonian multiphase model. Comput. Methods Biomech. Biomed. Engin. **18**, 760–768 (2015). https://doi.org/10.1080/10255842.2013.845880
18. Buradi, A., Mahalingam, A.: Effect of stenosis severity on wall shear stress based hemodynamic descriptors using multiphase mixture theory. J. Appl. Fluid Mech. **11**, 1497–1509 (2018). https://doi.org/10.29252/jafm.11.06.29062
19. Buradi, A., Mahalingam, A.: Numerical simulation of pulsatile blood flow in an idealized curved section of a human coronary. Int. J. Mech. Prod. Eng. **2016**, 15–19 (2016)
20. Carvalho, V., et al.: Hemodynamic study in 3D printed stenotic coronary artery models: experimental validation and transient simulation. Computer Methods in Biomechanics and Biomedical Engineering. 1–14 (2020). https://doi.org/10.1080/10255842.2020.1842377
21. Ansys, I.: ANSYS® Fluent Theory Guide, Release 2020 R2 (2020)
22. Muhammad, A., Zhang, N., Wang, W.: CFD simulations of a full-loop CFB reactor using coarse-grained eulerian-lagrangian dense discrete phase model: effects of modeling parameters. Powder Technol. **354**, 615–629 (2019). https://doi.org/10.1016/j.powtec.2019.06.016
23. Cristea, E.D., Conti, P.: Hybrid eulerian multiphase-dense discrete phase model approach for numerical simulation of dense particle-laden turbulent flows within vertical multi-stage cyclone heat exchanger. Am. Soc. Mech. Eng. Fluids Eng. Div. (Publication) FEDSM. **2**, 1–15 (2018). https://doi.org/10.1115/FEDSM2018-83058
24. Ariyaratne, W.K.H., Manjula, E.V.P.J., Ratnayake, C., Melaaen, M.C.: CFD approaches for modeling gas-solids multiphase flows - a review. In: Proceedings of The 9th EUROSIM Congress on Modelling and Simulation, EUROSIM 2016, The 57th SIMS Conference on Simulation and Modelling SIMS 2016, vol. 142, pp. 680–686 (2018). https://doi.org/10.3384/ecp17142680
25. Ansys, I.: ANSYS® Fluent User's Guide, Release 2020 R2 (2020)
26. Ostadfar, A.: Biofluid mechanics: principles and applications. Elsevier (2016). https://doi.org/10.1016/C2014-0-01583-3
27. Carvalho, V., Rodrigues, N., Lima, R.A., Teixeira, S.: Numerical simulation of blood pulsatile flow in stenotic coronary arteries: the effect of turbulence modeling and non-Newtonian assumptions. In: International Conference on Applied Mathematics & Computer Science, pp. 112–116. IEEE, Athens (2020). https://doi.org/10.1109/CSCC49995.2020.00027
28. Wu, W.T., Li, Y., Aubry, N., Massoudi, M., Antaki, J.F.: Numerical simulation of red blood cell-induced platelet transport in saccular aneurysms. Appl. Sci. **7**, 484 (2017). https://doi.org/10.3390/app7050484

29. Buradi, A., Morab, S., Mahalingam, A.: Effect of stenosis severity on shear-induced diffusion of red blood cells in coronary arteries. J. Mech. Med. Biol. **19**, 1950034 (2019). https://doi.org/10.1142/S0219519419500349
30. Gidaspow, D.: Multiphase Flow and Fluidization: Continuum and Kinetic Theory Descriptions. Academic Press, Cambridge (1994). https://doi.org/10.1016/C2009-0-21244-X
31. Caro, C.G., Pedley, T.J., Schroter, R.C., Seed, W.A., Parker, K.H.: The Mechanics of the Circulation. Cambridge University Press, Cambridge (2012). https://doi.org/10.1017/CBO9781139013406
32. Doutel, E., Carneiro, J., Oliveira, M.: Fabrication of 3D milis-scale channels for hemodynamic studies. J. Mech. Med. Biol. **12**, 367–374 (2015)
33. Huh, H.K., Ha, H., Lee, S.J.: Effect of non-newtonian viscosity on the fluid-dynamic characteristics in stenotic vessels. Exp. Fluids **56**(8), 1–12 (2015). https://doi.org/10.1007/s00348-015-2037-0
34. Duncker, D.J., Koller, A., Merkus, D., Canty, J.M.: Regulation of coronary blood flow in health and ischemic heart disease. Prog. Cardiovasc. Dis. **57**, 409–422 (2015). https://doi.org/10.1016/j.pcad.2014.12.002
35. Freidoonimehr, N., Chin, R., Zander, A., Arjomandi, M.: An experimental model for pressure drop evaluation in a stenosed coronary artery. Phys. Fluids **32** (2020). https://doi.org/10.1063/1.5139701
36. Gijsen, F., Katagiri, Y.: Expert recommendations on the assessment of wall shear stress in human coronary arteries: existing methodologies, technical considerations, and clinical applications. Eur. Heart J. **40**, 3421–3433 (2019). https://doi.org/10.1093/eurheartj/ehz551

Predicting the Efficacy of Stalk Cells Following Leading Cells Through a Micro-Channel Using Morphoelasticity and a Cell Shape Evolution Model

Q. Peng[1,2,3(✉)], F.J. Vermolen[3,2], and D. Weihs[4]

[1] Mathematical Institute, Leiden University, Niels Bohrweg 1, 2333 CA Leiden, The Netherlands
q.peng@math.leidenuniv.nl
[2] Delft Institute of Applied Mathematics, Delft University of Technology, Mekelweg 4, 2628 CD Delft, The Netherlands
[3] Computational Mathematics Group, Department of Mathematics and Statistics, Faculty of Science, Hasselt University, Campus Diepenbeek, Agoralaan Gebouw D, 3590 Diepenbeek, Belgium
fred.vermolen@uhasselt.be
[4] Faculty of Biomedical Engineering, Technion-Israel Institute of Technology, 3200003 Haifa, Israel
daphnew@technion.ac.il

Abstract. Cancer cell migration between different body parts is the driving force behind cancer metastasis, which causes mortality of patients. Migration of cancer cells often proceeds by penetration through narrow cavities in possibly stiff tissues. In our previous work [12], a model for the evolution of cell geometry is developed, and in the current study we use this model to investigate whether followers among (cancer) cells benefit from leading (cancer) cells during transmigration through micro-channels and cavities. Using Wilcoxon's signed-rank text on the data collected from Monte Carlo simulations, we conclude that the transmigration time for the stalk cell is significantly smaller than for the leading cell with a p-value less than 0.0001, for the modelling set-up that we have used in this study.

Keywords: Cell metastasis · Evolution of cell geometry · Morphoelasticity · Agent-based model

1 Introduction

Cells are complex viscoelastic objects, of which the shape is maintained by its boundaries. Generally speaking, the cell shape is an outcome of a local balance between reaction and protrusion [3]. Therefore, cells can deform as a result of external stimuli, for instance, cell-substrate adhesion, biochemical signaling and forces exerted on the cell surface [9,10].

J. M. R. S. Tavares et al. (Eds.): CMBBE 2021, LNCVB 38, pp. 112–122, 2023.
https://doi.org/10.1007/978-3-031-10015-4_10

When cells migrate, in particular, when a cancer cell invades and migrates through a narrow and stiff cavity, it has to deform to adapt to the obstacles. More invasive cancer cells appear to be more pliable and dynamic both internally [4] and externally [2,5,10,15], hence, they are able to adjust their cytoskeleton and geometry significantly, which could lead to possible diagnosis for cancer. Furthermore, it has been observed that cancer cells exert a significantly larger traction forces on their direct environment than benign cells do [8].

In our previous work [12], we developed a phenomenological formalism to model the evolution of the cell geometry, where we included the morphoelasticity model to describe the possibly permanent impact of cellular forces on the extracellular matrix. In this manuscript, we consider cells migrating through a microtube and the objective is to reproduce the experiments by Mak et al. [7]. In our computational setting and in the experiments [7], the width of the micro-channel is smaller than the cell size. Therefore, Hertz' contact model in combination with friction is used in order to account for the mechanical contact between channel wall and cell membrane (cell boundary).

In this study, the model by Peng et al. [12] is used and applied to a deformable channel that is subject to plastic deformation. The idea is that the first cells make the channel permanently wider, which is modelled by the use of morphoelasticity, so that cells that arrive later can transmigrate through the channel more easily and hence faster. The objective of this study is to quantify the decrease of transmigration of later cells with respect to the transmigration time of the early cells.

The manuscript is structured as follows: Sect. 2 summarizes the model developed in [12] regarding the evolution of the cell geometry. In Sect. 3, we present an application of the model where two cells consecutively penetrate through a flexible channel as well as the results from Monte Carlo simulations regarding the penetration time of the first and second cells through the channel. Afterwards, some conclusions are drawn in Sect. 4.

2 Mathematical Models

We present a brief summary of the model that has been developed in our previous work. For more details about the model, we refer to Peng et al. [12].

2.1 Cell Cytoskeleton

First we only consider the contribution from the cell stiffness to the movement of nodal points on the cell membrane. We assume that the cell maintains its geometry by a series of elastic springs that connect the cell centre and the nodal points on the cell boundary; see Fig. 1 for an illustration. Together with the rotation matrix, the displacement of the nodal point j on the cell boundary is given by

$$d\boldsymbol{x}_j = E_c(\boldsymbol{x}_c(t) + \boldsymbol{B}(\tilde{\phi})\hat{\boldsymbol{x}}_j - \boldsymbol{x}_j(t))dt. \tag{1}$$

Here, E_c is the stiffness of the spring that connects the nodal point and cell centre, $\hat{\boldsymbol{x}} = \boldsymbol{x}_c(t) - \tilde{\boldsymbol{x}}_j(t)$ is the vector connecting the equilibrium position of nodal point i on the cell membrane to the cell centre, $\boldsymbol{x}_c$ is the central position of the cell and $\tilde{\boldsymbol{x}}_j$ represents the equilibrium position of the nodal point j corresponding to the cell centre $\boldsymbol{x}_c$, and $\boldsymbol{B}(\tilde{\phi})$ describes the rotation of the cell, where $\boldsymbol{B}(\phi)$ is defined by

$$\boldsymbol{B}(\phi) = \begin{pmatrix} \cos(\phi) & -\sin(\phi) \\ \sin(\phi) & \cos(\phi) \end{pmatrix}, \tag{2}$$

such that ϕ can be computed from

$$\tilde{\phi} = \arg \min_{\phi \in [0, 2\pi)} \left(\sum_{i=1}^{N} \| B(\phi)\tilde{\boldsymbol{x}}_i(t) - \boldsymbol{x}_i(t) \|^2 \right). \tag{3}$$

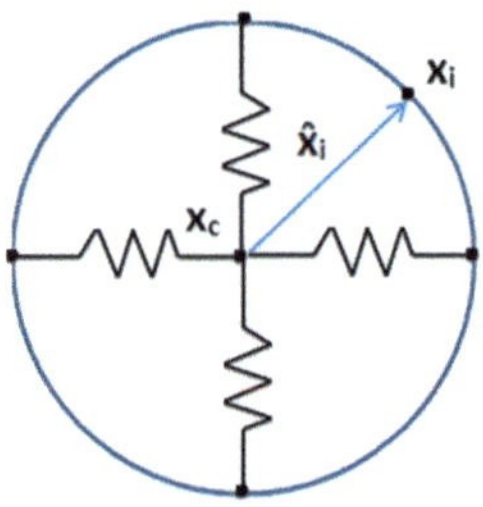

Fig. 1. A schematic of the cell cytoskeleton, which is kept by a series of springs.

2.2 Chemotaxis and Random Walk

There are various cues invoking cell migration, and one of them is chemotaxis, which provides the cell with a tendency to migrate towards (positive chemotaxis) or away from (negative chemotaxis) the gradient of a chemical. Just like in Peng et al. [12], we use a point source that secretes a chemical at the end of the microchannel. The concentration of the chemical is modelled by a reaction-diffusion equation with a point source in open domain Ω with boundary $\partial\Omega$:

$$\begin{cases} \dfrac{\partial c(\boldsymbol{x},t)}{\partial t} + \nabla \cdot (\boldsymbol{v}c(\boldsymbol{x},t)) - \nabla \cdot (D\nabla c(\boldsymbol{x},t)) = k\delta(\boldsymbol{x}(t) - \boldsymbol{x}_s), \text{ in } \Omega,\, t > 0, \\ c(\boldsymbol{x},0) = 0, \text{in } \Omega,\, t = 0, \\ \dfrac{\partial c}{\partial \boldsymbol{n}} + \kappa_s c = 0, \text{on } \partial\Omega, t > 0, \end{cases} \tag{4}$$

where $c(\boldsymbol{x},t)$ is the concentration of the signalling molecule; D is the diffusion rate which has been taken constant in the current study; k is the secretion rate of the signal source; $\boldsymbol{x}_s$ is the position of the source; $\delta(\boldsymbol{x}(t))$ is the Dirac Delta distribution, which is defined by

$$\delta(\boldsymbol{x}) = 0, \text{ for } \boldsymbol{x} \neq \mathbf{0},$$

and constrained to satisfy the identity that

$$\int_{\Omega} \delta(\boldsymbol{x}) d\Omega = 1, \text{ if } \mathbf{0} \in \Omega,$$

for any dimensionality (note that Ω is open); $\boldsymbol{v}$ is the displacement velocity of the substrate that results from the cellular forces exerted on their surroundings. The velocity is computed by solving the balance of the momentum, which will be discussed in the Sect. 2.3.

Together with the random displacement of the cell, which is described by a vector-Wiener process and the contribution from the cell stiffness, the displacement of a nodal point on the cell membrane is modelled by

$$d\boldsymbol{x}_j = E_c(\boldsymbol{x}_c(t) + \boldsymbol{B}(\tilde{\phi})\hat{\boldsymbol{x}}_j - \boldsymbol{x}_j(t))dt + \beta \frac{\nabla c(\boldsymbol{x}_j)}{\|\nabla c(\boldsymbol{x}_j)\| + \gamma} dt + \sigma_{rw} d\boldsymbol{W}(t), \quad (5)$$

where β is the chemotaxis sensitivity parameter, γ is a small positive constant to prevent division by zero, and σ_{rw} is the random walk parameter (in particular, the standard deviation of the random fluctuation, of which the square is proportional to the cell diffusion constant).

2.3 Force Balance and Friction

It has been documented in many studies, see for instance [8], that cancer cells exert forces on the immediate environment and that they are able to change biochemical properties of their immediate environment. We incorporate the forces that are exerted by the (migrating) cells on the walls of the micro-channel. We accommodate for the plastic nature of the resulting deformations by the use of a morphoelastic formalism. The morphoelastic model combines a mechanical balance, where viscoelasticity has been included, with an evolutionary equation for the effective Eulerian strain. This results into the following set of partial differential equations:

$$\begin{cases} \rho[\dfrac{D\boldsymbol{v}}{Dt} + \boldsymbol{v}(\nabla \cdot \boldsymbol{v})] - \nabla \cdot \boldsymbol{\sigma} = \boldsymbol{f}_m, \text{in } \Omega,\, t > 0, \\ \dfrac{D\boldsymbol{\varepsilon}}{Dt} + \boldsymbol{\varepsilon}\,\text{skw}(\boldsymbol{L}) - \text{skw}(\boldsymbol{L})\boldsymbol{\varepsilon} + [\text{tr}(\boldsymbol{\varepsilon}) - 1]\,\text{sym}(\boldsymbol{L}) = -\alpha\boldsymbol{\varepsilon}, \text{in } \Omega,\, t > 0, \\ \boldsymbol{v}(\boldsymbol{x}, t) = \mathbf{0}, \text{on } \partial\Omega,\, t > 0, \\ \boldsymbol{v}(\boldsymbol{x}, 0) = \mathbf{0}, \text{in } \Omega,\, t = 0, \\ \boldsymbol{\varepsilon}(\boldsymbol{x}, 0) = \mathbf{0}, \text{in } \Omega,\, t = 0, \end{cases} \quad (6)$$

where ρ is the density of the extracellular matrix, $\boldsymbol{L} = \nabla \boldsymbol{v}$ and α is a non-negative constant, and $\boldsymbol{\varepsilon}$ is the effective Eulerian strain tensor that is to be solved. Note that if $\alpha = 0$, then as soon as the force $\boldsymbol{f}_m = \boldsymbol{0}$, then the tissue will gradually recover to its original shape and volume. Here, $\frac{Dy}{Dt} = \frac{\partial y}{\partial t} + \boldsymbol{v} \cdot \nabla y$ is material derivative where y is any scalar field and $\boldsymbol{v}$ is the migration velocity of any point within the domain of computation. In order to have a fixed boundary, we use a homogeneous Dirichlet boundary condition for the velocity. From a mechanical point of view, we treat the computational domain as a continuous linear isotropic domain. Further, as a result of the presence of liquid phases in the tissue, the mechanical balance is also subject to viscous, that is friction, effects. Therefore, we use Kelvin-Voigt's viscoelastic dashpot model, of which the stress tensor reads as

$$\begin{aligned}\boldsymbol{\sigma} &= \boldsymbol{\sigma}_{elas} + \boldsymbol{\sigma}_{visco} \\ &= \frac{E}{1+\nu_s}\{\boldsymbol{\varepsilon} + \mathrm{tr}(\boldsymbol{\varepsilon})[\frac{\nu_s}{1-2\nu_s}]\boldsymbol{I}\} + \mu_1 \,\mathrm{sym}(\boldsymbol{L}) + \mu_2 \,\mathrm{tr}(\mathrm{sym}(\boldsymbol{L}))\boldsymbol{I},\end{aligned} \tag{7}$$

where ν_s is the Possion's ratio of the substrate, $\boldsymbol{\varepsilon}$ is the effective Eulerian strain tensor, μ_1 and μ_2 are the shear and bulk viscosities respectively. The morphoelasticity model solves two (nonlinear) partial differential equations and both velocity $\boldsymbol{v}$ and strain tensor $\boldsymbol{\varepsilon}$ are unknowns. The displacement of the domain can be approximated by numerically integrating the velocity over time: $\boldsymbol{u}(t) = \int_0^t \boldsymbol{v}(s)ds$.

We consider cells transmigrating through a deformable channel. The cell boundaries exert forces on the channel wall and the channel wall exerts reaction forces on the cell boundaries. As mentioned earlier, the cell boundary points are connected to the cell centre by springs and further these points are connected to their immediate neighboring points on the cell boundary. Similar to Peng et al. [12], we assume that $N_l = \{i_1, \ldots, i_m\}$ and $N_p = \{j_1, \ldots, j_{m+1}\}$, respectively, are the line segments and nodal points of the cell membrane that are in mechanical contact with the wall of the channel. Then the force $\boldsymbol{f}_m$ in Eq. (6) reads as

$$\boldsymbol{f}_m(\boldsymbol{x}, t) = \sum_{i \in N_l} Q\boldsymbol{n}(\boldsymbol{x}^i(t))\delta(\boldsymbol{x}(t) - \boldsymbol{x}^i(t))\Delta\Gamma^i, \tag{8}$$

where Q is the magnitude of the traction force per unit length that is exerted by the cell, $\boldsymbol{n}(\boldsymbol{x}(t))$ is the unit outward pointing vector at point $\boldsymbol{x}(t)$, $\boldsymbol{x}^i(t)$ is the midpoint of line segment i, $\delta(\boldsymbol{x}(t))$ is the Dirac delta distribution and $\Delta\Gamma^i$ is the length of line segment i.

Besides the pushing traction forces that the cell exerts on the channel, the wall exerts a friction and compressive force on the boundary of the cell. This friction force is assumed to be proportional to the forcing between the walls of the cell and the channel and hence both the spring force contribution of the cell and the force exerted on the channel wall are incorporated. Therefore, the friction $\boldsymbol{f}_f(\boldsymbol{x}, t)$, which is only at the nodal point $j \in N_p = \{j_1, \ldots, j_{m+1}\}$, is expressed as a sum of two aforementioned forces:

$$\boldsymbol{f}_f(\boldsymbol{x}_j, t) = \boldsymbol{f}_m(\boldsymbol{x}_j, t) + E_c(\boldsymbol{B}(\tilde{\phi})\hat{\boldsymbol{x}}_j - \boldsymbol{x}_j(t)). \tag{9}$$

Therefore, combining with the passive convection of the substrate and the effect of the friction, Eq. (1) is now given by

$$\begin{aligned} d\boldsymbol{x}_j &= E_c(\boldsymbol{x}_c(t) + \boldsymbol{B}(\tilde{\phi})\hat{\boldsymbol{x}}_j - \boldsymbol{x}_j(t))dt + \beta\frac{\nabla c(\boldsymbol{x}_j)}{\|\nabla c(\boldsymbol{x}_j)\| + \gamma}dt + \boldsymbol{v}(\boldsymbol{x}_j(t))dt \\ &- \mu_f\|\boldsymbol{f}_f(\boldsymbol{x}_j(t))\| \cdot (d\boldsymbol{x}_j(t), \boldsymbol{\tau}_{ob}(\boldsymbol{x}_j(t)))\boldsymbol{\tau}_{ob}(\boldsymbol{x}_j(t)) + \sigma_{rw}d\boldsymbol{W}(t), \end{aligned} \tag{10}$$

where μ_f is the friction coefficient, and $\boldsymbol{\tau}_{ob}(\boldsymbol{x})$ is the unit tangential vector of the obstacle boundary. We bear in mind that the forces are only exerted on the direct environment of the cell if both end points of the line segment are in direct contact with the channel wall; similarly, friction is only incorporated on a nodal point that is in direct contact with the channel wall.

3 Applications and Numerical Results

In our previous work [12], we presented several applications of the model, for instance, cell migration, cell differentiation and cell penetrating through a micro-tube. In this manuscript, we consider a flexible channel, which deforms due to the traction forces that a cell exerts on the channel wall. Figure 2(a) shows the initial settings of the simulation, where the cell is initially located in one side of the channel and migrates towards the gradient of the concentration of the signalling molecules through the channel. Once the first cell exits the channel, it will be discarded in further computations, and the second cell will start migrating from its initial position, i.e. the position where the red circle is, in Fig. 2(a). The parameter values are shown in Table 1. The horizontal channel is determined by

$$y = \pm(3.5 + \sin(0.5x))\mu\text{m}, x \in (-60, 60)\mu\text{m}.$$

Several screenshots are shown in Fig. 2, which shows the locations of the cells and the shape of the cell and the channel. In Fig. 2(b), the part of the channel that the first cell has penetrated, has clearly expanded. Once the first cell has entirely left the channel, it is disregarded in further calculations, and simultaneously the second cell appears at the initial position (the red circle), see Fig. 2(d) and 2(e). The plastic deformation that has been caused by the first cell makes the second cell transmigrate through the micro-channel more easily because less force needs to be applied by the cell. From Fig. 2, it is observed that the channel is slightly deformed in the direction of the migration of the cells.

Due to the significant deformation of the channel, we are interested in whether the second cell can benefit from the first cell expanding the flexible channel, that is, whether it takes less time for the second cell to penetrate through the channel than for the first cell, see Fig. 2(e)–(h). We carry out Monte Carlo simulations without taking any other input parameters as random variables. Hence, the only stochastic process in the model is the random walk contribution to migration of the cell. We collected 1130 samples regarding the penetration time of each cell through the channel. The box plot and density plot of the data are presented in Fig. 3, respectively. From these plots, it is hard to conclude whether

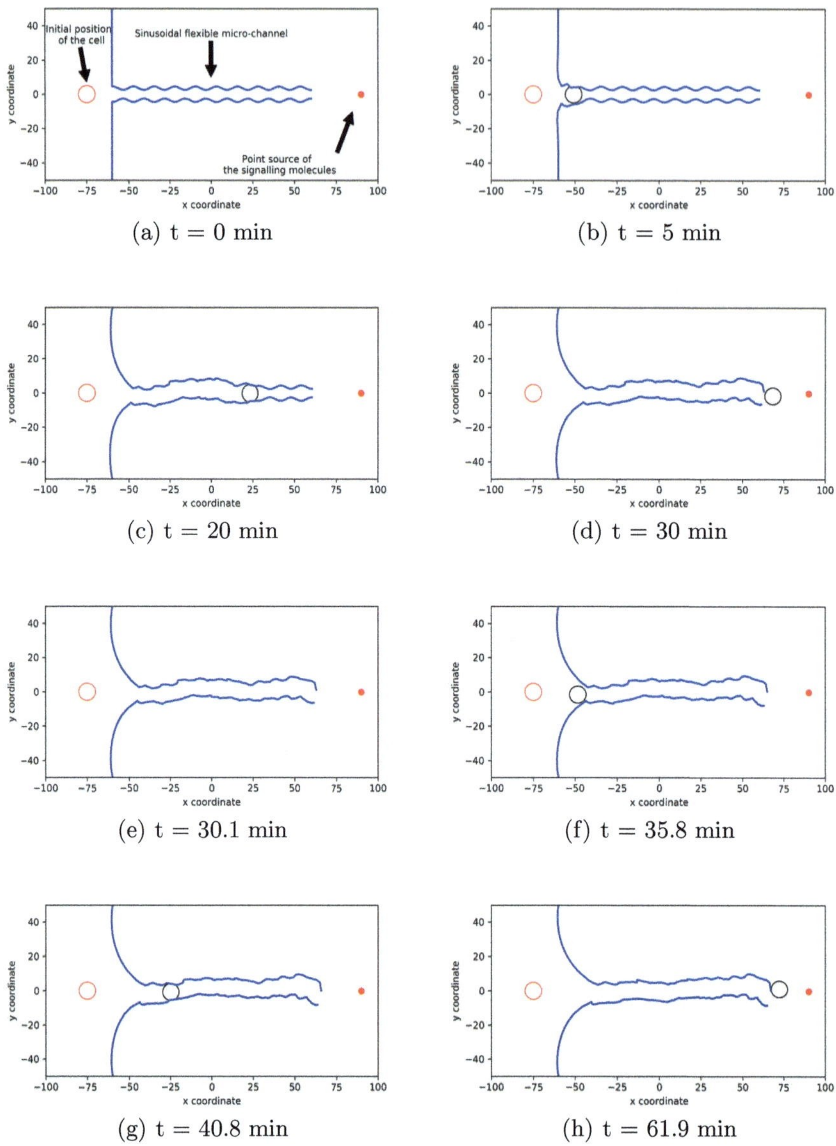

(a) t = 0 min
(b) t = 5 min
(c) t = 20 min
(d) t = 30 min
(e) t = 30.1 min
(f) t = 35.8 min
(g) t = 40.8 min
(h) t = 61.9 min

Fig. 2. Screenshots of the simulation at different moments when two cells penetrate the flexible channel sequentially. The red dot at the very right position with coordinates (90, 0) in the figure is the point source of the signalling molecules; blue curves are the deformed channel, which is initially sinusoidal; red circle represents the initial position and shape of the cells.

Table 1. Parameter values used in the computational simulations

Parameter	Description	Value	Units	Source
P_0	Maximal traction force exerted by cells	10.4	$kg \cdot \mu m/h^2$	[6]
μ_f	Cell friction coefficient	0.03	–	[1]
ν_s	Poisson's ratio of the substrate	0.49	–	[6]
k	Secretion rate of the cell-attractive signal	2.5	$kg/(\mu m^3 \cdot min)$	[11]
κ_s	Parameter in Robin's boundary condition to solve Eq. (4)	100	μm^{-1}	[11]
μ_1	Shear viscosity of the ECM	33.783	–	[11]
μ_2	Bulk viscosity of the ECM	22.523	–	[11]
Estimated parameter values				
$x0$	Length of the computational domain in x-axis	200	μm	
$y0$	Length of the computational domain in x-axis	100	μm	
Δt	Time step	0.1	min	
E_s	Substrate elasticity	50	$kg/(\mu m \cdot min^2)$	
E_c	Cell elasticity and the stiffness of the springs in the cell cytoskeleton	10	$kg/(\mu m \cdot min^2)$	
R	Cell radius	5	μm	
β	Maximal mobility of points on cell membrane	5	min^{-1}	
D	Diffusion rate of the signal	233.2	$\mu m^2/min$	
N	Number of nodal points on the cell membrane	40	–	
σ_{rw}	Weight of random walk	1	–	
α	Degree of permanent ECM deformation in Eq. (6)	2	–	

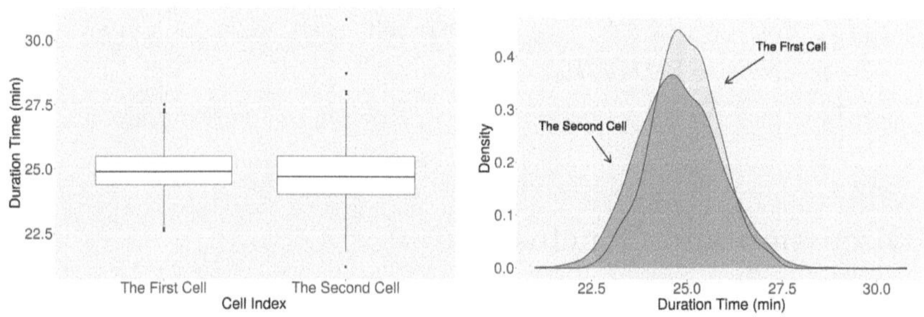

(a) Box Plot (b) Density Plot

Fig. 3. Empirical plots of the data collected from the Monte Carlo simulations. (a) Box plot shows the median (middle line), 25% (Q_1) and 75% (Q_3) percentile (lower and upper bound of the box), the minimum ($Q_1 - 1.5\frac{Q_3 - Q_1}{2}$) and maximum ($Q_3 + 1.5\frac{Q_3 - Q_1}{2}$) of the data excluding the outliers (verticle lines) and outliers (dots). (b) Estimated probability density of the penetration time for the first and the second cell, respectively.

the second cell takes significantly less time than the first cell to migrate through the channel. This seemingly small difference between the transmigration times of the second and first cells is probably the result of the thickness of the channel being comparable to the cell diameter. We expect that the differences in transmigration times increase if the channel width is much smaller than the cell size. Of course, this trend will depend on the amount of plasticity in the deformation of the channel. This will be a topic for further research. Furthermore, according to Shapiro's test [14], neither of the dataset is normally distributed with p-value less than 0.05. Therefore, instead of the t-test, we use Wilcoxon's signed-rank test for these two paired samples, since Wilcoxon's signed-rank test [13] does not require normality of the data.

Table 2 displays the statistical results from the Wilcoxon's signed-rank test of the two paired dataset regarding the penetration time in the channel. The p-value is much smaller than the significance level 0.05, therefore, based on the data collected from the Monte Carlo simulation, we conclude that the second cell does benefit from the first cell expanding the channel.

Table 2. Statistical results of wilcoxon's signed-rank test on the paired dataset of penetration time of each cell. T_1 and T_2 represent the penetration time of the first and second cell, respectively.

Null Hypothesis ($H_0 : T_1 = T_2$)	There is no difference in penetration time for both cells
Alternative Hypothesis ($H_1 : T_1 > T_2$)	The second cell has shorter penetration time than the first cell.
$p - value = \mathbb{P}(Data \vert H_0) = 1.158 \times 10^{-5}$	

4 Discussion and Conclusions

Our previously developed model for the evolution of cell geometry has been used to investigate whether followers among (cancer) cells benefit from leading (cancer) cells during transmigration through micro-channels and cavities. Since cell migration is subject to randomness, Monte Carlo simulations have been done to estimate the probability density of penetration times for both the leading and stalk cells. Using Wilcoxon's signed-rank test, it has been concluded that the transmigration time for the stalk cell is significantly smaller than for the leading cell with a p-value of less than 0.0001. The physical explanation is that the first (leading) cell deforms the micro-channel plastically so that the channel is wider when the second (stalk) cell transmigrates through the channel. The second cell has the advantage that it needs to exert fewer forces than the first cell. This effect could possibly even be amplified if metabolism would be taken into account.

We realize that we have shown the (Monte Carlo) results for one modelling set-up only. All input variables have been fixed and all differences in simulation runs were exclusively caused by the random nature of cell migration through the Wiener process (white noise – random walk). Nevertheless, we may conclude via Wilcoxon's signed-rank test that the current simple 'toy model' predicts a significant difference between the transmigration times of the second and first cell, from which it is concluded that the second cell transmigrates more easily than the first cell for the current modelling set-up. In our future Monte Carlo simulations, we will add uncertainties in the channel width, channel length, Young's modulus of the ECM, distance between the initial cell positions and the channel, distance between the consecutive cells, amount of plasticity in the deformation and possibly more (biophysical) parameters by the use of prior statistical distributions.

References

1. Angelini, T.E., Dunn, A.C., Urueña, J.M., Dickrell, D.J., Burris, D.L., Sawyer, W.G.: Cell friction. Faraday Discuss. **156**, 31 (2012)
2. Cross, S.E., Jin, Y.-S., Rao, J., Gimzewski, J.K.: Nanomechanical analysis of cells from cancer patients. Nat. Nanotechnol. **2**(12), 780–783 (2007)
3. Ebata, H., et al.: Persistent random deformation model of cells crawling on a gel surface. Sci. Rep. **8**(1) (2018). https://doi.org/10.1038/s41598-018-23540-x
4. Gal, N., Weihs, D.: Intracellular mechanics and activity of breast cancer cells correlate with metastatic potential. Cell Biochem. Biophys. **63**(3), 199–209 (2012)
5. Guck, J., et al.: Optical deformability as an inherent cell marker for testing malignant transformation and metastatic competence. Biophys. J. **88**(5), 3689–3698 (2005)
6. D. Koppenol. Biomedical implications from mathematical models for the simulation of dermal wound healing. PhD-thesis at the Delft University of Technology, The Netherlands (2017)
7. Mak, M., Reinhart-King, C.A., Erickson, D.: Elucidating mechanical transition effects of invading cancer cells with a subnucleus-scaled microfluidic serial dimensional modulation device. Lab Chip. **13**(3), 340–348 (2013)
8. Massalha, S., Weihs, D.: Metastatic breast cancer cells adhere strongly on varying stiffness substrates, initially without adjusting their morphology. Biomech. Model. Mechanobiology **16**(3), 961–970 (2016)
9. Mogilner, A., Keren, K.: The shape of motile cells. Current Biol. **19**(17), R762–R771 (2009)
10. Paluch, E., Heisenberg, C.-P.: Biology and physics of cell shape changes in development. Current Biol. **19**(17), R790–R799 (2009)
11. Peng, Q., Vermolen, F.: Agent-based modelling and parameter sensitivity analysis with a finite-element method for skin contraction. Biomech. Model. Mechanobiol. **19**(6), 2525–2551 (2020). https://doi.org/10.1007/s10237-020-01354-z
12. Peng, Q., Vermolen, F.J., Weihs, D.: A formalism for modelling traction forces and cell shape evolution during cell migration in various biomedical processes. Biomech. Model. Mechanobiol. **20**(4), 1459–1475 (2021). https://doi.org/10.1007/s10237-021-01456-2

13. Rey, D., Neuhäuser, M.: Wilcoxon-signed-rank test. In: International Encyclopedia of Statistical Science, pp. 1658–1659. Springer, Berlin (2011). https://doi.org/10.1007/978-3-642-04898-2_616
14. Shapiro, S.S., Wilk, M.B.: An analysis of variance test for normality (complete samples). Biometrika **52**(3/4), 591–611 (1965)
15. Swaminathan, V., Mythreye, K., O'Brien, E.T., Berchuck, A., Blobe, G.C., Superfine, R.: Mechanical stiffness grades metastatic potential in patient tumor cells and in cancer cell lines. Cancer Res. **71**(15), 5075–5080 (2011)

Simulation of Cell Proliferation Using a Meshless Tool

M. I. A. Barbosa[1(✉)], J. Belinha[2], R. M. Natal Jorge[3], and A. Carvalho[4]

[1] Institute of Science and Innovation in Mechanical and Industrial Engineering (INEGI), University of Porto, Porto, Portugal
up201405732@edu.fe.up.pt

[2] Department of Mechanical Engineering, School of Engineering Polytechnic of Porto, Porto, Portugal
job@isep.ipp.pt

[3] Department of Mechanical Engineering, University of Porto, Porto, Portugal
rnatal@fe.up.pt

[4] Institute for Research and Innovation in Health (I3S), University of Porto, Porto, Portugal
anacarvalho@ibmc.up.pt

Abstract. To preserve the haemostasis of the body, a balance between cell formation and cell death has to be maintained. Thus, when necessary, cells divide into two new ones through a complex and controlled process that is mainly dependent on the availability of oxygen and glucose. Numerical methods have been used to study and solve several problems in science and engineering. Regarding biomechanics and the study of cells, these methods were already combined with distinct formulations. Within these methods, it is possible to highlight the meshless methods, which are a novel and efficient approach that emerged as an alternative to traditional methods. However, the available literature is not yet as extensive as these last ones. The aim of this work is to simulate the process of cell proliferation resorting to a new algorithm. This was based on a phenomenological growth law proposed by the authors, dependent on the presence of oxygen and glucose and solved by a meshless method (Radial Point Interpolation Method). In the end, the algorithm leads to satisfactory and coherent results. It was capable of simulating cell proliferation, under different concentration of oxygen and glucose, as expected. Furthermore, the RPIM proved to be a viable option to solve this problem since the results obtained were adequate.

Keyword: Cell proliferation · Cell growth · Numerical simulation · Meshless method · RPIM

1 Introduction

One of the most important features of cells is their capacity to proliferate, which is crucial to maintain the haemostasis of the body [1, 2]. To proliferate, cells increase their volume and then divide into two equal cells, through an extremely organized set of reactions [1, 3]. To ensure that this process occurs without any issues, the presence

J. M. R. S. Tavares et al. (Eds.): CMBBE 2021, LNCVB 38, pp. 123–127, 2023.
https://doi.org/10.1007/978-3-031-10015-4_11

of some nutrients, such as oxygen and glucose, is required [4, 5]. In the case of oxygen, a balanced concentration of this molecule is indispensable for metabolic processes and cellular functions to occur correctly [6]. On the other hand, glucose is one of the main sources of energy of cells [7, 8].

Numerical methods have emerged and have been used to study and solve several problems in science and engineering. Regarding biomechanics, these methods are already used to solve different types of approaches [9–11]. Within these methods, it is possible to highlight the meshless methods, which are a novel discrete approach that emerged as an alternative to mesh-based discretization methods. However, the available literature does not present as many works as the ones solved with mesh-dependent discretization techniques [12].

Meshless methods are known to use a nodal set to discretize the domain, instead of using a mesh and so, these do not present any relation between them. Due to that, more accurate solutions are obtained [13, 14].

The aim of this work is to simulate the process of cell proliferation resorting to a new algorithm that is solved by the meshless methods Radial Point Interpolation (RPIM). In this, the cell grows due to the existence of oxygen and glucose, following a new proposed phenomenological law. At the same time, the viability of the method to solve this problem was also taken into account since, to the knowledge of the authors, it was never used in the study of this type of problem.

2 Radial Point Interpolation Method (RPIM)

To simulate the proliferation of a cell, the RPIM was used. In this section, the procedure of this method is described. To start the analysis, the geometry, the limits and the boundary conditions of the problem domain need to be defined. To discretize the domain, an unstructured nodal set is used, with no pre-established relation between the nodes. After that, integration points are distributed in the domain, in order to build the background mesh. This will be used to integrate the weak equation ruling the phenomenon under study. Note that, all integration points will have their own influence-domain and through that, nodal connectivity is imposed. The last step is the construction of the shape functions. These are used to obtain the global system of equations and, in the case of RPIM, built through the RPI technique. Moreover, it is important to highlight that the RPIM shape functions present the delta Kronecker property. Thus, the boundary conditions are directly imposed [14].

3 Cell Proliferation Algorithm

The proposed algorithm was created and ran in MATLAB®. It was developed to be solved by the RPIM and to simulate the growth of a cell until it divides, in the presence of oxygen and glucose. To do that, the first step is to introduce input data, such as the diameter of the cell, the number of maximum iterations, the cell and medium properties, the initial concentration and the diffusion coefficients of oxygen and glucose. After this, the cell starts to grow following the new proposed phenomenological growth law. This growth continues until the cell doubles its initial size and if the concentrations of oxygen and glucose are not 0. Note that, these two molecules are consumed as the cell grows. When the cell reaches the double of its volume, it divides into two new ones and the analysis ends.

4 Numerical Results

After developing the algorithm, different values of concentration of oxygen and glucose were used in order to study the viability of the algorithm. First, values of normal concentration were taking into account (3.5×10^{-16} mol/μm^2 for oxygen and 2.75×10^{-14} mol/μm^2 for glucose). Then, these concentrations were reduced in separated analyses. The remaining parameters were maintained constant.

Moreover, to perform the analyses, the colour maps of the evolution of the cell were assumed, as seen below (Fig. 1, Fig. 2 and Fig. 3). Note that, the figures presented do not represent sequential iterations.

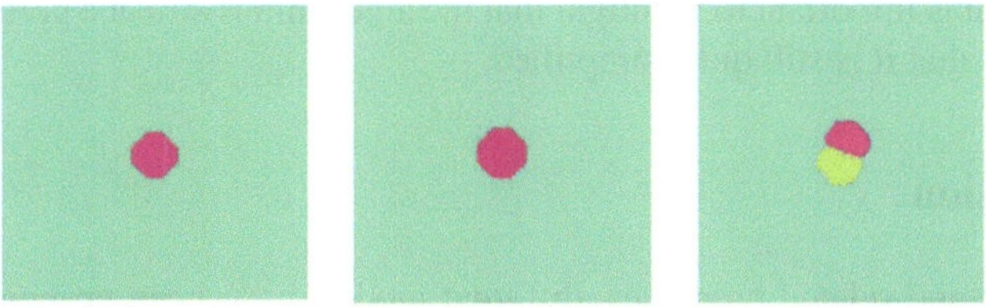

Fig. 1. Evolution of the volume of the cell when normal concentrations of oxygen and glucose are used. Dark green represents the medium, pink the first cell and light green the second cell.

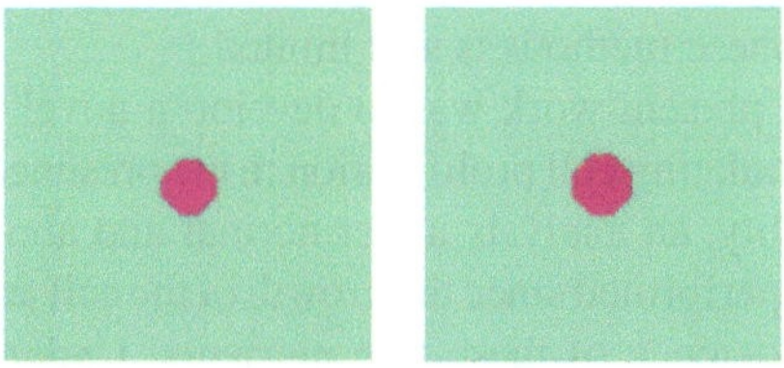

Fig. 2. Evolution of the volume of the cell when abnormal concentrations of oxygen are used. Dark green represents the medium and pink the cell.

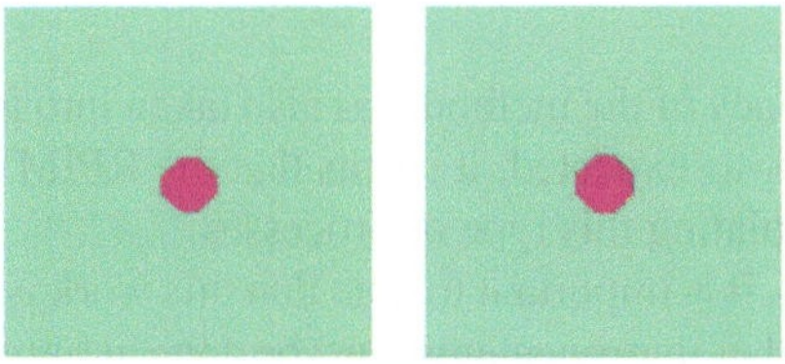

Fig. 3. Evolution of the volume of the cell when abnormal concentrations of glucose are used. Dark green represents the medium and pink the cell.

Considering Fig. 1, it is noticeable that the cell was able to grow until the double of its volume and then, to divide into two new cells, as defined in the algorithm. Since the concentrations of oxygen and glucose were set to normal values, the growth occurred without any unexpected error and so, the process of division occurred. This behaviour is in accordance with what occurs at a physiological environment since cells grow and divide correctly with a normal source of nutrients.

On the other hand, under abnormal lower concentrations of these two molecules, cellular proliferation is compromised, since cells tend to decrease their rate of proliferation, and it can even stop or not occur at all. In both Fig. 2 and Fig. 3, this behaviour is visible. In these two cases, the volume achieved by the cells is lower than the double of the initial volume. Moreover, the process stops before reaching the point of division and, for this reason, this process does not occur. Such occurred because all oxygen or glucose available was consumed and, according to what was defined in the algorithm, it is necessary that both molecules be present in the medium for the cell to grow.

Thus, the obtained results were coherent with what was expected and so, it seems that the algorithm is promoting adequate results regarding the process of cell proliferation. Even though, it is important to highlight that the algorithm is still at a preliminary stage, which implies that it is still quite simplified.

5 Conclusion

To study cells and related processes, different models have been proposed, as well as different discrete numerical methods (which are used to solve them). Regarding these methods, the interest in the use of meshless methods increased in the past years, such as the RPIM. These emerged as an alternative to mesh-based discretization methods, and are known for promoting more accurate results. However, despite their reliability, the number of works using these methods is still limited.

The objective of the present work was to develop a novel algorithm solved by the RPIM and capable of simulating cell proliferation in the presence of oxygen and glucose. Thus, to verify its viability, an analysis under normal and abnormal concentrations of oxygen and glucose was performed since the growth of the cell and the process of division are dependent on the existence of these two nutrients. Under normal concentrations, the two processes occurred without any unexpected problem, while under abnormal concentrations, this was not possible, as expected. Both for lower concentrations of oxygen or glucose, the process of division did not occur and the double of the initial volume of the cells was not achieved. Such is in accordance with what is described in the literature.

Moreover, the efficiency of the method was also taken into account. Since the results were according to what was expected, it seems that the RPIM is capable of generating accurate results and simulating this type of processes.

However, once again, it is important to note that this work is just a preliminary stage. Bearing this in mind, further improvements must be done in future work in order to obtain a more complex algorithm. For example, other parameters, such as the pH, ought to be added as a variable that affects cellular growth, other processes, as angiogenesis, could be included as well and more than one cell could be analysed. Such improvements might lead to a more realistic algorithm. Despite this, the algorithm seems to be a promising approach to simulate the process of cell proliferation.

Acknowledgments. The authors truly acknowledge the funding provided by Ministério da Ciência, Tecnologia e Ensino Superior – Fundação para a Ciência e a Tecnologia (Portugal), under grants: SFRH/BD/146272/2019. Additionally, the authors gratefully acknowledge the funding provided by LAETA, under project UIDB/50022/2020.

References

1. Tortora, G.J., Derrickson, B.H.: Introduction to the Human Body, Wiley, New York (2017)
2. King, K., Cidlowski, J.: Cell cycle regulation and apoptosis. Annu. Rev. Physiol. **60**(1), 601–617 (1998)
3. Standring, S.: Gray's Anatomy E-Book: the Anatomical Basis of Clinical Practice. Elsevier, Amsterdam (2015)
4. Sandal, T.: Molecular aspects of the mammalian cell cycle and cancer. Oncologist **7**(1), 73–81 (2002)
5. Saucedo, L.J., Edgar, B.A.: Why size matters: altering cell size. Curr. Opin. Genet. Dev. **12**(5), 565–571 (2002)
6. Carreau, A., et al.: Why is the partial oxygen pressure of human tissues a crucial parameter? Small molecules and hypoxia. J. Cell Mol. Med. **15**(6), 1239–1253 (2011)
7. Giri, B., et al.: Chronic hyperglycemia mediated physiological alteration and metabolic distortion leads to organ dysfunction, infection, cancer progression and other pathophysiological consequences: an update on glucose toxicity. Biomed. Pharmacother. **107**, 306–328 (2018)
8. Shao, D., et al.: Glucose promotes cell growth by suppressing branched-chain amino acid degradation. Nat. Commun. **9**(1), 1–17 (2018)
9. Alfieri, R., et al.: Modeling the cell cycle: from deterministic models to hybrid systems. Biosystems **105**(1), 34–40 (2011)
10. Getto, P., Marciniak-Czochra, A.: Mathematical modelling as a tool to understand cell self-renewal and differentiation, In: Vivanco, M. (eds.) Mammary Stem Cells. Methods in Molecular Biology, vol. 1293, pp 247–266. Humana Press, New York (2015). https://doi.org/10.1007/978-1-4939-2519-3_15
11. Youssef, B.B.: Visualization of spatial patterns of cells using a 3-D simulation model for multicellular tissue growth. In: 2014 International Conference on Multimedia Computing and Systems (ICMCS), IEEE (2014)
12. El Zahab, Z., Divo, E., Kassab, A.: A localized collocation meshless method (LCMM) for incompressible flows CFD modeling with applications to transient hemodynamics. Eng. Anal. Boundary Elem. **33**(8–9), 1045–1061 (2009)
13. Belinha, J., et al.: The analysis of laminated plates using distinct advanced discretization meshless techniques. Compos. Struct. **143**, 165–179 (2016)
14. Belinha, J.: Meshless Methods in Biomechanics. 1ed. Lecture Notes in Computational Vision and Biomechanics, vol. 16, p. 320, Springer, Cham (2014)

Explicit Non-linear Finite Element Analysis for Prediction of Primary Stability in Uncemented Total Hip Arthroplasty

Marzieh Ovesy(✉) and Philippe K. Zysset

ARTORG Centre for Biomedical Engineering Research, University of Bern, Freiburgstrasse 3, 3010 Bern, Switzerland
Marzieh.ovesy@gmail.com

Abstract. Primary stability in uncemented total hip arthroplasty is often assessed by insertion force, relative micromotions at the interface or subsidence of the femoral stem. This study aimed to simulate implantation process and evaluate the press-fit effect imposed at this stage on the primary stability outcome. The femur was modeled as an elasto-plastic material with element deletion based on cumulation of plastic strain represented as damage. The implantation phase was simulated using a displacement constraint that led to implant self-positioning by frictional contact. A subsequent force-controlled, loading-unloading cycle corresponding to the peak of a gait cycle was applied. Implantation force and work, implant subsidence and relative micromotions during the loading cycle were quantified. The modeling strategy was performed on two levels of hypothetical initial press-fit to capture the extreme influence of press-fit on the subsequent outcomes. The implantation force was 21.5% higher for the high press-fit approach. The implant remained more stable during the loading cycle for low press-fit model based on subsidence values. Micromotions showed also a significant dependency on press-fit. The median values of micromotions were below 150 μm for both press-fit models, which is in good agreement with the literature. This study demonstrates feasibility of simulating the implantation stage and investigating the effect of press-fit on primary stability of an individual total hip arthroplasty planning.

Keywords: Explicit finite element · Press-Fit · Implantation · Micromotions

1 Introduction

Total hip arthroplasty (THA) is divided into cemented and uncemented surgical techniques on both femoral and acetabular sides. The surgical procedure for an uncemented femoral stem consists of resecting the femoral head and broadening the medullary canal using broaches with smaller dimensions than the implant followed by press-fit of the implant into the prepared canal [1, 2]. Uncemented THA is successful when a high extent of primary stability is achieved for adequate osseointegration without leading to early aseptic loosening of the implant. For instance, subsidence values more than 1.5 mm in the first two years after surgery would result in aseptic loosening and therefore need

J. M. R. S. Tavares et al. (Eds.): CMBBE 2021, LNCVB 38, pp. 128–142, 2023.
https://doi.org/10.1007/978-3-031-10015-4_12

for revision surgery [3]. Primary stability depends on surgical technique (broaching, number, and energy of the insertion impacts), implant size, shape [4, 5], coating [6, 7] and critically also on bone quality [8]. Although it was originally believed that a higher amount of initial press-fit and implantation impact energy would result in better primary stability, studies have shown that the excessive stresses associated with higher initial press-fit lead to an increase in post-operative fracture risk [9, 10]. It has also been reported that a micromotion window of approximately 100–200 μm is required to ensure osseointegration at the bone-implant interface [11–13]. Values above this range induce formation of fibrous tissue at the interface, leading to loss of primary stability followed by implant migration, whereas lower value would not trigger the osseointegration process [14–16]. Consequently, a pre-operative control of these parameters would be desirable to ensure an even better surgical outcome of THA.

Once properly validated against experiments, finite element analysis (FEA) has the potential to assess primary stability by computing micromovements over the entire bone-implant interface for various loading conditions. In fact, the effect of implant design [17], implant position and surgical technique [18, 19] on micromotions and subsidence was already investigated [20]. In addition, FEA enables modeling of patient-specific THA [21] by testing different implants and coating on the same femoral anatomy [22] and optimizing a number of parameters before surgery.

In most FE simulations, the material behavior of bone was considered to be linear elastic, however, some studies have also accounted for the post-yield behavior [23–25]. Bone-implant interaction has either been simulated using frictionless contact [26, 27] or by modeling the interface as node to node or node to surface contact with friction [23]. Some studies have also compared fully-bonded and frictional contact interfaces [28]. In all these studies, bone damage during implantation was either ignored or simulated as a surrogate constant press-fit applied to the entire bone-implant interface [29]. It has been reported that the amount of press-fit plays a significant role on micromotions [30]. However, the above studies neglected the variations in press-fit at different locations due to damage occurred during implantation [31, 32]. The implant was positioned virtually in the femur before starting the simulation regardless of the implantation path and the substantial bone damage produced during surgery [33]. In addition, in many studies, loading was applied using displacement-controlled boundary conditions, which does not reflect the real physiological gait conditions, that are force-controlled by nature [34, 35]. In fact, capturing higher amounts of bone damage and large sliding contact conditions between bone and implant lead to convergence issues when using implicit solvers. An explicit integration scheme with an element deletion technique using force-controlled loading may, therefore, overcome these limitations [36].

Accordingly, the aim of this study is to demonstrate the ability of a previously validated explicit nonlinear FE approach [37] to simulate the implantation of a rigid stem into a human proximal femur and to quantify the subsequent response under physiological loading. In order to explore the influence of broaching, we started the simulation of femoral stem implantation from two different bone configurations leading to low and high press-fit models. Due to the chosen amount of press-fit in this study, the implantation stage could be considered as showing the outcome of the combined broaching and implantation procedure. First, we quantified the implantation force and work required to bring the stem in an anatomically correct position. Second, we evaluated the residual stresses and damage produced in the femoral bone after insertion of the femoral stem. Finally, implant subsidence and micromotions between bone and the entire femoral stem were computed under the peak load occurring during a gait cycle.

2 Methods

2.1 Geometry Construction

A computed tomography (CT) reconstruction of the right femur of a human donor (female, age 87) was obtained at an in-plane resolution of 0.48 × 0.48 mm and 0.40 mm slice thickness.

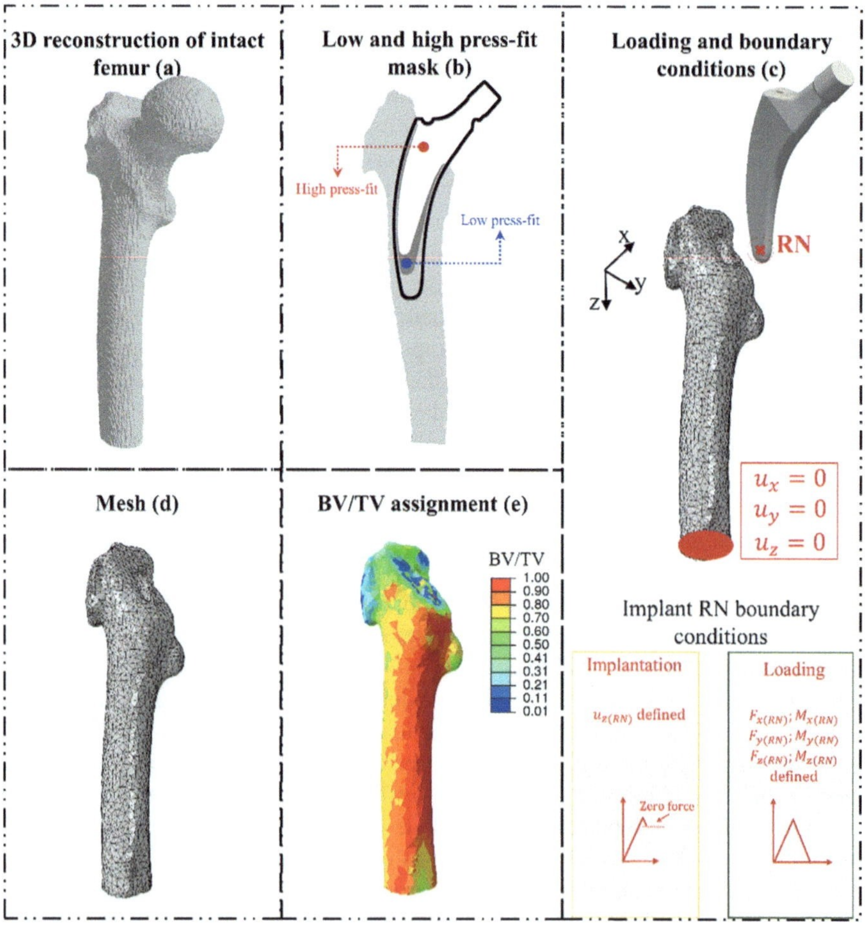

Fig. 1. Overview of FE model generation. Intact femur (a), two levels of press-fit indicated in red (high) and blue (low) (b), loading and boundary conditions (c), mesh (d) and BV/TV distribution (e). (Color figure online)

The image was resized to an isotropic resolution of 0.48 mm, calibrated in mg/cc of hydroxyapatite (BDC/6, QRM, Germany) and segmented with a U-Net machine learning approach [38]. A 2D planning approach was performed by first generating a mask from the outer surface of the intact femur (Fig. 1.a). Based on the pre-operative planning procedure on the 2D projection of the CT scan, an Optimys lateral implant with size 4 (length: 94 mm, offset: 42 mm, neck length: 36.5 mm) (Mathys, Bettlach, Switzerland) was selected.

The femoral head was removed virtually from the generated femoral mask according to the implant pre-operative position using 3D-slicer (Slicer 4.10.2, [39]). The femur shaft was cut transversely at a position of 60 mm from the implant distal tip. This value was chosen to eliminate the effect of boundary conditions on the results meanwhile reducing the model size [40]. Two distinct configurations were prepared that represent different amounts of broaching and press-fit occurring during the implant insertion. In order to reduce the modeling effort, the implant 3D file was shrunk in all directions by 2 (0.96 mm) and 4 voxels (1.92 mm) to produce extreme cases of low and a high press-fit model, respectively. In both cases, the resulting shape was subtracted from the femur geometry (Fig. 1.b). Later, a 3D model was generated from each mask and meshed using Cgal mesher [41] into 21381 and 22572 linear tetrahedral elements (C3D4) for low and high press-fit models, respectively. The element size was chosen to have the smallest element edge size equal to image isotropic resolution of 0.48 mm (Fig. 1.d). This is the smallest element size that could be applied to use image-based material property assignment. The previous analysis has shown that this choice of element size would have no significant influence on the force outcome [37]. Using Medtool 4.3 (Medtool, Dr. Pahr Ingenieurs, Vienna) BV/TV (bone volume per total volume, total volume here would be the sum of the volumes of both bone and marrow) was assigned to each element based on the averaged bone density around the element (Fig. 1.e). In order to eliminate the effort associated with segmentation of the femoral canal, the regions without bone were also meshed with very low BV/TV values (<0.01).

2.2 Material Modeling

The bone material was chosen to be similar to an already validated explicit representation of homogenized elasto-plastic behavior with damage [37]. The damage parameter reduced the stiffness of the model. The damage variable (D) was associated with cumulative plastic strain (κ) [39] and the highly damaged elements reaching a threshold of 95% of D_c were deleted to solve the issues related to existence of highly distorted elements (Eq. 1).

$$D(\kappa) = D_c(1 - e^{-k_p\kappa}) \tag{1}$$

The bone was modeled to be transversely isotropic in the direction of the femoral canal (using a fabric tensor with the eigen values being 0.9, 0.9, 1.2 [40]) and had elastic-perfect plastic behavior with a quadric yield surface [40]. The elastic and post-yield material properties of the bone were dependent on the BV/TV value and were scaled accordingly. This material model has been implemented through a user material subroutine (VUMAT) in the FE solver. The elastic and strength material parameters of bone were the same as a previous study (Table 1) [41].

Table 1. Elastic and post yield material properties.

Youngs modulus (E)	MPa	12000
Poisson ratio (ν)	–	0.249
k_p	–	8.0075
D_c	–	0.85
Ultimate tensile stress (σ^t)	MPa	81.6
Ultimate compression stress (σ^c)	MPa	111.6

The titanium implant was assumed to be rigid compared to the bone material being modeled by using a rigid body constraint. The implant rigid body reference node (RN) (Fig. 1.c) was set to the implant tip to ease the application of boundary conditions. The bone nodes at the bottom surface of the femoral shaft were fixed in all three degrees of freedom. The bone-implant interface was modeled as normal and tangential contact with friction. The friction coefficient was set to 0.3 [17].

2.3 Virtual Implantation

Initially, the implant was positioned out of the femur. Since one of the main outcomes of this stage is to measure the implantation force, the implantation procedure needed to be modeled by a displacement-based loading. In this study it was assumed that the implant is at its final position when the implant was fully placed in the bone. Accordingly, the implantation stage was modeled by applying a single translation in the z-direction (femoral canal direction) on the reference node to reach the final position of the implant. All the other degrees of freedom were free. This approach represented the implant insertion stage as the implant being pulled into the canal and allowed to self-position according to the surrounding bone material and contact conditions at the interface. The implantation has been simulated through two stages. First, inserting the implant until the final prescribed position (Fig. 1.c). Second, releasing the implant to a zero-force configuration (Fig. 1.c).

The work performed by the implant in this stage was calculated using:

$$W_{implant} = \sum\nolimits_{i=0}^{N} F_i.\Delta U_i \quad (2)$$

where F is the reaction force and ΔU is the displacement of the implant RN. The summation is performed over the time increments of the implantation stage.

The bone volume loss was calculated as the sum of the deleted elements multiplied by their level of BV/TV.

The von Mises stress, damage distribution and the final position of the implant at zero force were collected at the end of this implantation phase and reported for the coronal midplane view of the bone-implant system.

2.4 Loading-Unloading Cycles

After implantation, one loading-unloading cycle representing the maximum values achieved in a gait cycle during walking was applied to the implant RN. The force and moment values were chosen from Damm et al., 2015 in a manner to represent the closest age to the femur used in this study. The maximum values were all transferred to the RN of the implant rigid body as the values reported initially were measured at the center of the head of the prosthesis in the local Cartesian coordinate system ($F_x = 318N$; $F_y = -798N$; $F_z = 2100N$; $M_x = -0.89N.m$; $M_y = 41.97N.m$; $M_z = 16.02N.m$). The forces and moments were applied in one step starting from zero and reaching the maximum stated value (Fig. 1.c).

2.5 Implant Subsidence, Reversible and Irreversible Micromotions

Implant subsidence was reported as a rigid body translation of the implant by measuring the displacement of the implant RN from implantation stage (zero force) to end of the loading-unloading cycle. To extract the micromotion vectors, the deleted elements at the end of the cycle were omitted from the analysis. The closest bone node and implant node pairs at the beginning of the cycle were identified (interface nodes) by finding the nodes having the minimum distance to the implant surface nodes. The reversible micromotions were quantified as the relative displacement throughout the unloading section whereas the irreversible micromotions were calculated by the relative displacement throughout the whole cycle. Micromotion vectors were decomposed into normal and tangential components with respect to the surface of the implant. Box plots were used to visualize the distributions. Micromotions for all the nodes around the implant were also reported for the loading-unloading section. In addition, a global damage score was extracted by averaging the damage value of all the bone elements including the maximum damage achieved by the deleted elements.

All the simulations have been performed in Abaqus explicit finite element software (Abaqus 6.14 Dassault Systems, France) using a Linux cluster (16 Intel Xeon E5 cores, 256 GB RAM) for 72 h per simulation. In all simulations, variable mass scaling of 0.001 was used at the implantation stage to reduce the analysis time and was reset to initial values at the beginning of the loading stage. As the simulations were all performed in a quasi-static manner, the kinetic energy was ensured to remain less than 5% of the internal energy in the bone phase throughout the total simulation.

3 Results

The implantation force vs displacement curves of the RN node in the z-direction for two cases of low (0.96 mm) and high (1.92 mm) press-fits models are illustrated in Fig. 2. It is observed that the high amount of press-fit would result in overall higher forces especially at the beginning of the implantation stage. The force continues to increase by the further movement of the stem into the canal reaching forces up to 19.5 and 23.7 kN by the end of the implantation stage for low and high press-fit cases, respectively. The oscillations observed in the curves are due to varying contact between the stem and bone locations with changing densities combined with reduction of stiffness due to element deletion.

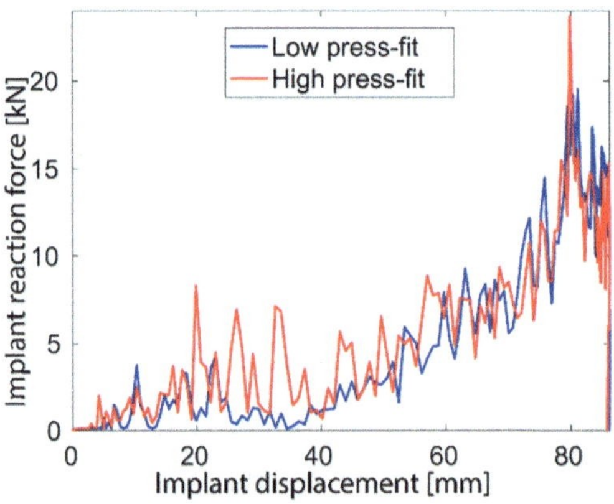

Fig. 2. Implant RN reaction force vs displacement for low (blue) and high (red) press-fit models. (Color figure online)

The implantation work performed by the low and high press-fit models is 367.87 J and 430.09 J, respectively, which translates into 17% higher work required for the high press-fit model.

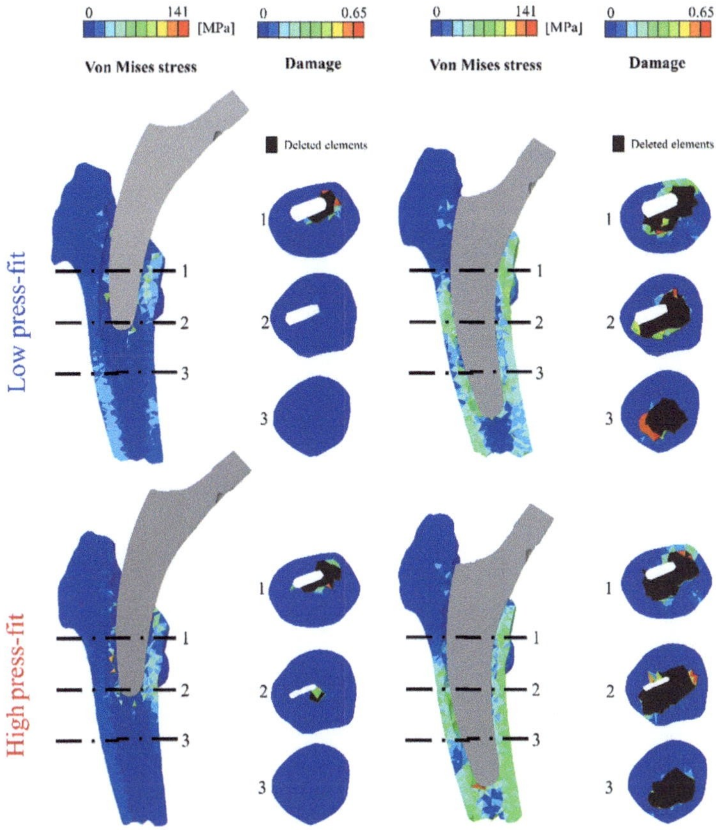

Fig. 3. von Mises stress, damage scores distribution and position of the deleted elements at the beginning and end of implantation stage (zero-force configuration) for low (top) and high (bottom) press-fit models.

According to Fig. 3, the von Mises and damage distribution at the end of implantation stage for the coronal mid-plane view of the femur varies for different press-fit models. The von Mises stress plot shows existence of residual stress in the femur after finishing the implantation procedure. The highest residual stresses are located in the medial and distal part of the femur shaft due to the bending and press-fit occurring during the implant insertion. It is observed that the high press-fit model induces more damage around the implant path, which is visible through the elements that are showing maximum damage. Damage is less at the distal part due to less bone elements in the canal. The bone volume loss is 32% higher for the high press-fit than the low press-fit case. The upward displacement of the implant between the stages where the implant reaches its prescribed position and zero force configuration is 0.081 mm and 0.135 mm for low and high press-fit cases, respectively.

The micromotion distributions around the implant are shown in Fig. 4 for an average of 440 interface nodes. As illustrated, the maximum amplitudes of reversible micromotions are located in the distal part, whereas the highest irreversible micromotions are mostly located in the medial and lateral parts. The maximum levels are 0.44 mm for the irreversible and 0.40 mm for the reversible micromotions.

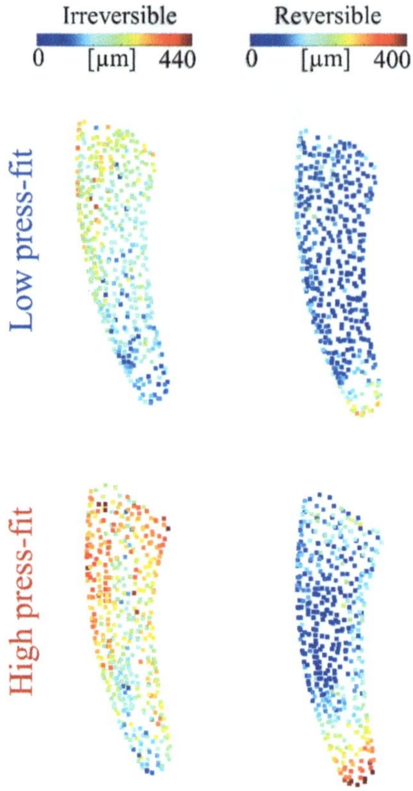

Fig. 4. Irreversible (left) and reversible (right) micromotions for low and high press-fit models.

Figure 5 illustrates the boxplots for the extracted micromotions. The amplitudes are mainly composed of the tangential part due to friction at the interface. The high press-fit model shows 39% and 12% higher values for the median of tangential and normal irreversible micromotions, respectively. The global damage score is 13% higher for the high press-fit model (Fig. 3). The subsidence of the high press-fit model is 23% higher than the low press-fit case.

The sensitivity analysis of this type of modeling on different parameters of mesh size, type, BV/TV and friction has been previously investigated by [37].

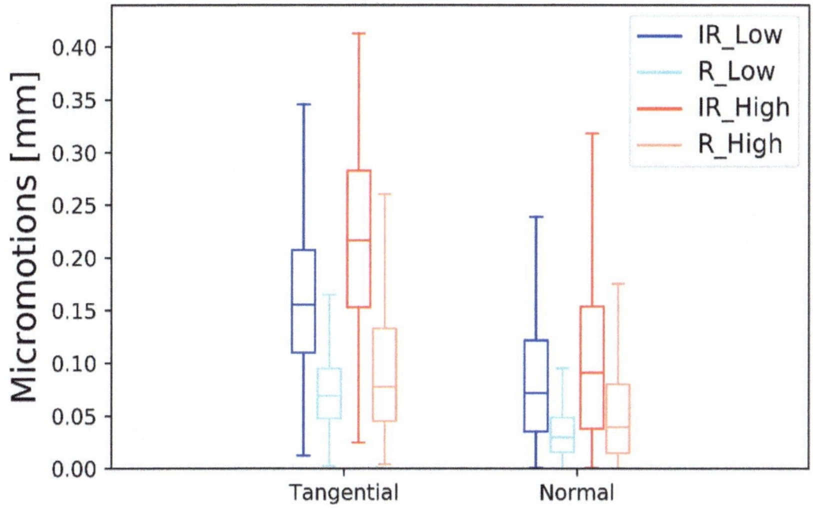

Fig. 5. Box plots of irreversible (IR) and reversible (R) tangential and normal micromotions for low (blue) and high (red) press-fit models. (Color figure online)

4 Discussion

The aim of this study was to demonstrate the ability of an explicit FE approach to quantify primary stability of an uncemented femoral stem by accounting for the bone damage produced during implantation for two levels of press-fit. The explicit FE approach allowed the implant to adjust its trajectory according to the femoral geometry and density. The insertion process was simulated, and evaluated in terms of implantation force, work and stress generated in the femur. Implant subsidence and the distribution of micromotions were extracted and compared for a force-controlled loading cycle.

To the authors' knowledge, this is the first model that allows simulating implant self-positioning in the femur according to contact and bone density variation at the bone-implant interface. The amount of press-fit chosen in this study is high compared to the level reported in the literature but represented a way to investigate implant positioning and micromotions at two relatively different interface conditions. This simple procedure mimics the combination of broaching and implantation and the maximum implantation force and work achieved by the simulation is within the range of the experimental forces (15–70 kN) reported in the literature [25, 42]. The values obtained in this study (19.5 and 23.7 kN) fall in the lower limit of reported force, which may be due to the quasi-static nature of the simulation and the rate-independent material behavior used in this study. This allows a first assessment of primary stability based on the implantation force magnitude [43]. The FE analyses showed slightly higher force and implantation work for the higher interference fit at the final stage of implantation, probably due to the larger amount of bone damage necessary to achieve the same final position. It should be emphasized here that the high press-fit configuration used in this study is higher than a typical clinical situation in order to explore a rather extreme case of both broaching and implantation.

High von Mises stresses produced during implantation can lead to bone fractures and should be avoided [44]. The reason for choosing von Mises stress is comparison with existing data in the literature. In fact, the spatial distribution of von Mises stresses is in good agreement with previous analyses that observed higher values at the medial, distal location [45–47]. The stress amplitudes range up to 140 MPa, which is close to the reported ultimate tensile stress of cortical bone of 150 MPa [44]. Since the amount of press-fit simulated in this study is larger than in a typical hip surgery where broaching precedes implantation, the obtained von Mises stresses should be interpreted as broaching stresses.

The modeling approach allowed to measure the subsidence for loading-unloading cycle. Accordingly, the variation of subsidence with press-fit and BV/TV at the boundaries could be captured. Subsidence of the implant during the loading cycle is 23% higher for the high press-fit configuration. Since no rate-dependent material properties were used in this study, the results would not change by repeating the same loading cycle. However, the implant subsidence associated with fatigue loading over a large number of cycles [3] could theoretically be estimated from a quasi-static damage analysis by increasing the applied load amplitude for a single cycle.

Computation of the reversible micromotions had three main outcomes. 1) The higher amplitudes of reversible micromotions are located at the distal part of the implant. This has been shown both experimentally [46] and by means of FE analysis [17] and is related to the conical stem design achieving the highest load transfer at the proximal part. 2) The reversible micromotions increase with increasing damage. Accordingly, the high press-fit achieved a higher level of micromotions, thus a lower primary stability for the same load. This effect is attenuated by the element deletion procedure adopted in this study as the elements removed at the interface disappear from the micromotion statistics. 3) The median of the reversible micromotions remained below 0.15 mm. This ensures primary stability, osseointegration and prevents therefore aseptic loosening [11, 13]. Ex-vivo μCT [46] measurements indicated a similar range of micromotions as the one observed in our study. The FE models described in the literature provided a broad spectrum of micromotions from below 0.04 mm [45] to values around 1 mm [18]. This is likely due to modeling assumptions such as press-fit, loading conditions and bone material properties. Specifically, our findings are consistent with [31, 47].

The irreversible micromotions also increased by increasing the extent of interference fit. The extent of damage and deleted elements resulted in a different behavior during the loading-unloading cycle for both models. The proportion between median micromotions of low and high press-fit models is similar to the subsidence observed for the implant. Probably due to the friction associated with high normal stresses, the maximum values of irreversible micromotions were observed at the medial and lateral aspects of the femoral stem. The values obtained are similar to the ranges reported in the literature with medians below 0.25 mm [48, 49].

Despite the interesting findings, there are several limitations associated with this study. First, drawing clinical conclusions is too early as no experimental validation was performed directly that included both implantation and loading stages. However, the overall results are in a fair agreement with those reported in the literature. In addition, the amount of press-fit simulated is not clinically relevant and was modeled to illustrate the extreme effects of press-fit and the ability of the model to simulate such conditions. Second, the insertion of a rigid implant at the implantation phase was modeled by translation of the distal tip instead of loading the head of the implant. This made it possible to achieve a stable implantation with a maximum number of degrees of freedom and was confirmed by the level of implantation force. Third, simplified loading and boundary conditions representing in vivo contact forces reported by [50] were applied. The fixation of the femur in the distal part was following the approach proposed by [29] and no muscle forces were transferred to the femur. The consideration of one loading cycle and simplified boundary conditions limited the already high computational costs of the model. Fourth, element deletion does not capture bone compaction. This is the compromise consented to overcome the convergence problems associated to highly distorted elements appearing in implicit FE analyses. This technique influences micromotions at the bone-implant interface as the stiffness and therefore the normal stresses driving frictional forces would drop as an element disappears, especially for high BV/TV values. This could be corrected by adding virtual forces to the adjacent elements to compensate the loss of stiffness [51] or by using meshless methods that would not encounter convergence difficulties due to element distortion [52, 53]. Finally, the use of simplified transversely isotropic bone material with a global alignment along the femoral canal does not reflect rigorously the curvilinear orthotropy of compact and trabecular bone. Nevertheless, BV/TV remains the main determinant of bone stiffness, yield strength and post-yield properties.

5 Conclusion

The explicit nonlinear FE approach developed in this study can simulate the implantation process and its effect on a subsequent loading-unloading gait cycle. The model captures the interface behavior due to the contact and density variation for two distinct levels of initial press-fit. The force and work required for the implantation stage are computed. The subsidence level, amount and location of the micromotions are quantified and the influence of press-fit on the FE results is investigated. Although extensive experimental validation will be necessary to gain confidence in the quantitative predictions, this explicit, non-linear FE model is a proof of concept and represents the first attempt to investigate numerous facets of primary stability of the femoral stem by accounting for damage produced during implantation and post-operative loading.

In the long term, such models may help improve personalized preoperative planning of total hip arthroplasty with implant and press-fit selection designed to achieve the highest primary and long-term stability for a specific femur morphology.

Acknowledgement. This work was co-funded by the RMS Foundation (grant no E16_0001/HOM-FEM). We also gratefully thank Mathys (Bettlach, Switzerland) for providing the CAD files of the Optimys stem and Benjamin Voumard for the CT scan of the femur.

References

1. Levine, B.R., Klein, G.R., Di Cesare, P.E.: Surgical approaches in total hip arthroplasty. Bull. NYU Hosp. Jt. Dis. **65**, 5–18 (2007)

2. Gheduzzi, S., Miles, A.W.: A review of pre-clinical testing of femoral stem subsidence and comparison with clinical data. In: Proceedings of the Institution of Mechanical Engineers, Part H: Journal of Engineering in Medicine, vol. 221, pp. 39–46 (2007)
3. Krismer, M., Biedermann, R., Stöckl, B., Fischer, M., Bauer, R., Haid, C.: The prediction of failure of the stem in THR by measurement of early migration using EBRA-FCA. J. Bone Joint Surg. Br. **81**, 273–280 (1999)
4. Fottner, A., et al.: Influence of undersized cementless hip stems on primary stability and strain distribution. Arch. Orthop. Trauma Surg. **137**(10), 1435–1441 (2017). https://doi.org/10.1007/s00402-017-2784-x
5. Heinecke, M., Rathje, F., Layher, F., Matziolis, G.: The proximal and distal femoral canal geometry influences cementless stem anchorage and revision hip and knee implant stability. Orthopedics **41**, e369–e375 (2018)
6. Bätz, J., et al.: Effect of cavity preparation and bone mineral density on bone-interface densification and bone-implant contact during press-fit implantation of hip stems. J. Orthop. Res. **37**, 1580–1589 (2019)
7. Giebaly, D.E., Twaij, H., Ibrahim, M., Haddad, F.S.: Cementless hip implants: an expanding choice. Hip Int. **26**, 413–423 (2016)
8. Søballe, K., et al.: Fixation of titanium and hydroxyapatitecoated implants in arthritic osteopenic bone. J. Arthroplasty. **6**, 307–316 (1991)
9. Lindberg-Larsen, M., Jørgensen, C.C., Solgaard, S., Kjersgaard, A.G., Kehlet, H., for Fast-Track Hip, L.F.C., Group, K.R.C.: Increased risk of intraoperative and early postoperative periprosthetic femoral fracture with uncemented stems: 7,169 total hip arthroplasties from 8 Danish centers. Acta Orthop. **88**, 390–394 (2017)
10. Meek, R.M.D., Garbuz, D.S., Masri, B.A., Greidanus, N.V, Duncan, C.P.: Intraoperative fracture of the femur in revision total hip arthroplasty with a diaphyseal fitting stem. JBJS. **86**, 480–485 (2004)
11. Pilliar, R.M., Lee, J.M., Maniatopoulos, C.: Observations on the effect of movement on bone ingrowth into porous-surfaced implants. Clin. Orthop. Relat. Res. **208**, 108–113 (1986)
12. Ramamurti, B.S., Orr, T.E., Bragdon, C.R., Lowenstein, J.D., Jasty, M., Harris, W.H.: Factors influencing stability at the interface between a porous surface and cancellous bone: a finite element analysis of a canine in vivo micromotion experiment. J. Biomed. Mater. Res. **36**, 274–280 (1997)
13. Engh, C.A., O'connor, D., Jasty, M., McGovern, T.F., Bobyn, J.D., Harris, W.H.: Quantification of implant micromotion, strain shielding, and bone resorption with porous-coated anatomic medullary locking femoral prostheses. Clin. Orthop. Relat. Res. **285**, 13–29 (1992)
14. Maloney, W.J., et al.: Biomechanical and histologic investigation of cemented total hip arthroplasties. A study of autopsy-retrieved femurs after in vivo cycling. Clin. Orthop. Relat. Res. **249**, 129–140 (1989)
15. Sugiyama, H., Whiteside, L.A., Kaiser, A.D.: Examination of rotational fixation of the femoral component in total hip arthroplasty. A mechanical study of micromovement and acoustic emission. Clin. Orthop. Relat. Res. 122–128 (1989)
16. Freeman, M.A., Plante-Bordeneuve, P.: Early migration and late aseptic failure of proximal femoral prostheses. J. Bone Joint Surg. Br. **76**, 432–438 (1994)
17. Pancanti, A., Bernakiewicz, M., Viceconti, M.: The primary stability of a cementless stem varies between subjects as much as between activities. J. Biomech. **36**, 777–785 (2003)
18. Viceconti, M., Cristofolini, L., Baleani, M., Toni, A.: Pre-clinical validation of a new partially cemented femoral prosthesis by synergetic use of numerical and experimental methods. J. Biomech. **34**, 723–731 (2001)
19. Bah, M.T., Nair, P.B., Taylor, M., Browne, M.: Efficient computational method for assessing the effects of implant positioning in cementless total hip replacements. J. Biomech. **44**, 1417–1422 (2011)

20. Gabarre, S., Herrera, A., Ibarz, E., Mateo, J., Gil-Albarova, J., Gracia, L.: Comparative analysis of the biomechanical behaviour of two cementless short stems for hip replacement: Linea Anatomic and Minihip. PLoS One. **11**, e0158411 (2016)
21. Sas, A., Pellikaan, P., Kolk, S., Marty, P., Scheerlinck, T., Lenthe, G.H. Van: effect of anatomical variability on stress-shielding induced by short calcar-guided stems : automated finite element analysis of 90 femora. **37**(3), 681–688 (2019). https://doi.org/10.1002/jor.24240
22. Taylor, M., Prendergast, P.J.: Four decades of fi nite element analysis of orthopaedic devices : where are we now and what are the opportunities ? J. Biomech. 1–12 (2015). https://doi.org/10.1016/j.jbiomech.2014.12.019
23. Rubin, P.J., Rakotomanana, R.L., Leyvraz, P.F., Zysset, P.K., Curnier, A., Heegaard, J.H.: Frictional interface micromotions and anisotropic stress distribution in a femoral total hip component. J. Biomech. **26**, 725–739 (1993)
24. Taylor, M., Tanner, K.E., Freeman, M.A.R., Yettram, A.L.: Cancellous bone stresses surrounding the femoral component of a hip prosthesis: an elastic-plastic finite element analysis. Med. Eng. Phys. **17**, 544–550 (1995)
25. Shultz, T.R., Blaha, J.D., Gruen, T.A., Norman, T.L.: Cortical bone viscoelasticity and fixation strength of press-fit femoral stems: a finite element model. J. Biomech. Eng. **128**, 7–12 (2006). https://doi.org/10.1115/1.2133765
26. Verdonschot, N.J.J., Huiskes, R., Freeman, M.A.R.: Pre-clinical testing of hip prosthetic designs: a comparison of finite element calculations and laboratory tests. Proc. Inst. Mech. Eng. Part H J. Eng. Med. **207**, 149–154 (1993)
27. Van Rietbergen, B., Huiskes, R., Weinans, H., Sumner, D.R., Turner, T.M., Galante, J.O.: The mechanism of bone remodeling and resorption around press-fitted THA stems. J. Biomech. **26**, 369–382 (1993)
28. Dopico-González, C., New, A.M., Browne, M.: Probabilistic finite element analysis of the uncemented hip replacement—effect of femur characteristics and implant design geometry. J. Biomech. **43**, 512–520 (2010)
29. Abdul-Kadir, M.R., Hansen, U., Klabunde, R., Lucas, D., Amis, A.: Finite element modelling of primary hip stem stability: the effect of interference fit. J. Biomech. **41**, 587–594 (2008). https://doi.org/10.1016/j.jbiomech.2007.10.009
30. Harrigan, T.P., Harris, W.H.: A finite element study of the effect of diametral interface gaps on the contact areas and pressures in uncemented cylindrical femoral total hip components. J. Biomech. **24**, 87–91 (1991)
31. Viceconti, M., Brusi, G., Pancanti, A., Cristofolini, L.: Primary stability of an anatomical cementless hip stem: a statistical analysis. J. Biomech. **39**, 1169–1179 (2006). https://doi.org/10.1016/j.jbiomech.2005.03.024
32. Park, Y., Choi, D., Hwang, D.S., Yoon, Y.-S.: Statistical analysis of interfacial gap in a cementless stem FE model. J. Biomech. Eng. **131**, 021016 (2009)
33. Al-Dirini, R., Huff, D., Taylor, M.: How does surgical variation in stem placement affect the primary stability of cementless femoral stems? In: Orthopaedic Proceedings, p. 6 (2017)
34. Halloran, J.P., Petrella, A.J., Rullkoetter, P.J.: Explicit finite element modeling of total knee replacement mechanics. J. Biomech. **38**, 323–331 (2005)
35. Godest, A.C., Beaugonin, M., Haug, E., Taylor, M., Gregson, P.J.: Simulation of a knee joint replacement during a gait cycle using explicit finite element analysis. J. Biomech. **35**, 267–275 (2002)
36. Abdel-Wahab, A.A., Maligno, A., Silberschmidt, V.: V: Dynamic properties of cortical bone tissue: Izod tests and numerical study. Comput. Mater. Contin. **19**, 217 (2010)
37. Ovesy, M., Aeschlimann, M., Zysset, P.K.: Explicit finite element analysis can predict the mechanical response of conical implant press-fit in homogenized trabecular bone. J. Biomech. **107**, 109844 (2020)

38. Ronneberger, O., Fischer, P., Brox, T.: U-net: Convolutional networks for biomedical image segmentation. In: International Conference on Medical image computing and computer-assisted intervention, pp. 234–241 (2015)
39. Philippe, W.U.W.H.J.Z.: Damage accumulation in vertebral trabecular bone depends on loading mode and direction (2011)
40. Schwiedrzik, J.J., Wolfram, U., Zysset, P.K.: A generalized anisotropic quadric yield criterion and its application to bone tissue at multiple length scales. Biomech. Model. Mechanobiol. **12**(6), 1155–1168 (2013). https://doi.org/10.1007/s10237-013-0472-5
41. Luisier, B., Dall'Ara, E., Pahr, D.H.: Orthotropic HR-pQCT-based FE models improve strength predictions for stance but not for side-way fall loading compared to isotropic QCT-based FE models of human femurs. J. Mech. Behav. Biomed. Mater. **32**, 287–299 (2014). https://doi.org/10.1016/j.jmbbm.2014.01.006
42. Preutenborbeck, M., Reuter, J., Ferrari, E.: Quantitative characterisation of impaction events during femoral broaching in total hip arthroplasty. Med. Eng. Phys. **76**, 13–19 (2020). https://doi.org/10.1016/j.medengphy.2019.12.004
43. Tijou, A., et al.: Monitoring cementless femoral stem insertion by impact analyses: an in vitro study. J. Mech. Behav. Biomed. Mater. **88**, 102–108 (2018). https://doi.org/10.1016/j.jmbbm.2018.08.009
44. Wolfram, U., Schwiedrzik, J.: Post-yield and failure properties of cortical bone. Bonekey Rep. **5**, 829 (2016)
45. Abdul Kadir, M.R., Kamsah, N.: Interface micromotion of cementless hip stems in simulated hip arthroplasty. Am. J. Appl. Sci. **6**, 1682–1689 (2009). https://doi.org/10.3844/ajassp.2009.1682.1689
46. Malfroy Camine, V., Rüdiger, H.A., Pioletti, D.P., Terrier, A.: Full-field measurement of micromotion around a cementless femoral stem using micro-CT imaging and radiopaque markers. J. Biomech. (2016). https://doi.org/10.1016/j.jbiomech.2016.10.029
47. Camine, V.M., Rüdiger, H.A., Pioletti, D.P., Terrier, A.: Full-field measurement of micromotion around a cementless femoral stem using micro-CT imaging and radiopaque markers. J. Biomech. **49**, 4002–4008 (2016)
48. Britton, J.R., Lyons, C.G., Prendergast, P.J.: Measurement of the relative motion between an implant and bone under cyclic loading. Strain **40**, 193–202 (2004)
49. Gortchacow, M., Wettstein, M., Pioletti, D.P., Müller-Gerbl, M., Terrier, A.: Simultaneous and multisite measure of micromotion, subsidence and gap to evaluate femoral stem stability. J. Biomech. **45**, 1232–1238 (2012)
50. Heller, M.O., Bergmann, G., Kassi, J.-P., Claes, L., Haas, N.P., Duda, G.N.: Determination of muscle loading at the hip joint for use in pre-clinical testing. J. Biomech. **38**, 1155–1163 (2005)
51. Harrison, N.M., McDonnell, P., Mullins, L., Wilson, N., O'Mahoney, D., McHugh, P.E.: Failure modelling of trabecular bone using a non-linear combined damage and fracture voxel finite element approach. Biomech Model Mechanobiol. **12**, 225–241 (2013). https://doi.org/10.1007/s10237-012-0394-7
52. Nguyen, V.P., Rabczuk, T., Bordas, S., Duflot, M.: Meshless methods: a review and computer implementation aspects. Math. Comput. Simul. **79**, 763–813 (2008)
53. Belytschko, T., Krongauz, Y., Organ, D., Fleming, M., Krysl, P.: Meshless methods: an overview and recent developments. Comput. Methods Appl. Mech. Eng. **139**, 3–47 (1996)

Hemodynamic Effects of Entry Versus Exit Tear Size and Tissue Stiffness in Simulations of Aortic Dissection

Kathrin Bäumler[1(✉)], Judith Zimmermann[1,4], Daniel B. Ennis[1,5], Alison L. Marsden[2,3], and Dominik Fleischmann[1]

[1] Department of Radiology, Stanford University, Stanford, CA, USA
baeumler@stanford.edu
[2] Department of Pediatrics, Stanford University, Stanford, CA, USA
[3] Department of Bioengineering, Stanford University, Stanford, CA, USA
[4] Department of Computer Science, Technical University of Munich, Munich, Germany
[5] Division of Radiology, Veterans Affairs Health Care System, Palo Alto, CA, USA

Abstract. Based on the patient-specific model of a Type B aortic dissection we created a second model with reduced entry and exit tear size. Two sets of simulations were performed for each model: (i) fluid structure interaction (FSI) and (ii) rigid wall simulations. In both simulation modalities we found that alterations in tear size had substantial impact on true to false lumen flow ratios, true and false lumen pressure differences, and loss of systolic pressure along the dissection. Compared to rigid wall simulations, FSI simulations yielded decreased true lumen flow ratios, increased dampening of flow waveforms along the aorta, smaller negative pressure differences in the distal dissection, and smaller systolic pressure drops across entry and exit tears. These results underline the sensitivity of simulation-based quantitative hemodynamics in aortic dissections to tear size and tissue stiffness.

Keywords: Computational hemodynamics · Fluid-structure interaction · Aortic dissection

1 Introduction

The arterial wall consists of three layers, the innermost intima, the media layer with 40–60 lamellae, and the outermost adventitia. In an aortic dissection the lamellae in the media layer are separated and the intima layer possesses one or more tears (the most proximal one being called the *primary intimal tear* or *entry tear*), allowing blood to penetrate into the arterial wall, the *false lumen*. In most cases of aortic dissection, the blood flow in the false lumen extends to a distal *exit tear* where it merges back into the original flow channel - the *true lumen*.

According to the Stanford classification scheme, aortic dissections are *type A* if the ascending aorta is involved, and *type B* aortic dissections (TBAD) otherwise. TBADs are furthermore considered *complicated*, if signs of rupture, rapid

J. M. R. S. Tavares et al. (Eds.): CMBBE 2021, LNCVB 38, pp. 143–152, 2023.
https://doi.org/10.1007/978-3-031-10015-4_13

expansion, or branch vessel malperfusion are observed [1]. While type A aortic dissections generally require immediate surgery, uncomplicated TBADs are often medically managed and surveilled via noninvasive imaging. However, up to 20%–55% of medically treated patients develop aneurysmal degeneration after 5 years [2].

To date, it is challenging to predict which patients with uncomplicated TBADs will benefit from thoracic endovascular aortic repair (TEVAR), and for which patients the procedural risks outweigh benefits [1–3]. Multiple morphological features have been suggested as prognostic factors of adverse events. More recently also hemodynamic features, such as false lumen pressurization, wall shear stress, false lumen ejection fraction, and others are being discussed. 4D-flow MRI and computational fluid dynamics (CFD) have been proposed as means to investigate hemodynamic parameters in aortic dissections [3–7].

In this study we performed CFD simulations in two patient-specific TBAD models that differ regarding entry and exit tear size. For each model, we compare two simulation modalities, fluid-structure interaction (FSI) and rigid wall simulations, to quantify the effects of tissue stiffness and changes in tear size on hemodynamic parameters in TBAD.

2 Methods

2.1 Model Construction and Mesh Generation

Geometric Model. We obtained computed tomography (CT) imaging data of a 25 year old female TBAD patient with Marfan syndrome in compliance with HIPAA regulations and IRB processes at our institution. In this patient-specific case, the Type B aortic dissection originates distal to the left subclavian artery and ends proximal to the celiac trunk, with no major arteries branching off in the dissected region of the aorta. The lumen of the thoracic aorta was segmented using the active contour algorithm with manual refinements (itk-SNAP v3.4) and Meshmixer (Autodesk, Inc.), Fig. 1. For rigid simulations we built a single model of the fluid domain, for FSI simulations we additionally built a model of the structural domain, comprising the outer vessel wall and the intimal flap. To reduce geometric uncertainties due to variable tissue thickness we followed the description in [7] and prescribed an outer wall and intimal flap thickness of $h = 2$ mm.

Model Augmentation. Based on the patient-specific geometric model, we reduced the entry and exit tear size to 27% of their original size, using the software Meshmixer (Autodesk, Inc.). The entry tear size was reduced from originally 228 mm^2 to 62 mm^2. The exit tear size was reduced from originally 227 mm^2 to 61 mm^2.

The resulting models are denoted as (i) model O with the original entry and exit tear, (ii) model R with the reduced entry and exit tear, Fig. 1.

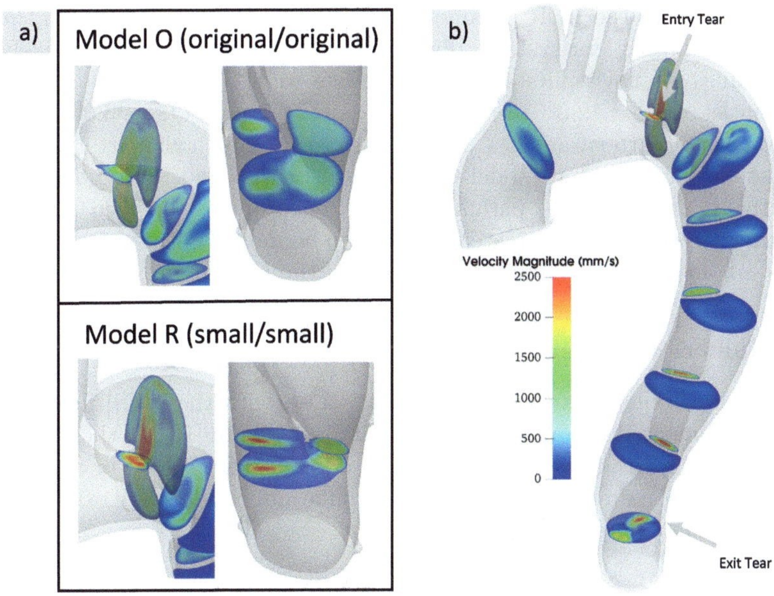

Fig. 1. a) Models O and R with the respective configuration of entry and exit tear sizes. b) Full view of patient-specific model.

Meshing. We created the corresponding unstructured tetrahedral meshes with the TetGen mesh generator which is embedded in SimVascular [8,9]. The target mesh size was set to 1.3 mm which was found sufficiently small in both previous simulations [7], and a mesh study targeted to the present geometry and parameters (results not shown). The final meshes consisted of 1M elements for rigid simulations, and of 1.5M elements for FSI simulations.

2.2 Governing Equations

The fluid was governed by the Navier-Stokes equations for incompressible Newtonian fluids (μ_f = 0.00 392 Pa s, ϱ_f = 1100 kg/m^3). In the rigid simulations these equations were solved in Eulerian formulation on a fixed grid. In FSI simulations, they were solved in arbitrary Lagrangian-Eulerian (ALE) formulation on a moving mesh [7,9]. In the FSI simulations the structural domain represented the arterial wall and intimal flap, which we modeled as a homogeneous, isotropic, Neo-Hookean material (E =1.3 MPa, ϱ_s = 1450 kg/m^3) [7].

2.3 Boundary Data

Boundary conditions for the inlets and outlets of the 3D domain were obtained from Zimmermann et al. [10]. At the inlet of the fluid domain we prescribed a pulsatile flow waveform with a parabolic velocity profile. The outlet boundaries were modeled as three element Windkessel boundaries [7,11]. The pressure targets for tuning the RCR parameters were $P_s = 119\,\text{mmHg}$, $P_d = 42\,\text{mmHg}$, and $P_P = 77\,\text{mmHg}$, for systolic, diastolic, and pulse pressure, respectively. The target values for flow splits across the outlets were $Q_O = 78.4\%$ at the distal aortic outlet, $Q_{A_1} = 12.3\%$ at the brachiocephalic trunk, Q_{A_2} =3.0% at the left common carotid artery, and $Q_{A_3} = 5.2\%$ at the left subclavian artery. The RCR parameters were tuned in model O for the FSI setup and the rigid wall setup, respectively, and the same parameters were subsequently applied in model R. The outer arterial wall was treated as a no-slip boundary in the rigid wall setting, and as a homogeneous Neumann condition in the FSI setting.

2.4 Numerical Solver

The governing equations were numerically solved with the svFSI finite element solver, implemented in SimVascular [9] and described in [10]. In the FSI setting, the equations for the fluid and structural domain were solved in a monolithic fashion [12]. To assure periodicity of the simulation, results were extracted from the fifth cardiac cycle.

3 Results

We evaluated simulation results at pre-defined cross-sectional locations, Fig. 2a. Location S_{BCT} and $S_{\text{EntryTear}}$ were placed proximal to the dissected aorta and through the entry tear, respectively. Locations DAo1 - DAo5 were placed along the dissected aorta with the true lumen slices S_{TL1} - S_{TL5} and the false lumen slices S_{FL1} - S_{FL5}, respectively. Location S_{ExitTear} is located distal to the exit tear.

3.1 Flow Waveforms and False Lumen Flow Ratio

FSI simulations resulted in dampened flow waveforms (i) from $S_{\text{EntryTear}}$ to S_{ExitTear} and (ii) along the dissected false lumen. No dampening of flowrate waveforms was observed in rigid simulations. These observations were similar in model O and model R, results shown for model R in Fig. 2b.

The false lumen flow ratio Q_{FL} was defined as the false lumen net flow rate divided by the true lumen net flow rate, Fig. 2c. In model R the false lumen flow rate was 24% and 23% lower compared to model O in the FSI and rigid simulations respectively. The rigid wall assumption decreased Q_{FL} by 7% in model O and by 6% in model R.

3.2 Pressure

Systolic Pressure Along the Aorta. We extracted the systolic pressure P_{sys} at each cross-section and normalized it with the upstream systolic pressure P^* at S_{BCT} to obtain the dimensionless, normalized systolic pressure $P^*_{sys} = \frac{P_{sys}}{P^*}$, Fig. 4. In FSI simulations, following a path from the ascending aorta through

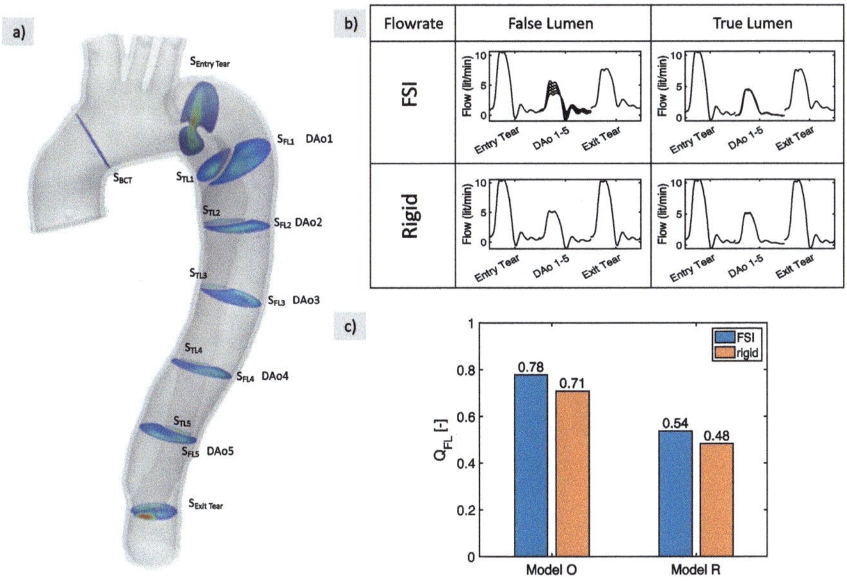

Fig. 2. a) Pre-defined cross-sectional locations along the aorta. b) Transient flowrates in model R (small entry tear/small exit tear) in rigid and FSI simulations. c) False lumen flow ratio.

the entry tear into the false lumen, P_{sys} dropped 2% between S_{BCT} and S_{FL2} in model O with the original entry tear. A reduced entry tear increased the pressure drop to 11% in model R. P^*_{sys} remained constant along the false lumen and no second pressure drop was observed across the exit tear (from S_{FL5} to S_{ExitTear}) in model O with the original sized exit tear. In model R, P^*_{sys} dropped by 3% across the exit tear. Along the true lumen, P_{sys} experienced the majority of pressure loss between S_{TL2} and S_{TL5}, Fig. 3. Rigid wall simulations showed an identical pattern of systolic pressure loss as described above with consistently larger pressure drops compared to the FSI simulations. In the rigid simulations, P^*_{sys} at S_{ExitTear} was 4% smaller in model O and 9% smaller in model R compared to the FSI simulations.

Table 1. Pressure differences between true and false lumen $\Delta P_{\mathrm{TL-FL}} := P_{\mathrm{TL}} - P_{\mathrm{FL}}$.

	Model O FSI	Model O rigid	Model R FSI	Model R rigid
$\Delta P^{\mathrm{MAX}}_{\mathrm{TL-FL}}$	11.9 mmHg	11.1 mmHg	27.7 mmHg	27.3 mmHg
$\Delta P^{\mathrm{MIN}}_{\mathrm{TL-FL}}$	−1.9 mmHg	−3.4 mmHg	−6.2 mmHg	−15.5 mmHg

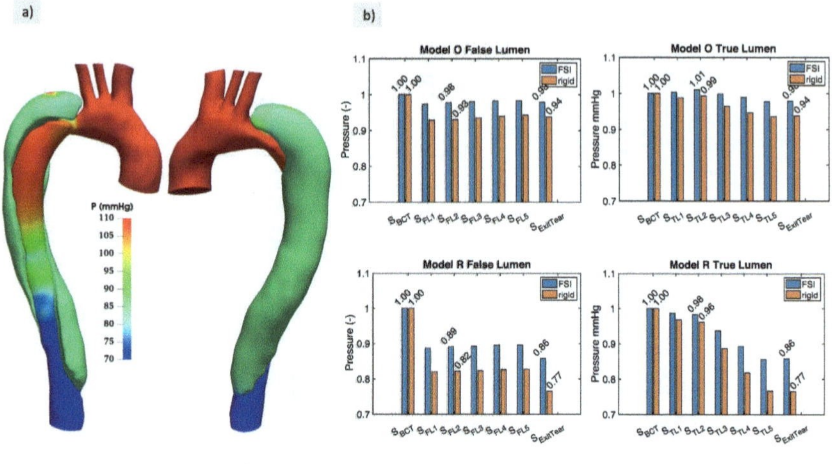

Fig. 3. a) Pressure in model R at $t = 0.2$ s , FSI simulation. b) Dimensionless, normalized systolic pressure P^*_{sys} along aorta in true and false lumen in models O, R.

Pressure Difference Between True and False Lumen. We defined the interluminal pressure difference $\Delta P_{\mathrm{TL-FL}} := P_{\mathrm{TL}} - P_{\mathrm{FL}}$. In all cases, the systolic pressure difference in the proximal dissection was positive ($\Delta P_{\mathrm{TL-FL}} > 0$), and incrementally decreased towards the distal dissection. At DAo 5 minimal negative values $P_{\mathrm{TL-FL}}$ were observed, Fig. 4, quantitative values in Table 1.

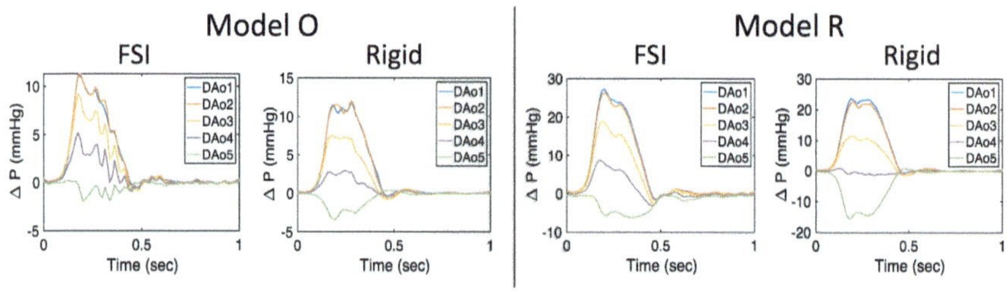

Fig. 4. Pressure differences $\Delta P_{\mathrm{TL-FL}} := P_{\mathrm{TL}} - P_{\mathrm{FL}}$ between true lumen pressure and false lumen pressure at pre-defined cross-sectional locations along the aorta.

3.3 Changes in Cross-Sectional Area

Absolute areas in the false lumen were approximately constant within 6% across all cross-sections. Absolute areas in the true lumen monotonically decreased from 235 mm at DAo 1 to 57 mm at DAo 5 in the undeformed model. We defined the relative area change over the cardiac cycle as $A^* := 100 * (\frac{A_{\text{max}}}{A_{\text{dia}}} - 1)$, where A_{max} denotes the maximal area during the cardiac cycle and A_{dia} denotes the end-diastolic area. In FSI simulations, the relative area changes during the cardiac cycle along the dissected false lumen cross-sections were within 10.4%–12.9%. Along the true lumen, the relative area decreased from 7.2% and 9.2% at S_{TL1} in model O and R respectively to 2.2% and 1.7% at S_{TL5} in model O and R respectively, Fig. 5.

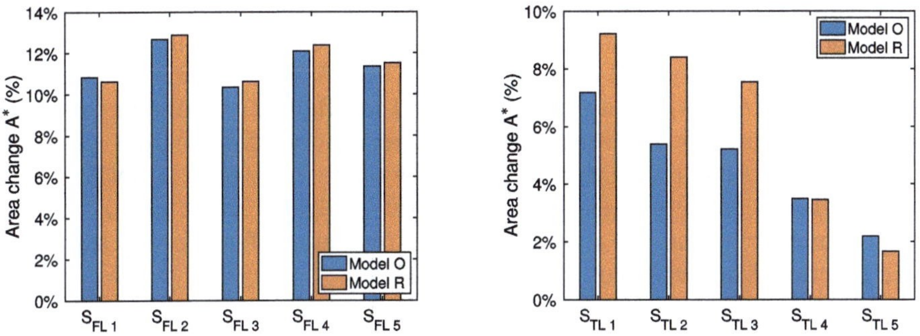

Fig. 5. Relative area change A^* in the false lumen (left) and the true lumen (right) in model O and R in the FSI simulation.

4 Discussion

We first discuss the effect of entry and exit tear size on computational results and subsequently the effect of the rigid wall assumption.

4.1 Effect of Entry and Exit Tear Size on Hemodynamic Results

Effect of Tear Size on Flow Rates and Flow Waveforms. Not surprisingly, the reduced entry and exit tear size limit the amount of blood that flows under otherwise identical flow conditions into the false lumen and consequently decreases the false lumen flow rates Q_{FL} by 24% (FSI simulation) and 23% (rigid simulation), Fig. 2c. Qualitative changes in flow waveforms based on changes in tear size were not observed.

Effect of Tear Size on P^*_{sys}. Altering the size of an entry or exit tear in the model substantially affects peak pressures throughout the model. As one might expect, a small entry or exit tear results in an increased systolic pressure drop across the entry and exit tear, respectively. In our simulations, a reduction of the entry tear size by 73% decreases the normalized, dimensionless, systolic pressure P^*_{sys} by 12% in the FSI simulations and by 17% in the rigid simulation. This comparison of large versus small tear size suggests that a small entry or exit tear can affect blood flow in aortic dissections similar to an aortic coarctation or stenosis. Similarly, peak systolic pressures mirror the changes in vessel cross-sectional areas along the true and false lumen: Along the false lumen, the cross-sectional areas are fairly constant (all cross-sectional areas are within 6% of each other) and correspondingly P^*_{sys} is approximately constant along the false lumen. In the true lumen, cross-sectional areas incrementally decrease from 235 mm at DAo 1 to 57 mm at DAo 5, and correspondingly P^*_{sys} incrementally decreases along the true lumen, Fig. 3.

Effect of Tear Size on $\Delta P_{\mathrm{TL-FL}}$. Pressure differences between true and false lumen are affected by changes in entry and exit tear size, Fig. 4. In the FSI simulations in the proximal dissected aorta (DAo 1), a small entry tear and exit tear increase $\Delta P_{\mathrm{TL-FL}}$ from approximately 11 mmHg to 27 mmHg, and decrease negative pressure differences in the distal dissected aorta (DAo 5) from −2 mmHg to −6 mmHg, Table 1. Negative pressure differences have been hypothesized to promote late adverse events, and in a recent study by Xu et al. [6] the first longitudinal occurrence of a negative mean pressure difference was associated with functional improvement after TEVAR.

4.2 Effect of Rigid Wall Assumption on Hemodynamic Results

Effect of Rigid Wall Assumption Flow Rates and Flow Waveforms. Q_{FL} is 6%–7% smaller in FSI simulations compared to rigid wall simulations. This is consistent with the observation that the false lumen area dilates substantially more than the true lumen area, both in relative and absolute numbers, Fig. 5. Consequently, the relative volume of the false lumen to the true lumen increases in the FSI simulations, potentially reducing resistance to blood flow through the false lumen. Additionally, we observe that FSI simulations and the corresponding distensibility of the arterial wall dampens flow waveforms along the aorta, while no damping was observed in rigid simulations, Fig. 2.

Effect of Rigid Wall Assumption on P^*_{sys}. Deformable vessel walls in FSI simulations reduce the amount of systolic pressure loss across sites of stenosis and along the tapered true lumen. In FSI simulations we observe not only a pulsatile increase in cross-sectional area during the cardiac cycle, but also an inflation of the end-diastolic area (as no prestress of the vessel wall was applied in our simulations). This results in a larger volume of the 3D model during simulation, and potentially mitigates the amount of systolic pressure loss along the aorta when compared to rigid wall simulations.

Effect of Rigid Wall Assumption on $\Delta P_{\mathrm{TL-FL}}$. Maximum values of $\Delta P_{\mathrm{TL-FL}}$ are reached in the proximal dissected aorta (DAo 1 and DAo 2) and agree well between FSI and rigid simulations (<1 mmHg difference in model O and model R, respectively). Minimal values of $\Delta P_{\mathrm{TL-FL}}$ are reached in the distal dissected aorta (DAo 5) and smaller in rigid wall simulations compared to the FSI simulations, especially in model R. In model O at DAo 5 the minimal negative values of $\Delta P_{\mathrm{TL-FL}}$ differ 1.5 mmHg between rigid and FSI simulation, in model R at

DAo 5 the minimal negative values of $\Delta P_{\mathrm{TL-FL}}$ differ 9.3 mmHg. This finding suggests, that while a rigid wall assumption may result in only small to moderate changes in interluminal pressure differences in certain patient-specific cases, it may substantially alter pressure differences in other patient-specific cases.

4.3 Limitations

Our simulations were only performed in a single patient-specific model and patient-specific effects of geometric features such as the tapered true lumen may affect the simulations results. Similarly, all material parameters such as tissue stiffness are subjected to uncertainties. We did not incorporate prestress of the arterial wall in the FSI simulations, and therefore the end-diastolic volume in FSI simulations is larger than the corresponding volume in rigid simulations which may affect simulation results.

5 Conclusions

We find that a reduction of entry and exit tear size to 27% of its original size (i) decrease false lumen flow rates, (ii) substantially increase the loss in nondimensionalized, normalized systolic pressure P^*_{sys}, (iii) amplify maximal and minimal interluminal pressure differences $\Delta P_{\mathrm{TL-FL}}$. These results suggest that small entry and exit tears have the potential to substantially alter hemodynamics in aortic dissection (under otherwise identical flow and morphological conditions).

In the present model, the rigid wall assumption (i) qualitatively changes the dampening of flow waveforms along the aorta, (ii) decreases FL flow rates, (iii) amplifies the systolic pressure loss across sites of stenosis and the tapered true lumen, (iv) amplifies minimal negative pressure differences in the distal dissection, especially in model R.

In general, the effects of varying tear size are captured similarly in the rigid and the FSI simulation. However, interestingly and perhaps surprisingly, we find that the assumption of rigid walls may affect hemodynamic results differently depending on the tear size (and thus morphology of the patient). Our findings suggest that especially when computing interluminal pressure differences ($\Delta P_{\mathrm{TL-FL}}$) and pressure losses across entry and exit tears, different simulation modalities may yield substantially different results.

Validation experiments with measured *in vivo* or *in vitro* data are warranted to further investigate the effects of simulation modalities or morphology in aortic dissection.

Acknowledgment. This work used the Stanford Research Computing Center (SRCC). Additionally, we acknowledge the open-source projects Paraview at www.paraview.org, Meshmixer at www.meshmixer.com, itk-SNAP at www.itksnap.org, and the open-source SimVascular project at www.simvascular.org.

References

1. Erbel, R., et al.: 2014 ESC guidelines on the diagnosis and treatment of aortic diseases. Eur. Heart J. **35**(41), 2873–2926 (2014). https://doi.org/10.1093/eurheartj/ehu281
2. Yuan, X., et al.: Conservative management versus endovascular or open surgery in the spectrum of type B aortic dissection. J. Visualized Surg. **4**, 59 (2018). https://doi.org/10.21037/jovs.2018.02.15
3. Zilber, Z.A., et al.: Noninvasive morphologic and hemodynamic evaluation of type b aortic dissection: state of the art and future perspectives. Radiol. Cardiothorac. Imaging **3**(3) (2021). https://doi.org/10.1148/ryct.2021200456
4. Chen, H.Y., et al.: Editor's choice - fluid-structure interaction simulations of aortic dissection with bench validation. Eur. J. Vasc. Endovasc. Surg. **52**(5), 589–595 (2016)
5. Burris, N.S., et al.: Retrograde flow in the false lumen: marker of a false lumen under stress ? J. Thorac. Cardiovasc. Surg., 1–5 (2018). ISSN 0022–5223. https://doi.org/10.1016/j.jtcvs.2018.06.092
6. Huanming, X., et al.: Computed tomography-based hemodynamic index for aortic dissection. J. Thorac. Cardiovasc. Surg. **162**, e165–e176 (2020)
7. Bäumler, K., et al.: Fluid–structure interaction simulations of patient-specific aortic dissection. Biomech. Model. Mechanobiol. **19**(5), 1607–1628 (2020). https://doi.org/10.1007/s10237-020-01294-8
8. Si, H.: Adaptive tetrahedral mesh generation by constrained Delaunay renement. Int. J. Numer. Methods Eng. **75**, 856–880 (2008). ISSN: 00295981. eprint: 1010.1724. https://doi.org/10.1002/nme.2318
9. Updegrove, A., Wilson, N.M., Merkow, J., Lan, H., Marsden, A.L., Shadden, S.C.: SimVascular: an open source pipeline for cardiovascular simulation. Ann. Biomed. Eng. **45**(3), 525–541 (2016). https://doi.org/10.1007/s10439-016-1762-8
10. Ennis, D.B., Perotti, L.E., Wang, V.Y. (eds.): FIMH 2021. LNCS, vol. 12738. Springer, Cham (2021). https://doi.org/10.1007/978-3-030-78710-3
11. Vignon-Clementel, I.E., Alberto Figueroa, C., Jansen, K.E., Taylor, C.A.: Outflow boundary conditions for three-dimensional finite element modeling of blood flow and pressure in arteries. Comput. Methods Appl. Mech. Eng. **195**(29–32), 3776–3796 (2006). https://doi.org/10.1016/j.cma.2005.04.014
12. Esmaily-Moghadam, M., Bazilevs, Y., Marsden, A.L.: A bipartitioned iterative algorithm for solving linear systems arising from incompressible ow problems. Comput. Methods Appl. Mech. Eng. **286**, 40–62 (2015). https://doi.org/10.1016/j.cma.2014.11.033

Reproducibility of in Vivo Constitutive Parameter Identification Based on 4D Ultrasound Strain Imaging

Andreas Wittek[1,2](✉), Claus-Peter Fritzen[2], Armin Huß[1], and Christopher Blase[1]

[1] Frankfurt University of Applied Sciences, Frankfurt/Main, Germany
{wittek,huss,cblase}@fb2.fra-uas.de
[2] University of Siegen, Siegen, Germany
Claus-Peter.Fritzen@uni-siegen.de

Abstract. Time-resolved three-dimensional ultrasound combined with speckle tracking algorithms (4D ultrasound) is a noninvasive medical imaging technique that provides full field displacement and strain data of aortic wall motion. We have developed a Finite Element Model Updating (FEMU) approach to determine the individual nonlinear and anisotropic constitutive behavior of healthy and diseased human aortae in vivo based on 4D ultrasound data, diastolic and systolic blood pressure, only.

In an in vitro study by our group, the 95% confidence interval of the random error of local strain measurement by 4D ultrasound was determined to be ±2.1%. The uncertainty of the mean of the local strains was slightly smaller (± 1.6%). However, the random error of the mean in repeated measurements is the possible systematic error of local strain values in a single measurement.

The uniqueness of the parameter identification and the effect of the measurement uncertainty on the identified constitutive behavior were examined in numerical experiments. The results indicate that the solution of the inverse problem is unique with regard to the identified stretch-CAUCHY stress curves ($R^2 \geq 0.978$), but non-unique with regard to the parameters of the constitutive equation. These results were confirmed by parameter identification based on strain field data with overlaid random error ($R^2 \geq 0.912$). In contrast, a systematic error of the mean strain (+1.6%) resulted in the identification of a decisively softer material. The numerical verification shows that it is feasible to identify the constitutive behavior of the geometrically irregular aortic wall based on just two load-cases if heterogeneous strain fields are available.

Keywords: In vivo constitutive parameter identification · Human aortic wall · 4D ultrasound · Uncertainty

1 Introduction

Abdominal Aortic Aneurysms (AAA) are a permanent dilation of the human abdominal aorta to a diameter of more than 30 mm, compared to a healthy diameter between 15 mm

J. M. R. S. Tavares et al. (Eds.): CMBBE 2021, LNCVB 38, pp. 153–169, 2023.
https://doi.org/10.1007/978-3-031-10015-4_14

and 20 mm [1]. Since the increase of the aortic diameter is associated with degenerative (micro-) structural changes of the aortic wall, the disease may result in the rupture of the wall, which is associated with high mortality rates between 30% and 80% due to heavy blood loss in short time [2–4]. The threat of immediate rupture is the criterion for surgical or endovascular deactivation of an AAA, which consists of mechanical stabilization of the wall by means of prostheses which carry the load of pulsatile blood pressure instead of the weakened wall. Clinically established indications of immediate rupture risk are either a symptomatic AAA, a diameter ≥5 cm and ≥5.5 cm in women and men, respectively, a growth rate >0.5 cm per year, and saccular shape, i.e. a geometrical configuration with large variation of local radii of curvature [1]. These criteria correspond to a rupture risk of about 10%. This threshold is chosen for clinical intervention because it exceeds the complication rate which is associated with both, open surgery and endovascular treatment [3, 5–7]. It is widely acknowledged that more precise individual rupture risk predictors would help to avoid ruptures in small AAA, which are currently not treated, as well as unnecessary treatment of large but stable AAA: in some patient collectives up to 25% of AAA with a diameter <5 cm ruptured whereas up to 64% of AAA ≥7 cm remained stable [8, 9].

Because the rupture of the AAA wall represents its mechanical failure, the engineering concept of mechanical strength verification has been adopted in order to obtain additional patient-individual rupture risk predictors, cf. [10–12] and citations therein. Mostly, the individual wall stress that results from the loading of the wall by pulsatile blood pressure and residual stresses is estimated by nonlinear Finite Element (FE) analyses whereas statistical models have been proposed to estimate the tensile strength of the individual wall noninvasively [11, 13]. All parameters that determine the mechanical boundary value problem to be solved, i.e. irregular geometry, load and boundary conditions as well as constitutive behavior, show large interindividual variations [11, 14]. Therefore, appropriate FE models for computing AAA wall stresses and estimating rupture risk have to be patient-individual with regard to these parameters. This is still a challenge because all in vivo data that are acquired for research purposes, only, on which the models are based have to be acquired noninvasively and radiation free. Thus, most of the proposed FE models are patient-specific with regard to the geometrical configuration which is obtained from medical imaging data and with regard to blood pressure as the most relevant load, only. In contrast, the individual "pseudo-elastic" [15] nonlinear and orthotropic constitutive behavior of the AAA wall usually is unknown in a clinical setting, cf. e.g. [11, 12, 16].

The inverse identification of constitutive parameters tends to be an ill-posed problem, since few load cases can be observed noninvasively in vivo, only. Usual medical imaging techniques are either too slow (MRI) to capture the dynamics of the cyclic deformation of the human aorta or do not provide the complete irregular 3D configuration (2D ultrasound). Moreover, minimum (diastolic) and maximum (systolic) blood pressure can be measured noninvasively in vivo, only. Therefore, several approaches to in vivo material identification have restricted themselves to linear elastic tangent moduli of the nonlinear material, e.g. [17–19]. Other approaches have assumed a regular cylindrical geometry, which prevents their application to irregularly shaped aneurysms, cf.

[13, 20] and citations therein. It has been shown that the use of heterogeneous deformation fields improves the reliability of constitutive parameter identification in cases where only few load cases can be observed [21]. Several groups have presented approaches to the identification of the nonlinear and anisotropic constitutive behavior of geometrically irregular aneurysms that are based either on full field deformation data that were obtained in laboratory experiments [21–24] or on virtual, numerically generated full field data [20, 25–27]. Recently, Liu et al. have applied the multi-resolution direct search approach to in vivo CT and blood pressure data of ascending thoracic aortic aneurysm patients [28]. They compared the constitutive parameters that were identified based on in vivo data to those that were obtained from biaxial tensile tests on excised tissue from the same patients in vitro. However, deviations between the constitutive parameters that were identified based on in vivo and on in vitro data were large.

In collaboration with the Heart Center of the University Hospital Marburg and the Department of Vascular and Endovascular Surgery of the Goethe University Hospital Frankfurt am Main, our group has established a method for noninvasive in vivo full-field measurement of all components of the in-plane strain tensor using time-resolved three-dimensional ultrasound imaging combined with speckle tracking algorithms (4D ultrasound) [29–31]. Based on these data, we have developed a Finite Element Model Updating (FEMU) approach to the identification of the individual constitutive behavior of the aortic and aneurysmal wall that was applied to in vivo data of volunteers and patients of different age and suffering from various cardiovascular diseases. Reasonable and characteristically different material properties were identified [32, 33].

In the current paper, we present the results of a numerical verification experiment by which we have investigated the reproducibility of the proposed approach and the uniqueness of the identified constitutive behavior and load-free configuration of the aortic wall. In further numerical experiments, we have evaluated the effect of the uncertainty of 4D ultrasound strain measurement on the identified constitutive behavior and load-free configuration.

2 Methods

2.1 Real-Time 3D Ultrasound Imaging Combined with Speckle Tracking (4D Ultrasound)

In vivo full field deformation measurements of aortic wall segments were performed using a customized commercial real-time 3D ultrasound device (Artida®, Toshiba Medical Systems, Otawara, Japan) that was equipped with a 3D transthoracic probe (Toshiba, PST-25SX, 1–4 MHz phased array matrix transducer). Combined with speckle tracking analysis using a commercial algorithm (Advanced Cardiac Package (ACP), Toshiba Medical Systems, Otawara, Japan), this imaging technique does not only allow for the measurement of the global change of shape of the imaged aortic segments but for the tracking of the motion of specific tissue regions (Fig. 1).

A customized data interface allows for the export of position vector fields of up to 1188 measuring points on the arterial wall for each of between 15 and 40 temporal frames throughout an entire cardiac cycle as ASCII files. These describe the trajectories

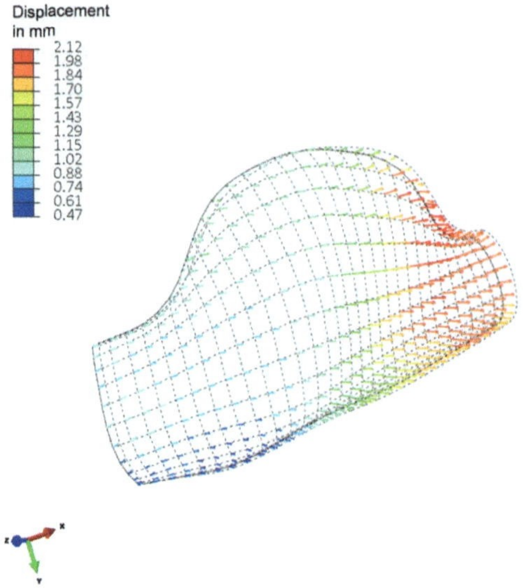

Fig. 1. Two configurations of a small AAA with a maximum diameter of 31.8 mm and displacement field of material points on the surface of the AAA wall as measured by 4D ultrasound.

of distinct tissue regions and can be understood as motion functions of *discrete* material points X of the aortic wall in a continuum mechanical sense:

$$\boldsymbol{x}_{i,k} = \boldsymbol{x}_i\left(\boldsymbol{X}_{i,ref}, t_k\right) = \boldsymbol{X}_{i,ref} + \boldsymbol{u}_i(t_k) \tag{1}$$

where $\boldsymbol{X}_{i,ref}$ is the position vector of a material point X_i in the deformed end-diastolic reference configuration, $\boldsymbol{x}_{i,k}$ is the position vector of the same material point at the time t_k of the cardiac cycle, and $\boldsymbol{u}_i$ is the displacement between both.

In a recent study, we have determined the uncertainty of 4D ultrasound displacement and strain measurement. Using a custom-built inflation-extension setup, a tubular specimen of porcine aorta was loaded physiologically by axial prestretch that was held constant and a cyclic pressure amplitude that was applied with a frequency of 1 Hz. The locally varying deformation of the specimen was measured in parallel by 4D ultrasound and optically. In repeated measurements, 4D ultrasound did not show any systematic error compared to the optical reference. Random error of displacement measurement was characterized by an interquartile range (IQR) of 0.21 mm in propagation direction of the ultrasound signal and IQR = 0.18 mm orthogonal to the propagation direction. The 95% confidence interval of the measurement of local and mean strain was ±2.1% and ±1.6%, respectively. It must be noted that the random error of the mean of local strain values that was determined from repeated measurements, may indicate the systematic error of a single measurement. In clinical examinations, single ultrasound measurements are performed and evaluated for diagnostic purposes, usually. Therefore, a systematic error of 1.6% strain must be expected for single 4D ultrasound measurements.

2.2 Continuum Mechanical Framework

The deformation of the aortic wall and its passive elastic properties are described in this work based on the theory of finite non-linear elastic deformations. The right stretch tensor $\boldsymbol{U}$ and the *nominal* or BIOT's strain tensor $\boldsymbol{\varepsilon}$ are used for the representation of the local deformation of aortic wall segments. These rotation-free 2nd order tensors are suited for large deformations. They can be obtained from the measured discrete material point fields as follows: According to the polar decomposition theorem, the local deformation gradient $\boldsymbol{F}_{i,k} = \partial \boldsymbol{x}_{i,k}/\partial \boldsymbol{X}_{i,ref}$ can be decomposed uniquely into the positive definite symmetric second-order right stretch tensor $\boldsymbol{U}_{i,k} = \lambda_{i,k}\boldsymbol{N}_i \otimes \boldsymbol{N}_i$ and a proper orthogonal second-order tensor $\boldsymbol{R}_{i,k}$ such that $\boldsymbol{F}_{i,k} = \boldsymbol{R}_{i,k} \cdot \boldsymbol{U}_{i,k}$. The principal stretches, i.e. the ratios of current length and reference length in the principal directions of the current and the reference configuration, respectively, $\lambda_{i,k} = l_i(t_k)/l_{i,ref}$, $i = 1, 2, 3$ are the Eigenvalues of $\boldsymbol{F}_{i,k}$. The BIOT's strain tensor is defined as $\boldsymbol{\varepsilon}_{i,k} = \boldsymbol{U}_{i,k} - \boldsymbol{I}$. The CAUCHY- GREEN strain tensor that is used as argument of constitutive equations in this paper may be defined by means of the right stretch tensor: $\boldsymbol{C}_{i,k} = \boldsymbol{U}^2_{i,k}$.

The invariant-based Holzapfel-Gasser-Ogden model [34] that is available as built-in feature of Abaqus® (Dassault Systèmes Simulia Corp., Vélizy-Villacoublay, France) was chosen as constitutive equation. As the aortic wall is usually considered to be incompressible [35], the constitutive equation is represented here without the distinction of a deviatoric and a volumetric term that is relevant for compressible materials, only:

$$\Psi(\boldsymbol{C}, \boldsymbol{a}_{0i} \otimes \boldsymbol{a}_{0i}) = \mu(C_1 - 3) + \frac{k_1}{2k_2} \sum\nolimits_{\alpha=1}^{2} \left(e^{k_2 E_\alpha^2} - 1\right) \tag{2}$$

where $E_\alpha = \kappa(C_1 - 3) + (1 - 3\kappa)(C_{0\alpha} - 1)$, and $C_1 = \mathrm{Tr}(\boldsymbol{C})$ is the first invariant of the CAUCHY- GREEN strain tensor $\boldsymbol{C}$. $C_{0\alpha} = \boldsymbol{C} : \boldsymbol{a}_{0\alpha} \otimes \boldsymbol{a}_{0\alpha}$ ($\alpha = 1, 2$) denotes a mixed tensor invariant of $\boldsymbol{C}$ and the structural tensor $\boldsymbol{a}_{0\alpha} \otimes \boldsymbol{a}_{0\alpha}$ that governs the anisotropic response of the constitutive equation. $\boldsymbol{a}_{01}$ and $\boldsymbol{a}_{02}$ are spatial vectors within the wall plane that denote the main directions of anisotropy in the reference configuration. They are assumed to be symmetric to the longitudinal and circumferential directions of the vessel and depend on the direction angle $0° \le \varphi \le 90°$. The strict convexity of the constitutive equation restricts the parameter space to $\mu > 0$, $k_1 > 0$ and $k_2 > 0$. The parameter κ $(0 \le \kappa \le 1/3)$ determines the degree of the anisotropy of the material.

2.3 Finite Element Model Updating Approach to the Identification of the Individual Constitutive Behavior

The developed Finite Element Model Updating (FEMU) approach to in vivo constitutive parameter identification is described in detail in [32, 33]. A flowchart of the FEMU is shown in Fig. 2. The idea of this approach is to minimize the difference between the measured strain fields and those that are estimated by the FE analysis under the assumption of a set of constitutive parameters.

Measured Strain Fields: The left-hand side of the flow chart indicates the heterogeneous fields of local in-plane strains (longitudinal, circumferential, and shear) that were obtained from in vivo 4D ultrasound measurements. These are no "true" strains with

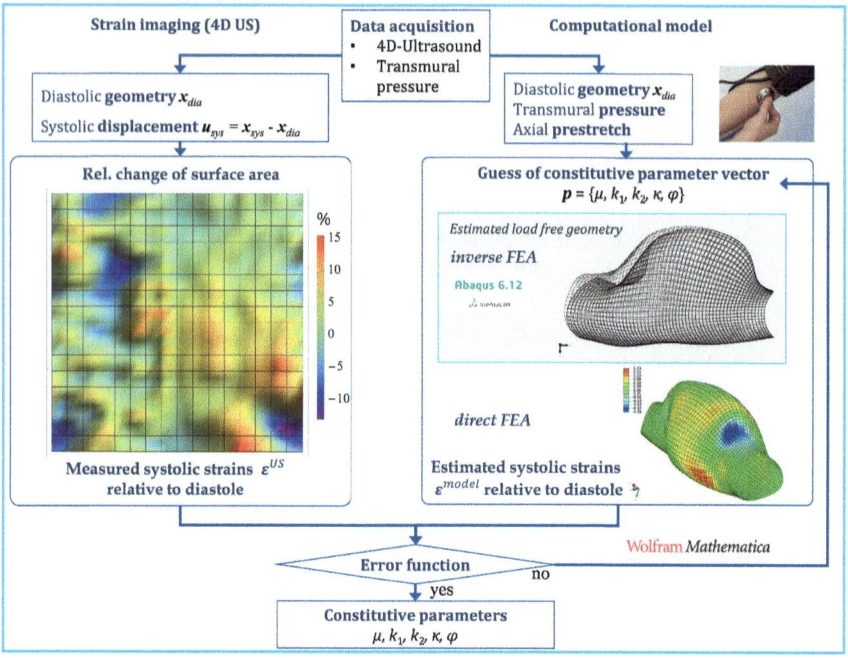

Fig. 2. Flow chart of the Finite Element Model Updating approach.

regard to a load- and stress-free reference configuration (which cannot be accessed in vivo) but give the systolic deformation with regard to the chosen diastolic reference configuration (cf. [17, 29] for more detailed information).

FE Model: The right-hand side of the flow chart indicates the computational model, which estimated the strain distribution based on the assumption of a set of constitutive parameters. Using in-house software that was written in Visual Basic for Applications (Microsoft Corp., Redmond, WA, USA),

- the deformed diastolic reference configuration was reconstructed from 4D ultrasound data,
- discretized into 4-node shell elements in such a way that the measured diastolic position vectors $\boldsymbol{x}_{dia}$ were preserved as nodes in the mesh, and
- local cartesian nodal and element coordinate systems were defined on the irregular geometries indicating the physiologically relevant longitudinal (11), circumferential (22), and radial (33) directions of the vessel.

In Abaqus Standard finite element analysis finite-strain shell elements of type S4R with reduced integration and hourglass control were used. Since the wall thickness cannot be measured in vivo, a reference wall thickness was assumed for the imaged segment according to literature data. Population averaged data were used here that, however, could be personalized with regard to age and type of pathology. The model was loaded by a hydrostatic transmural pressure and a reduced axial force $\boldsymbol{F}_{red}$ that was estimated based on literature data for age-dependent axial prestrech [36] and held constant in all load steps. According to [37] the abdomen behaves as a hydraulic system. Thus, the influence of surrounding organs and tissue was represented in the model by subtracting the intra-abdominal pressure - that acts on the wall from the outside - from the blood pressure that was measured individually.

The geometry of the FE model was derived from 4D ultrasound imaging data of a deformed configuration that corresponded to a loading of the wall by diastolic transmural pressure and $\boldsymbol{F}_{red}$. However, the mesh of the reconstructed model was stress and strain free. The application of physiological loads and boundary conditions would have resulted in artificial deformation of the model. Therefore, an iterative optimal design algorithm [38] was used to estimate the load-free configuration of the vessel segment and update the reference configuration of the FE model accordingly. The solution of this inverse problem depended on the assumed constitutive behavior. Physiological loading of the estimated load-free geometry resulted in the imaged diastolic configuration including the corresponding prestresses and prestretches. A Hausdorff distance of 0.01 mm between the measured and estimated deformed diastolic configuration was chosen as convergence criterion.

In a two-step direct FEA, diastolic and systolic transmural pressures and the constant reduced axial force $\boldsymbol{F}_{red}$ were applied to the estimated load-free configuration. The respective strain fields with regard to the load-free configuration were computed. Finally, systolic deformation with regard to the chosen diastolic deference was obtained by post-processing of the results.

Optimization Procedure: The fitting of the model to the experimental data was a non-linear, multivariate, and multi-objective optimization problem: The chosen constitutive Eq. (2) and, thus, the error function that was to be minimized in the optimization procedure, depended nonlinearly on five parameters. The identified parameters of the constitutive equation had to characterize all three independent components of the orthotropic elastic behavior. Multi-objectivity was taken into account by the construction of the error function: it was composed of the weighted sums of squared residuals of all in-plane strain components predicted by the FE model ($\varepsilon_{i,j}^{model}$) compared to measured strain values ($\varepsilon_{i,j}^{US}$):

$$f_{err} = \frac{1}{n}\sum_{i}\sum_{j=1}^{n} w_i\left(\varepsilon_{i,j}^{model}(\mu, k_1, k_2, \kappa, \varphi) - \varepsilon_{i,j}^{US}\right)^2, \tag{3}$$

where i marks the in-plane components of the BIOT's strain tensor with regard to local element coordinates (11 - axial, 22 - circumferential, 12 - shear). n is the number of elements in the model, w_i is the weighting factor for the squared residuals of the axial (w_{11} = 2/5), circumferential (w_{22} = 2/5) and in plane shear (w_{12} = 1/5) components, respectively.

In an iterative process, the error function was minimized with regard to the constitutive parameters. In order to find a solution as close as possible to the global minimum of the error function, a mixed stochastic-deterministic approach was chosen: The six best out of 300 constitutive parameter guesses that were obtained using a Monte-Carlo algorithm were used as initial values for a deterministic nonlinear Nelder and Mead simplex algorithm [39]. For each of the six initial parameter guesses, the deterministic optimization was performed three times with randomized construction of the start simplex. Because the determination of the load-free configuration from the current reference configuration and corresponding loads depends on the chosen constitutive parameters, this inverse problem was solved for each guess of parameters, i.e. in each iteration of the optimization process. The optimal design algorithm, Monte-Carlo and Nelder and Mead simplex algorithms were coded in-house using Mathematica (Wolfram Research, Champaign, IL., USA).

2.4 Numerical Verification Experiment

The numerical experiment performed to verify the parameter estimation procedure was designed as follows: A reasonable but arbitrarily chosen parameter set (master parameter vector, see Table 1) was assigned to a likewise chosen geometry. This constituted the load-free reference configuration (master model). Using Simulia Abaqus® 6.12 the direct elastostatic problem was solved for two load steps:

Table 1. Comparison of the constitutive parameters of the master material and the parameter values that were identified based on numerically generated data. HD – Hausdorff distance.

	μ_1 in MPa	k_1 in MPa	k_2	κ	φ in°	HD in mm
Master	0.072	0.567	353.64	0.261	68.8	
Estimates based on input data without noise						
Estimate 1	0.072	0.593	331.92	0.260	68.5	0.02
Estimate 2	0.071	0.576	352.12	0.261	68.7	0.06
Estimate 3	0.074	0.500	353.00	0.260	68.3	0.08
Estimate 4	0.071	0.649	323.10	0.261	68.7	0.06
Estimate 5	0.073	0.482	367.10	0.262	68.8	0.10
Max. Deviation in %	3.2	– 15.0	– 8.6	±0.4	– 0.5	
Estimate based on data with superimposed error						
Random Error	0.068	0.883	277.87	0.258	70.2	0.30
Deviation in %	– 4.2	55.7	– 21.4	– 0.8	2.0	
Systematic Error	0.066	0.307	157.97	0.252	66.2	1.35
Deviation in %	– 7.2	– 45.9	– 55.3	– 3.4	– 3.9	

(i) 55 mmHg transmural pressure and a reduced axial force F_{red} inducing axial prestrain of $\varepsilon_{11} = 12\%$, and
(ii) 110 mmHg, F_{red} was held constant.

Resulting in-plane deformation fields were computed for these load steps. Out of these assumed and numerically generated data, only those that were available by measurements in the in vivo situation were used as input data for constitutive parameter identification with the FEMU approach: transmural pressures of 55 and 110 mmHg, the diastolic configuration $\boldsymbol{x}_{dia}$ that corresponded to load case (i) and the displacement field $\boldsymbol{u}_{dia-sys}$ that mapped this configuration onto the systolic configuration $\boldsymbol{x}_{sys}$ that corresponded to load case (ii). These three vector fields described the motion of the aortic wall with regard to the deformed reference configuration $\boldsymbol{x}_{dia}$, not with regard to the load-free geometry. The distribution of in-plane longitudinal and circumferential stretches that corresponded to this motion is shown in Fig. 3.

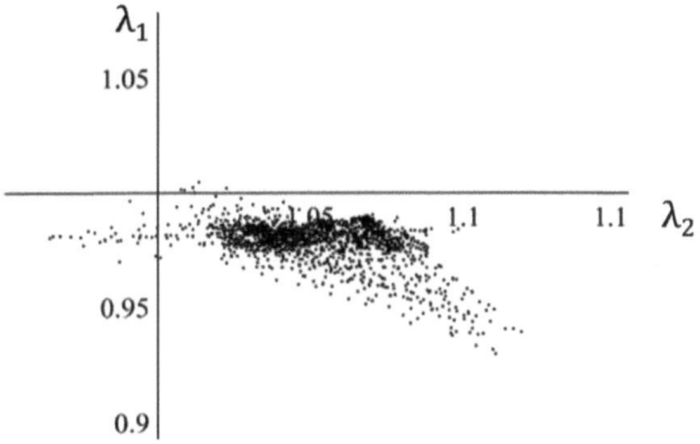

Fig. 3. Field of local deformations between the numerically generated diastolic and systolic configurations of an aortic segment. λ_1 and λ_2 are the longitudinal and circumferential stretches, respectively.

This numerical experiment was repeated for five times in order to assess the repeatability and uniqueness of the inverse constitutive parameter identification. In a second experiment, the numerically generated systolic strain field data were corrupted by a random error of $\pm 2.1\%$ that was identified as 95% confidence interval of local strain measurement by 4D ultrasound in the validation study. In a third numerical experiment, a systematic error was superimposed on the strain field data that were assumed to represent the ground truth: the corrupted data were obtained from the ground truth strain field by multiplication with a constant factor c. In this way a systematic error could be induced that was not constant in all elements but varied smoothly depending on the spatial location. c was chosen so that the mean circumferential strain deviated from the true mean by $+1.6\%$, which was the upper limit of the possible systematic error of circumferential strain in a single measurement:

$$c = 1 + \frac{0.016}{\bar{\varepsilon}_{22}}, \tag{4}$$

where $\bar{\varepsilon}_{22}$ is the arithmetic mean of local circumferential strain values. The same systematic error was applied to all in-plane strain components. Figure 4 shows numerically generated in-plane strain fields with and without superimposed noise.

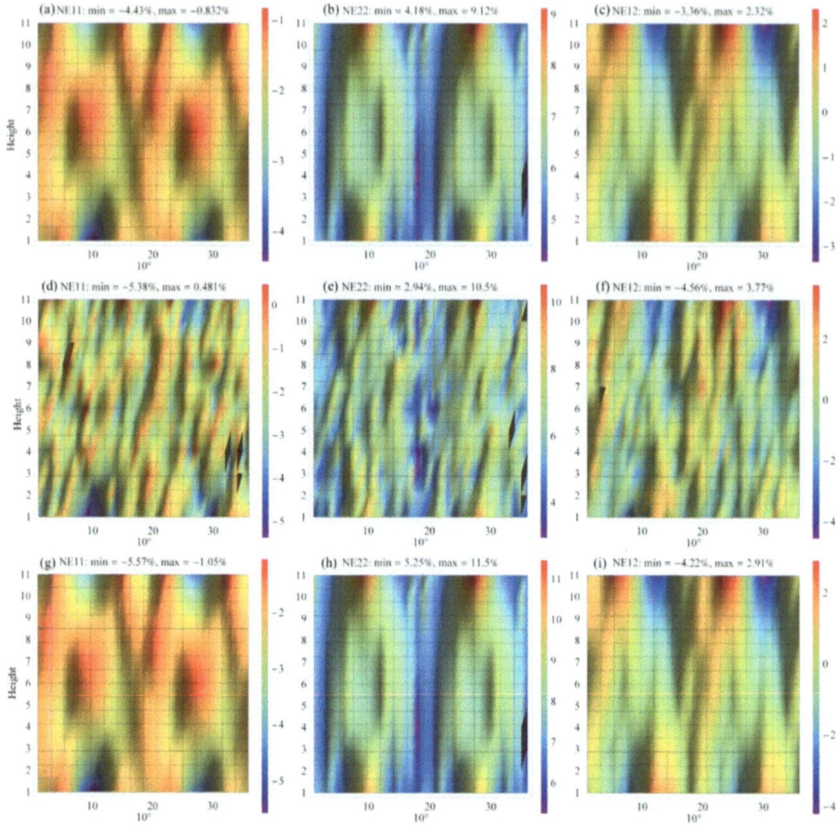

Fig. 4. Numerically generated in-plane strain fields without superimposed error (a, b, c), with superimposed random error (d, e, f), and with superimposed systematic error (g, h, i). NE11, NE22, and NE12 indicate the longitudinal, circumferential, and shear strain in the wall plane, respectively. 'Height' and '10°' indicate the position of the tissue segment in longitudinal and circumferential direction of the imaged wall segment, respectively. 0° and 18° correspond to the anterior and posterior face of the vessel wall. Height 1 and 11 indicate the cranial and caudal endings of the imaged wall segment, respectively.

In order to compare the resulting material properties to the known master material, the principal in-plane Cauchy stresses σ_1 and σ_2 were plotted for both parameter sets along three deformation paths in the λ_1, λ_2-plane: uniaxial longitudinal (I) and circumferential stretch (III), and equi-biaxial stretch (II). The coefficient of determination R^2 was calculated to judge the quality of fit of the stretch-stress curves of the identified material along each of the stretch paths to the respective curves of the given master material.

As already stated above, the inverse identification of the load-free vessel wall configuration depended on the chosen constitutive parameter vector. The Hausdorff distance between the inversely estimated and the true load-free configurations was calculated for each identified constitutive parameter set as a measure for the accuracy of the recovery of the load free master configuration.

3 Results

A comparison of the constitutive parameters characterizing the master material and the five estimates that were identified based on strain field data without superimposed error is given in Table 1. Maximum deviations of parameter values between −15% and +3.2% were observed for the three parameters that govern the stiffness of the material, μ, k_1, and k_2. In contrast, very small deviations <1% were observed for κ and φ, the two parameters that characterize the degree and the preferred directions of the anisotropy.

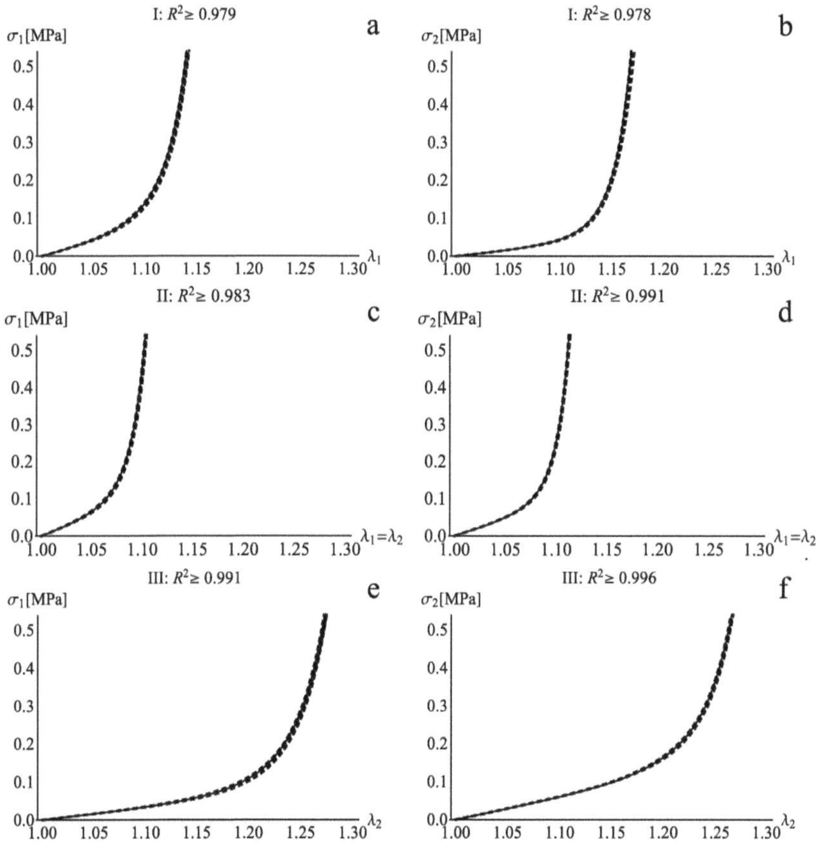

Fig. 5. Comparison of the CAUCHY stress functions for the master parameter set (solid line) and the five estimates (dashed lines) of the constitutive parameter vector along selected stretch paths. a and b: uniaxial longitudinal stretch (I), c and d: equi-biaxial stretch (II), e and f: uniaxial circumferential stretch (III). σ_1 and σ_2 are the principal in-plane CAUCHY stresses. The coefficient of correlation for the master stress-stretch curve and the worst estimate is given for each stretch path.

Despite the observed deviation of the parameter values, the principal CAUCHY stress-stretch curves of the five estimates along the different deformation paths are almost identical to the stress-stretch curves characterizing the master material, which is indicated by R^2-values ≥ 0.978 (Fig. 5). In all cases, the load-free configuration was recovered with a Hausdorff distance ≤ 0.10 mm (cf. Fig. 6).

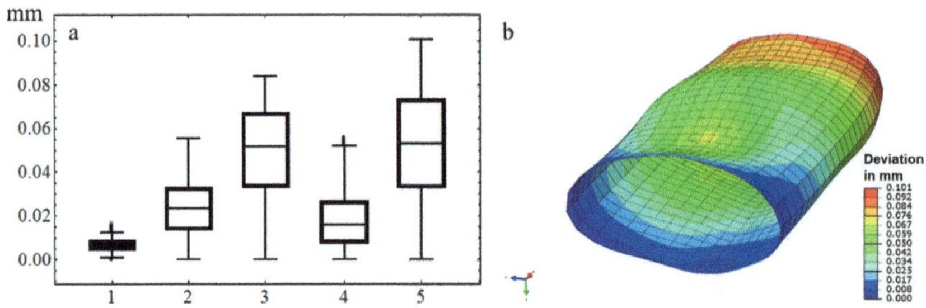

Fig. 6. Deviation between the load-free configuration of the master model and the five configurations that were inversely identified based on strain distibutions without superimposed error. a) Box-Whisker plots of the distributions of the nodal deviations of the five estimates from the initially given configuration. Upper and lower boundaries of the boxes indicate the 1st and 3rd quartile, respectively, the solid line at the center of the boxes indicates the median, upper and lower endings of the whiskers indicate the maximum and minimum values, respectively. Outliers are indicated by dots. The maximum value of each distribution marks the Hausdorff distance between given and estimated configuration. b) Nodal deviations of the worst estimate (cf. Table 1) plotted on the initially given configuration.

Based on strain-field data with superimposed random error, the constitutive parameters $\mu = 0.068$ MPa, $k_1 = 0.883$ MPa, $k_2 = 277.87$, $\kappa = 0.258$, and $\varphi = 70.2°$ were identified. This corresponded to deviations of –4.2%, 55.7%, –21.4%, –0.8%, and 2.0%, respectively. Despite this considerably increased deviation, the longitudinal and circumferential stretch-Cauchy stress curves of the identified material showed coefficients of determination $R^2 > 0.9$ along all stretch pathes (Fig. 7). This still indicates a good agreement with the master material. The load-free configuration was recovered with a median deviation of 0.18 mm per material point and a Hausdorff distance of 0.3 mm.

In contrast, the use of strain field data with superimposed systematic error resulted in a larger deviation of all constitutive parameters, but k_1 (see Table 1). Moreover, the parameters μ_1, k_1, and k_2 that govern the stiffness of the material, consistently showed smaller values compared to the master material. This indicates that the identified constitutive behavior was too distensible in all spatial directions. This was confirmed by the comparison of the stress-stretch plots that are given in Fig. 7 (dashed-dotted lines). Also, the accuracy of the identification of the load-free configuration decreased significantly: it was recovered with a median deviation of 0.81 mm and a Hausdorff distance of 1.5 mm, only.

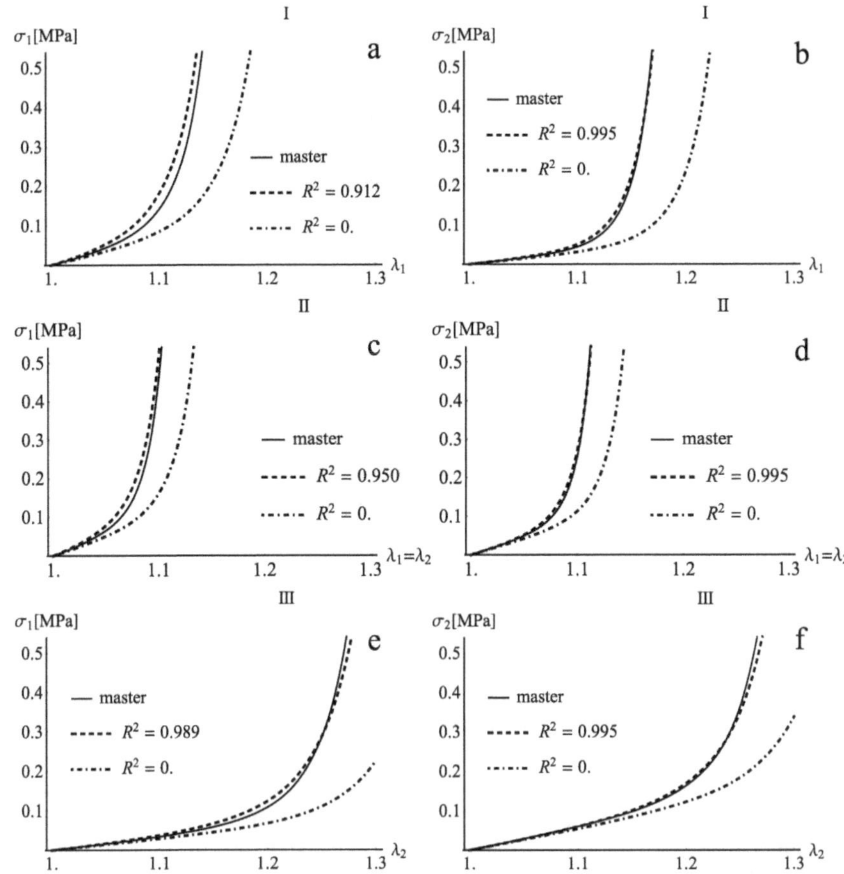

Fig. 7. Comparison of the CAUCHY-stress response of the master material (solid line) and the materials that were identified using the inverse FEMU approach based on numerically generated data with overlaid random (dashed line) and systematic (dashed-dotted line) measurement error. σ_1 and σ_2 indicate the 1st and 2nd principal CAUCHY stresses, respectively. Stress-stretch curves are plotted along selected stretch paths: a and b – uniaxial longitudinal stretch (I), c and d – equi-biaxial stretch (II), e and f – uniaxial circumferential stretch (III).

4 Discussion and Conclusions

The performed numerical verification experiment shows that it is feasible to identify the non-linear and orthotropic constitutive behavior of an aorta-like geometrically irregular structure based on the observation of just two load-cases if heterogeneous fields of all components of the in-plane strain tensor are available. The repeated identification of the constitutive behavior based on strain data that are not corrupted by measurement uncertainty indicates that the solution of the inverse problem is unique with regard to the identified stretch-CAUCHY stress curves, i.e. the elastic behavior at finite deformations, and the load-free configuration, but non-unique with regard to the parameters of the chosen constitutive equation. This applies in particular to the parameters μ_1, k_1, and k_2

which jointly determine the "stiffness" of the material so that changes in the value of one parameter can be compensated by changes of the others, thus resulting in similar constitutive behavior in the physiological range of finite deformation [32]. In contrast, unique parametrization was observed in case of the parameters that determine the degree and main directions of the orthotropy, κ and φ. Therefore, the observed non-uniqueness of the numerical values of μ_1, k_1, and k_2 seems to be a problem of the applied constitutive model [34] rather than the proposed parameter identification strategy. Moreover, it does not impair the possible clinical use of the identified constitutive parameters for wall stress computation in the physiological range: despite the observed variation of the numerical parameter values almost identical wall stresses and corresponding deformations will be computed for given physiological loads.

These results were confirmed by the parameter identification based on strain field data with superimposed random error in the scale of the measurement uncertainty of 4D ultrasound that was determined in a validation study: the deviation of the identified parameters μ_1, k_1, and k_2 from the master parameters is considerably increased, whereas the agreement of the estimated values of κ and φ as well as the agreement of the stretch-CAUCHY stress curves are only slightly affected. Thus, we conclude that the approach is robust with regard to random measurement uncertainty. In contrast, a deviation of the mean of the strain distribution by $+1.6\%$ marks a systematic overestimation of the cyclic deformation and results in the identification of a too distensible elastic behavior and large deviations of the recovered load free configuration.

The data that are required for the proposed approach to the identification of the constitutive behavior, i.e. two deformed configurations, the local displacement field that maps one of these onto the other, and the loads corresponding to both configurations, are available noninvasively in vivo by 4D ultrasound imaging and measurement of diastolic and systolic blood pressure. The use of patient-individual material properties in biomechanical modelling of AAAs is feasible and represents a step towards more personalized rupture risk assessment.

The investigation of the effect of measurement uncertainty on the identification of the constitutive behavior is a first step to the analysis of the sensitivity of the approach to necessary model assumptions. The effects of deviating and locally varying wall thickness and of inhomogeneous material properties will have to be addressed in future work.

Acknowledgements. AAA and aortic geometries that are shown in this paper in Figs. 1, 2, and 6 and those that were used to construct the numerical model for the verification experiment were derived from 4D ultrasound data sets that were already published earlier. The data sets were acquired by W. Derwich, MD (Department of Vascular and Endovascular Surgery of the Goethe University Hospital, Frankfurt am Main) and by K. Karatolios (Department of Cardiology, Universitätsklinikum Marburg und Gießen).

References

1. Deutsche Gesellschaft für Gefäßchirurgie: Leitlinien zum Bauchaortenaneurysma und Beckenarterienaneurysma (2011)

2. Kühnl, A., Erk, A., Trenner, M., Salvermoser, M., Schmid, V., Eckstein, H.-H.: Incidence, treatment and mortality in patients with abdominal aortic aneurysms. Dtsch. Arztebl. Int. **114**, 391–398 (2017). https://doi.org/10.3238/arztebl.2017.0391
3. Frömke, J.: Standardoperationen in der Gefäßchirurgie, 1st edn. Steinkopff-Verlag, Darmstadt (2006)
4. Torsello, G., Debus, E.S., Schmitz-Rixen, T., Grundmann, R.T.: Ultrasound screening for abdominal aortic aneurysms - a rational measure to prevent sudden rupture. Dtsch. Med. Wochenschr. **141**, 1030–1034 (2016). https://doi.org/10.1055/s-0041-108998
5. Collin, J.: UK small aneurysms trial. Lancet **353**, 407–408 (1999). https://doi.org/10.1016/S0140-6736(05)74979-5
6. Patel, R., Sweeting, M.J., Powell, J.T., Greenhalgh, R.M.: Endovascular versus open repair of abdominal aortic aneurysm in 15-years' follow-up of the UK endovascular aneurysm repair trial 1 (EVAR trial 1): a randomised controlled trial. Lancet **388**, 2366–2374 (2016). https://doi.org/10.1016/S0140-6736(16)31135-7
7. Karthikesalingam, A., et al.: The shortfall in long-term survival of patients with repaired thoracic or abdominal aortic aneurysms: retrospective case-control analysis of hospital episode statistics. Eur. J. Vasc. Endovasc. Surg. **46**, 533–541 (2013). https://doi.org/10.1016/j.ejvs.2013.09.008
8. Darling, R.C., Messina, C.R., Brewster, D.C., Ottinger, L.W.: Autopsy study of unoperated abdominal aortic aneurysms. The case for early resection. Circulation **56**, II161–4 (1977)
9. Cronenwett, J.L., et al.: Actuarial analysis of variables associated with rupture of small abdominal aortic aneurysms. Surgery **98**, 472–483 (1985)
10. Vorp, D.A.: Biomechanics of abdominal aortic aneurysm. J. Biomech. **40**, 1887–1902 (2007). https://doi.org/10.1016/j.jbiomech.2006.09.003
11. Humphrey, J.D., Holzapfel, G.A.: Mechanics, mechanobiology, and modeling of human abdominal aorta and aneurysms. J. Biomech. **45**, 805–814 (2012). https://doi.org/10.1016/j.jbiomech.2011.11.021
12. Gasser, T.C.: Biomechanical rupture risk assessment: a consistent and objective decision-making tool for abdominal aortic aneurysm patients. Aorta (Stamford) **4**, 42–60 (2016). https://doi.org/10.12945/j.aorta.2015.15.030
13. Farotto, D., Segers, P., Meuris, B., Vader Sloten, J., Famaey, N.: The role of biomechanics in aortic aneurysm management: requirements, open problems and future prospects. J. Mech. Behav. Biomed. Mater. **77**, 295–307 (2018). https://doi.org/10.1016/j.jmbbm.2017.08.019
14. Kassab, G.S.: Biomechanics of the cardiovascular system: the aorta as an illustratory example. J. R. Soc. Interface **3**, 719–740 (2006). https://doi.org/10.1098/rsif.2006.0138
15. Fung, Y.C., Fronek, K., Patitucci, P.: Pseudoelasticity of arteries and the choice of its mathematical expression. Am. J. Physiol. **237**, H620–H631 (1979). https://doi.org/10.1152/ajpheart.1979.237.5.H620
16. Avanzini, A., Battini, D., Bagozzi, L., Bisleri, G.: Biomechanical evaluation of ascending aortic aneurysms. Biomed. Res. Int. **2014**, 820385 (2014). https://doi.org/10.1155/2014/820385
17. Wittek, A., Derwich, W., Fritzen, C.-P., Schmitz-Rixen, T., Blase, C.: Towards non-invasive in vivo characterization of the pathophysiological state and mechanical wall strength of the individual human AAA wall based on 4D ultrasound measurements. ZAMM J. Appl. Math. Mech./Zeitschrift für Angewandte Mathematik und Mechanik **98**, 2275–2294 (2018). https://doi.org/10.1002/zamm.201700353
18. Farzaneh, S., Trabelsi, O., Avril, S.: Inverse identification of local stiffness across ascending thoracic aortic aneurysms. Biomech. Model. Mechanobiol. **18**(1), 137–153 (2018). https://doi.org/10.1007/s10237-018-1073-0

19. van Disseldorp, E.M.J., Petterson, N.J., Rutten, M.C.M., van de Vosse, F.N., Sambeek, M.R.H.M., Lopata, R.G.P.: Patient-specific mechanical characterization of abdominal aortic aneurysms using 4D ultrasound. Ultraschall Med. **37** (2016). https://doi.org/10.1055/s-0036-1587983
20. Liu, M., Liang, L., Sun, W.: Estimation of in vivo constitutive parameters of the aortic wall using a machine learning approach. Comput. Methods Appl. Mech. Eng. **347**, 201–217 (2019). https://doi.org/10.1016/j.cma.2018.12.030
21. Avril, S., et al.: Overview of identification methods of mechanical parameters based on full-field measurements. Exp. Mech. **48**, 381–402 (2008). https://doi.org/10.1007/s11340-008-9148-y
22. Avril, S., Badel, P., Duprey, A.: Anisotropic and hyperelastic identification of in vitro human arteries from full-field optical measurements. J. Biomech. **43**, 2978–2985 (2010). https://doi.org/10.1016/j.jbiomech.2010.07.004
23. Badel, P., Avril, S., Lessner, S., Sutton, M.: Mechanical identification of layer-specific properties of mouse carotid arteries using 3D-DIC and a hyperelastic anisotropic constitutive model. Comput. Methods Biomech. Biomed. Eng. **15**, 37–48 (2012). https://doi.org/10.1080/10255842.2011.586945
24. Avril, S., Bersi, M.R., Bellini, C., Genovese, K., Humphrey, J.D.: Regional identification of mechanical properties in arteries. Comput. Methods Biomech. Biomed. Eng. 1–2 (2015). https://doi.org/10.1080/10255842.2015.1070577
25. Kroon, M., Holzapfel, G.A.: Estimation of the distributions of anisotropic, elastic properties and wall stresses of saccular cerebral aneurysms by inverse analysis. Proc. Roy. Soc. A Math. Phys. Eng. Sci. **464**, 807–825 (2008). https://doi.org/10.1098/rspa.2007.0332
26. Liu, M., Liang, L., Sun, W.: A new inverse method for estimation of in vivo mechanical properties of the aortic wall. J. Mech. Behav. Biomed. Mater. **72**, 148–158 (2017). https://doi.org/10.1016/j.jmbbm.2017.05.001
27. Liu, M., Liang, L., Sun, W.: Estimation of in vivo mechanical properties of the aortic wall: a multi-resolution direct search approach. J. Mech. Behav. Biomed. Mater. **77**, 649–659 (2018). https://doi.org/10.1016/j.jmbbm.2017.10.022
28. Liu, M., et al.: Identification of in vivo nonlinear anisotropic mechanical properties of ascending thoracic aortic aneurysm from patient-specific CT scans. Sci. Rep. **9**, 109 (2019). https://doi.org/10.1038/s41598-019-49438-w
29. Karatolios, K., et al.: Method for aortic wall strain measurement with three-dimensional ultrasound speckle tracking and fitted finite element analysis. Ann. Thorac. Surg. **96**, 1664–1671 (2013). https://doi.org/10.1016/j.athoracsur.2013.06.037
30. Wittek, A., Derwich, W., Fritzen, C.-P., Schmitz-Rixen, T., Blase, C.: Validation of full field measurement of aortic wall motion by 4D ultrasound. In: Nithiarasu, P., Robertson, A.M. (eds.) CBME17: 5th International Conference on Computational & Mathematical Biomedical Engineering, vol. 2, pp. 656–659. Zeta Computational Resources Ltd., Swansea (2017)
31. Derwich, W., Wittek, A., Blase, C., Schmitz-Rixen, T.: Dynamische Analyse der biomechanischen Eigenschaften der infrarenalen Aorta. Gefässchirurgie **22**(6), 392–399 (2017). https://doi.org/10.1007/s00772-017-0308-1
32. Wittek, A., Karatolios, K., Bihari, P., Schmitz-Rixen, T., Moosdorf, R., Vogt, S., Blase, C.: In vivo determination of elastic properties of the human aorta based on 4D ultrasound data. J. Mech. Behav. Biomed. Mater. 167–183 (2013). https://doi.org/10.1016/j.jmbbm.2013.03.014
33. Wittek, A., et al.: A finite element updating approach for identification of the anisotropic hyperelastic properties of normal and diseased aortic walls from 4D ultrasound strain imaging. J. Mech. Behav. Biomed. Mater. **58**, 122–138 (2016). https://doi.org/10.1016/j.jmbbm.2015.09.022

34. Gasser, T.C., Ogden, R.W., Holzapfel, G.A.: Hyperelastic modelling of arterial layers with distributed collagen fibre orientations. J. R. Soc. Interface **3**, 15–35 (2006). https://doi.org/10.1098/rsif.2005.0073
35. Humphrey, J.D.: Cardiovascular Solid Mechanics: Cells, Tissues, and Organs. Springer, New York (2002)
36. Horny, L., et al.: Correlations between age, prestrain, diameter and atherosclerosis in the male abdominal aorta. J. Mech. Behav. Biomed. Mater. **4**, 2128–2132 (2011). https://doi.org/10.1016/j.jmbbm.2011.07.011
37. De Keulenaer, B.L., De Waele, J.J., Powell, B., Malbrain, M.L.N.G.: What is normal intra-abdominal pressure and how is it affected by positioning, body mass and positive end-expiratory pressure? Intensive Care Med. **35**(6), 969–976 (2009). https://doi.org/10.1007/s00134-009-1445-0
38. Sellier, M.: An iterative method for the inverse elasto-static problem. J. Fluids Struct. **27**, 1461–1470 (2011). https://doi.org/10.1016/j.jfluidstructs.2011.08.002
39. Nelder, J.A., Mead, R.: A simplex method for function minimization. Comput. J. **7**, 308–313 (1965). https://doi.org/10.1093/comjnl/8.1.27

A Systematic Review of the Uses and Benefits of 3-D Printing in Orthopaedic Surgery

Firas Nasr[1](✉) and Caroline Hing[2]

[1] St George's University of London, London, UK
M1906887@sgul.ac.uk
[2] St George's Hospital NHS Foundation Trust, London, UK

Abstract. Introduction: The objective of this systematic review was to analyse 3-D printing in orthopaedic surgery, focussing on pre-operative planning, patient specific implants, instruments, and orthoses.

Method: The PRISMA methodology was followed and literature searches were conducted on Medline, Embase and the Cochrane library. MeSH search terms and Boolean operators included '3-D printing', 'orthopaedic', 'pre-operative plan', 'implants', 'patient specific instruments' AND 'orthosis'.

Results: Searches resulted in 36 studies included in the review. The increasing interest in 3-D printing in orthopaedics is reflected in the rise in publications between 2015 and 2020. The most common application, reported by 75% of the studies was the use of 3-D printed anatomical models to aid in pre-operative planning. The models were also utilized for surgical simulation (31%), intraoperative navigation (8%), and patient/family and surgical education (8%). The use of 3-D printing to manufacture patient specific orthoses, implants and instruments was reported in 14%, 11% and 8% of the studies respectively. The advantage of 3-D printing reported most (56%) was the educational and training opportunities the models provided for junior surgeons. Doctor-patient communication and improved consenting was a reported benefit in 28% of the studies. Objective benefits of 3-D printing included reduced operating time (42%), instrumentation time (11%), fluoroscopy time (31%) and intraoperative blood loss (33%).

Conclusion: The literature shows 3-D printing has improved pre-operative planning, allowed for surgical simulation, training and education. These benefits have led to improved operating metrics. There are currently no studies which demonstrate these reported benefits led to improved patient outcomes.

Keywords: Three-dimensional · Printing · Orthopeadic

1 Introduction

Three-dimensional (3-D) printing was developed by Charles Hull in 1986 [1]. Since its introduction, 3-D printing has renovated the manufacturing industry. The first application of 3-D printing in medicine was documented by Mankovich et al. [2] in 1990, when the group used the technique to produce a model of cranial bone anatomy. Since then, the utilisation of 3-D printing in medicine and surgery has increased over the last decade. In

J. M. R. S. Tavares et al. (Eds.): CMBBE 2021, LNCVB 38, pp. 170–182, 2023.
https://doi.org/10.1007/978-3-031-10015-4_15

2014, the medical 3-D printing industry was worth $11 million dollars and was projected to become a $1.9 billion industry by 2024 [3].

In healthcare, 3-D printers can use magnetic resonance imaging (MRI) or computerised tomography (CT) scan images to provide the instructions for the printhead to build a 3-D construct one layer at a time. The process of generating a 3-D printed anatomical model is complex and is comprised of several phases [4]. Firstly, image acquisition and denoising is conducted where the CT scan or MRI Digital Imaging and Communication in Medicine (DICOM) data is acquired, and image artefacts are removed. Secondly, segmentation of the image takes place where the anatomical area of interest is labelled and separated from the rest of the image. This segmented image is then used to generate the computer aided design (CAD) model, which is the blueprint for the 3-D printer to produce the final 3-D model (Fig. 1).

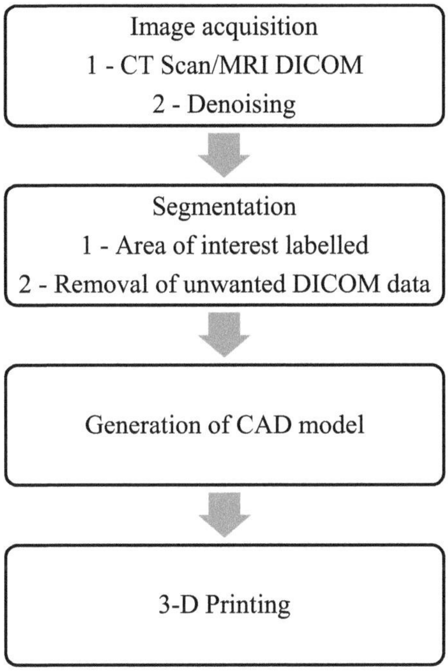

Fig. 1. Flowchart for generating a 3-D printed anatomical model.

Traditionally, surgeons would rely solely on the CT scan and MRI images to formulate a surgical plan preoperatively. In recent years, as radiological imaging software has improved, 3-D projections and reconstructions on the two-dimensional computer screen were made possible to assist the surgeon. The ability to produce 3-D printed models of the patient's anatomy has added new dimensions to surgical planning, multi-team discussions and even allowed for surgical rehearsal to be conducted [5].

In orthopaedic surgery, the use of 3-D printing has expanded to areas such as surgeon and patient education, construction of patient specific implants, instrumentation, and orthoses [6]. The objective of this systematic review was to analyse the uses and benefits of 3-D printing specifically in orthopaedic surgery, focusing on the areas of surgical pre-operative planning, surgeon and patient education, manufacture of patient specific implants and instruments, and orthoses. As the field of medical 3-D printing has advanced rapidly only references from 2015 to 2020 were analysed in this review to ensure the most up-to-date applications of 3-D printing were recorded.

2 Method

The 'Preferred Reporting Items for Systematic Reviews and Meta-Analyses' (PRISMA) methodology was used to conduct the systematic review [7]. The eligibility criteria for the review included papers that only discussed the use of 3-D printing in orthopaedic surgery. The literature searches were carried out on 28th December 2020 on Medline, Embase and the Cochrane library to ensure a thorough search was conducted. The MeSH search terms and Boolean operators used were "orthopaedic" OR "orthopedic", AND "three-dimensional printing" OR "3D printing" OR "stereolithography". To maintain focus on the areas of interest separate searches were also conducted which included the terms, "pre-operative plan", "implants", "patient specific instruments" and "orthosis". All the results were grouped together before going through the stages of screening and ensuring they were eligible to be included in the systematic review (Fig. 2). The exclusion criteria apart from studies that were both related to 3-D printing in orthopaedics included studies that were veterinary medicine related, studies that were not in the English language and conference abstracts that no extended papers could be identified.

The identification stage involved conducting the searches on the named databases. Screening involved removing the duplicates, filtering for date of publication (2015–2020). A search of the 'grey literature' was conducted by reviewing the titles and abstracts of each article referenced within the publication list to determine if the article met the criteria of being related to 3-D printing in orthopaedic surgery in human subjects. From the list of references that met these criteria, the full text was reviewed to ensure it could be included in the systematic review. The final list of references that were included in this review were the clinical case reports, case series, clinical trials that have been published discussing the direct application of 3-D printing in orthopaedic surgery. Articles that the full text could not be accessed or obtained through the Athens portal or other web searches were excluded from the final list. The application of 3-D printing in orthopaedic surgery, and the benefits its use provided to the orthopaedic procedure were recorded from the articles that were included in the systematic review and analysed. There was no funding sourced or required for this systematic review.

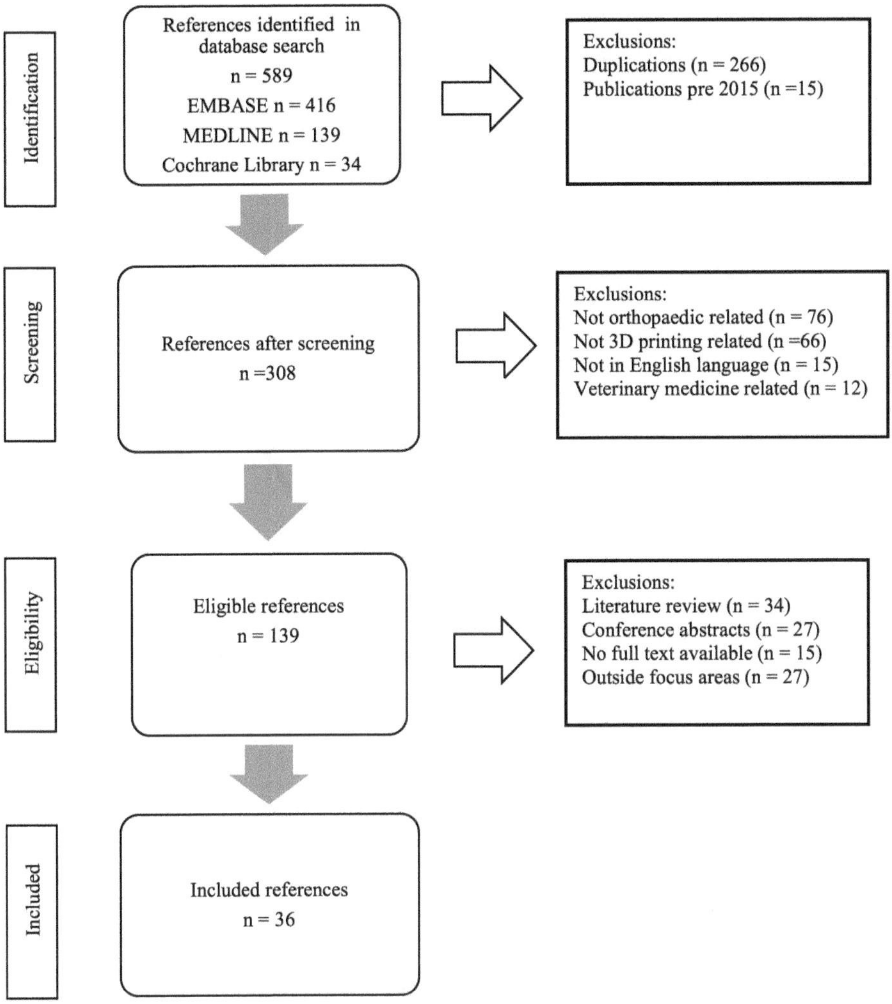

Fig. 2. PRISMA flow chart of included studies.

3 Results

The literature searches uncovered 36 references for inclusion in this systematic review [4, 8–42]. Table 1 shows that most of the references were comparison studies (12 randomised control studies and nine case-control studies) which separated their patients into '3-D printing used' groups and control groups (no 3-D printing used). The remaining references included were either case reports or case series detailing their experiences using 3-D printing in orthopaedics.

The popularity of 3-D printing in orthopaedics appears to be increasing based on the rising number of publications between 2015 (n = 1) and 2019 (n = 12), and only decreasing slightly in 2020 (n = 11) (Fig. 3).

Table 1. Distribution of types of studies in the systematic review.

Type of study	References included (n = 36)	(%)
Randomised control	12	33.3
Case-control	9	25.0
Case report	8	22.2
Case series	7	19.4

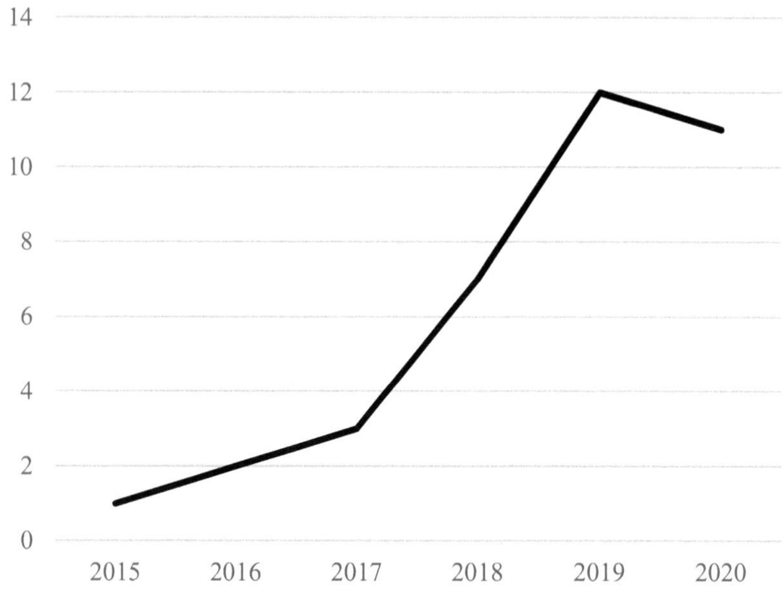

Fig. 3. Number of studies in the systematic review between 2015 and 2020.

The applications of 3-D printing in orthopaedic surgery outlined in the studies included in the systematic review were varied (Fig. 4).

The use of 3-D printing to create patient anatomical models to aid pre-operative planning was the most common application and was highlighted in 27 (75%) of the references. Eleven articles (31%) mentioned using the 3-D printed patient models to undertake surgical simulations. Three articles (8%) referenced using the 3-D printed patient models to aid intra-operative navigation and referencing. The application of 3-D printing for education refers to junior surgeon, patient and/or family education to assist in explaining the orthopaedic condition and assisting in gaining consent. There were three articles (8%) that highlighted education as one of the uses of 3-D printing.

Another application of 3-D printing in orthopaedics was in manufacturing patient specific orthoses (five articles, 14%). These included three articles referencing wrist/ankle orthoses, one article references 3-D printed insoles and one article detailing a 3-D printed cervical orthosis.

There were four articles (11%) that described the use of 3-D printing to manufacture patient specific implants. Luenam et al. [8] reported the use of a 3-D printed implant to reconstruct the lateral column of a patient's distal humerus following a traumatic injury. Han et al. [21] and Wong et al. [30] both specified the use of 3-D printing to produce patient specific pelvic implants to reconstruct a pelvis following tumour resection. So et al. [26] also referenced using 3-D printed patient specific implants to reconstruct large osseous defects in the foot and ankle.

Three articles detailed the use of 3-D printing to create patient specific instrumentation such as guide plates for core decompression and rod placement for treatment of femoral head osteonecrosis.

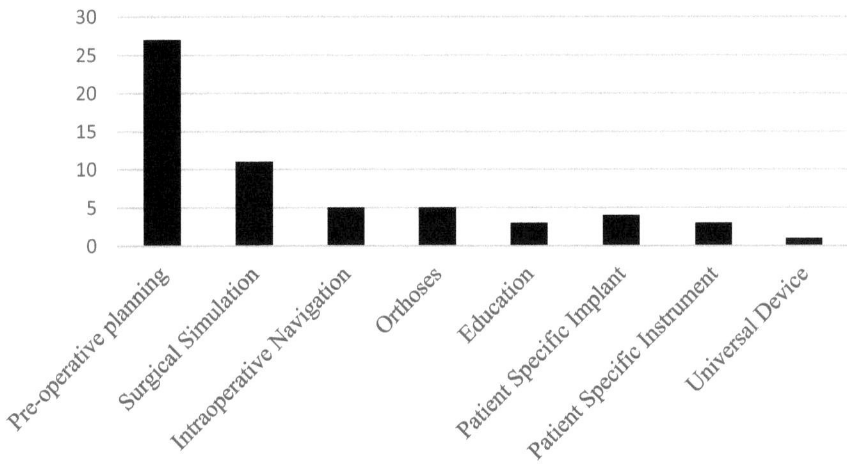

Fig. 4. Number of studies detailing applications of 3-D printing in orthopaedic surgery in the systematic review.

The benefits provided by 3-D printing that were reported in were wide-ranging so were therefore grouped into subjective benefits and objective benefits (Fig. 5).

When looking at the subjective benefits, the most common advantage of using 3-D printing was for the surgeon's education and training and was described in 20 articles (56%). This included the surgeons using the 3-D models to fully conceptualise and understand the patients' orthopaedic conditions as well as, using the models for the junior surgeons to undertake as surgical simulations, providing training opportunities. There were 10 articles (28%) that highlighted the use of 3-D printed models aiding the doctor-patient/family communication and assisting in the consenting process. The applications of 3-D printing in orthoses and manufacturing of the device designed to prevent excessive drilling [33] reported cost effectiveness (n = 3, 8%) and the ability to easily reproduce the 3-D printed item if it is lost (n = 5, 14%) as advantages provided by the technology. Four of the orthosis studies (11%) reported that the 3-D printed orthosis provided the patients with increased comfort when compared to traditional orthosis.

All the objective benefits of 3-D printing in orthopaedics surgery analysed were detailed in references that used the technology for pre-operative planning and measured comparable outcomes such as operation time, intraoperative blood loss, fluoroscopy time and instrumentation time. Fifteen articles (42%) stated using 3-D printed led to reduced operation time. Twelve papers (33%) showed that the patients that were assigned to the 3-D printed groups had a significant reduction in intraoperative blood loss. Eleven references (31%) reported 3-D printing reduced fluoroscopy time. Four articles (11%) highlighted 3-D printing reduced operative instrumentation time.

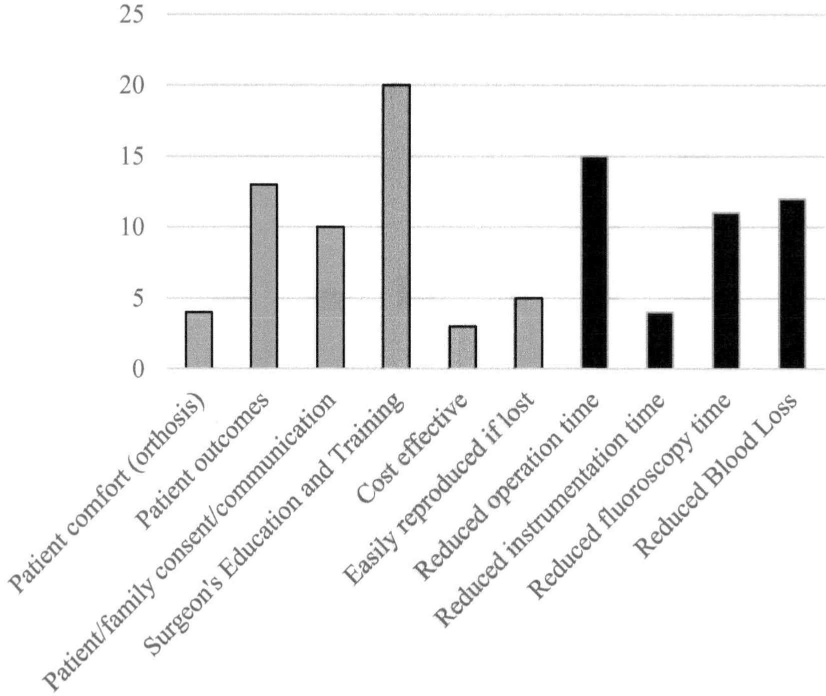

Fig. 5. Number of studies detailing subjective (grey) and objective (black) benefits of 3-D printing in orthopaedic in the systematic review.

The quality of the studies reporting the objective outcomes was measured based on study type, sample size and the National Institute of Health study quality assessment tools [43] (Table 2). Of the randomised control trials (RCTs), Chen et al. (2019) [36] had the highest quality paper as the study scored 12.5 of 14 marks (89.3%). The case-control studies by Li et al. [18, 32] were ranked lower in quality than the RCTs as both assorted the patients into the respective groups according to the decision of the chief surgeons, thus increasing the risk of selection bias in the papers. There were four case series studies that claimed a reduction in operation time as a benefit of utilising 3-D printing but were ranked lowest as they did not have a control group to compare against to verify the claims.

Table 2. Quality ranking of references reporting objective benefits of 3-D printing.

Quality ranking	Quality assessment (%)	Reference	Type of study	Sample size (3-D printed group = 3DPG, Control Group = CG)
1	89.3	Chen et al. [36]	Randomised control	48 (3DGP = 23, CG = 25)
2	85.7	Chen et al. [24]	Randomised control	107 (3DGP = 52, CG = 55)
3	82.1	Kong et al. [42]	Randomised control	32 (3DGP = 16, CG = 16)
4	78.6	Huang et al. [15]	Randomised control	40 (3DGP = 20, CG = 20)
5	78.6	Ozturk et al. [34]	Randomised control	37 (3DGP = 18, CG = 19)
6	71.4	Zheng et al. [35]	Randomised control	75 (3DGP = 35, CG = 40)
7	71.4	Yang et al. [41]	Randomised control	40 (3DGP = 20, CG = 20)
8	64.3	Wan et al. [40]	Randomised control	96 (3DGP = 48, CG = 48)
9	64.3	Zheng et al. [37]	Randomised control	91 (3DGP = 43, CG = 48)
10	64.3	You et al. [38]	Randomised control	66 (3DGP = 34, CG = 32)
11	57.1	Yang et al. [28]	Randomised control	30 (3DGP = 15, CG = 15)
12	58.3	Li et al. [18]	Case control	16 (3DGP = 7, CG = 9)
13	50	Li et al. [32]	Case-control	40 (3DGP = 20, CG = 20)
14	100	Samaila et al. [13]	Case series	52
15	100	Wong et al. [39]	Case series	10
16	77.8	Mishra et al. [20]	Case series	91
17	77.8	Bagaria and Chaudhary [4]	Case series	50

4 Discussion

4.1 Three-Dimensional Printed Models

The technology of 3-D printing to create the anatomical models from the patient's radiological images has been especially useful in orthopaedics, traumatology as well as other specialities such as maxillofacial surgery [44]. The anatomical models have proven to aid pre-operative planning, allow for surgical simulation, and assist in the education of both surgeons and patients.

Pre-operative planning was the most common application of 3-D printing in orthopaedic surgery in this review. Complex fracture patterns and revision arthroplasty surgeries are areas of orthopaedics that have found the application of 3-D printed models to be particularly useful.

Two-dimensional imaging and 3-D radiologic reconstruction on the computer screen are normally used for these complex procedures, but Wei et al. (2019) [10] were able to demonstrate how using 3-D printed models added to the pre-operative planning capabilities. Surgeons were able to use the models to plan for precise angle corrections, pre-bend plates, run a full simulation of the surgery to reduce the risk of any unanticipated complications as well as offering training opportunities for junior surgeons [13].

This extra dimension to the pre-operative planning process also translated to significant benefits to the patients as shown by Chen et al. (2018) [24] and several other studies. In these studies, the patients in the 3-D printed groups had reduced operation times, fluoroscopy time and intraoperative blood loss when compared the patients that did not have 3-D printed model utilised. Reductions in these measures can also lead to reduced risks associated with anaesthetic time, radiation exposure, intraoperative infections, and other post-operative complications. Furthermore, the cumulative effect of these will also lead to reduced healthcare costs by reducing both the theatre costs and potentially post-operative hospital stay.

The quality of the papers that recorded the objective operative metrics (Table 2) was generally high. The three studies with sample sizes over 90 patients [24, 37, 40] all highlighted small sample size as a limitation of the study. Notable limitations in the randomised control studies included the use of a coin toss as the method of randomisation [36], and the inability to include blinding to the surgeon or the patients in the randomisation process might have influenced the results.

It must be stated that some studies have that shown that using 3-D printed models showed no significant difference in operation time, estimated blood loss, fluoroscopy time and over all post-operative outcomes [17] but this might be explained by the small sample size of 20 patients.

The 3-D printed models can also be utilised to assist in the doctor-patient communication where it is difficult for them to understand the associated complexities of their conditions. Downey et al. [17], detailed the positive effect the 3-D printed models had on the communication which led to improved overall patient satisfaction. Bizzotto et al. [45] were able to demonstrate the positive impact of 3-D printed models in the consenting process due to this improved doctor patient communication. The benefits of the 3-D printed models are not limited to pre-operative planning. The model can also be sterilised and used during the operation to aid in intra-operative referencing and navigation as shown by Bagaria and Chaudhary's study [4].

Kong et al. [42] highlighted that the time needed for the 3-D printer to create an anatomical model is hours and therefore was not a feasible option for the emergency patients that required immediate operation despite showing the clinical benefits its utility.

4.2 Patient Specific Implants and Instrumentation

The use of 3-D printing to create non biologic patient specific implants has been widely reported and has benefited the treatment of orthopaedic tumours [21]. As well as being able to create complex anatomical structures particular to the patient's needs following resections of their pelvic tumour, Wong et al. [30] discussed the use of materials and porosity of the implant to promote osteointegration with the surrounding native bone following the reconstruction. Luenam et al. [8] were also able to demonstrate the role of patient specific implants in treating patients following traumatic injury leading to considerable bone loss. In this article the surgeons were able to reconstruct the lateral column of the distal humerus with the 3-D printed implant leading to the restoration of full function of the elbow joint.

Patient specific 3-D printed instruments are also commonly used particularly as sterilisable surgical guides to assist with screwing and drilling in orthopaedic spinal surgery [46]. These guides have increased surgical accuracy when comparing to the use of expensive surgical navigation systems and therefore open potential healthcare cost savings opportunities without compromising on patient safety.

4.3 Limitations of 3-D Printing

This review has focused on the benefits that 3-D printing provide to orthopaedic surgery, but the most common limiting factor for its routine use was the costs associated the technology. Heunis et al. [19] highlights the costs of purchasing the printer itself and the ongoing costs of technicians to convert the CT imaging data as the major disadvantage.

Crawford [5] discussed two main reasons as to why 3-D printing still is not being used routinely despite all the benefits it provides. Firstly, getting access to the raw imaging data and finding out what the surgeon needs to be able to see in the final printed model has proven to be logistically quite challenging. Secondly, the process of segmentation of the raw data to extract the anatomical area of interest is massively time-consuming. The process of segmentation can take about four hours per case if conducted manually as it requires anatomical, medical physics and 3-D printing systems knowledge and is therefore not feasible for everyday routine use if done in individual hospitals.

The limitations regarding the scalability of 3-D printing have largely been solved by using artificial intelligence (AI) and machine learning algorithms to streamline the segmentation process. AI is used to extract the relevant data in minutes by using a database of 100,000s labelled data images created by experts to automatically identify the required anatomical structures [47]. By automating this process, AI could allow for point of care 3-D printing without the ongoing costs of staffing to manually convert images.

5 Conclusion

The literature shows that applications of 3-D printing have resulted in improved operating metrics due to enhanced pre-operative planning. The use of 3-D printing to provide patient specific implants remains limited to specific cases. The application of 3-D printing in biologics with advancement in tissue engineering opens exciting treatment options for orthopaedic conditions such as osteoarthritis and ligament repair.

References

1. Hull, C.: Apparatus for production of three-dimensional object by stereolithography, U.S. Patent 4,575,330. https://patents.google.com/patent/US4575330A/en. Accessed 16 Nov 2021
2. Mankovich, N.J., Cheesman, A.L., Stoker, N.G.: The display of three-dimensional anatomy with stereolithographic models. J. Digit. Imaging **3**, 200–203 (1990)
3. Schubert, C., van Langeveld, M.C., Donoso, L.A.: Innovations in 3D printing: a 3D overview from optics to organs. Br. J. Ophthalmol. **98**(2), 159–161 (2014)
4. Bagaria, V., Chaudhary, K.: A paradigm shift in surgical planning and simulation using 3Dgraphy: experience of first 50 surgeries done using 3D-printed biomodels. Int. J. Care Injured **48**, 2501–2508 (2017)
5. Crawford, D.: 3D printing in surgical planning: Benefits for surgeons and patients [Lecture]. Innovation in surgery: Future Surgery Virtual 2020. Royal College of Surgeons of England, 22 October 2020
6. Wong, T.M., et al.: The use of three-dimensional printing technology in orthopaedic surgery: a review. J. Orthop. Surg. **25**(1), 1–7 (2017)
7. Moher, D., Liberati, A., Tetzlaff, J., Altman, D.G.: The PRISMA group, preferred reporting items for systematic reviews and meta-analyses: the PRISMA statement. PLoS Med. **6**(7), e1000097 (2009). https://doi.org/10.1371/journal.pmed.1000097
8. Luenam, S., Kosiyatrakul, A., Phakdeewisetkul, K., Puncreobutr, C.: The patient-specific implant created with 3D printing technology in treatment of a severe open distal humerus fracture with complete loss of the lateral column. J. Orthop. Surg. **28**(3), 1–8 (2020)
9. Wang, K., et al.: Research and development of 3D printing orthotic insoles and preliminary treatment of leg length discrepancy patients. Technol. Health Care **28**, 615–624 (2020)
10. Wei, Y.-P., Lai, Y.-C., Chang, W.-N.: Anatomic three-dimensional model-assisted surgical planning for treatment of pediatric hip dislocation due to osteomyelitis. J. Int. Med. Res. **48**(2), 1–9 (2019)
11. Chae, D.-S., et al.: The functional effect of 3D-printing individualized orthosis for patients with peripheral nerve injuries: Three case reports. Medicine **99**(16), 1–6 (2020)
12. Zheng, Y., et al.: Effects of a 3D-printed orthosis compared to a low-temperature thermoplastic plate orthosis on wrist flexor spasticity in chronic hemiparetic stroke patients: a randomized controlled trial. Clin. Rehabil. **34**(2), 194–204 (2019)
13. Samaila, E.M., et al.: Value of three-dimensional printing of fractures in orthopaedic trauma surgery. J. Int. Med. Res. **48**(1), 1–9 (2019)
14. Horas, K., Hoffmann, R., Faulenbach, M., Heinz, S.M., Langheinrich, A., Schweigkofler, U.: Advances in the preoperative planning of revision trauma surgery using 3D printing technology. J. Orthop. Trauma **34**, e181–e186 (2020)
15. Huang, J.-H., et al.: Surgical treatment for both-column acetabular fractures using pre-operative virtual simulation and three-dimensional printing techniques. Chin. Med. J. **133**(4), 395–401 (2020)

16. Liawrungrueang, W., Sornsa-ard, T., Boonchieng, E.: Bilateral pure facet joint dislocation in thoracolumbar junction (T11–T12) without facet fracture using a 3D digital printing model for surgical planning: a case report. Trauma Case Rep. **25**, 100273 (2020). https://doi.org/10.1016/j.tcr.2019.100273
17. Downey, C., McCarrick, C., Fenelon, C., Murphy, E.P., O'Daly, B.J., Leonard, M.: A novel approach using 3-D printing in the Irish National Centre for pelvic and acetabular surgery. Irish J. Med. Sci. **189**(1), 219–228 (2019). https://doi.org/10.1007/s11845-019-02055-y
18. Li, Y.-T., et al.: Surgical treatment for posterior dislocation of hip combined with acetabular fractures using preoperative virtual simulation and three-dimensional printing model-assisted precontoured plate fixation techniques. BioMed Res. Int. (2019). https://doi.org/10.1155/2019/3971571. Article ID 3971571
19. Heunis, J.C., Cheah, J.W., Sabnis, A.J., Wustrack, R.L.: Use of three-dimensional printing and intraoperative navigation in the surgical resection of metastatic acetabular osteosarcoma. BMJ Case Rep. **12**, e230238 (2019). https://doi.org/10.1136/bcr-2019-230238
20. Mishra, A., Verma, T., Vaish, A., Vaish, R., Vaishya, R., Maini, L.: Virtual preoperative planning and 3D printing are valuable for the management of complex orthopaedic trauma. Chin. J. Traumatol. **22**, 350–355 (2019)
21. Han, Q., et al.: Individual resection and reconstruction of pelvic tumor with three-dimensional printed customized hemi-pelvic prosthesis: a case report. Medicine **98**(36), e16658 (2019). https://doi.org/10.1097/MD.0000000000016658
22. Nie, W., Gu, F., Wang, Z., Wu, R., Yue, Y., Shao, A.: Preliminary application of three-dimension printing technology in surgical management of bicondylar tibial plateau fractures. Injury Int. J. Care Injured **50**, 476–483 (2019)
23. Coote, J.D., et al.: Three-dimensional printed patient models for complex pediatric spinal surgery. Ochsner J. **19**, 49–53 (2019)
24. Chen, C., Cai, L., Zhang, C., Wang, J., Guo, X., Zhou, Y.: Treatment of die-punch fractures with 3D Printing technology. J. Invest. Surg. **31**, 285–392 (2018)
25. Thayaparan, G.K., Owbridge, M.G., Thompson, R.G., D'Urso, P.S.: Designing patient-specific 3D printed devices for posterior atlantoaxial transarticular fixation surgery. J. Clin. Neurosci. **56**, 192–198 (2018)
26. So, E., Mandas, V.H., Hlad, L.: Large osseous defect reconstruction using a custom three-dimensional printed titanium truss implant. J. Foot Ankle Surg. **57**, 196–204 (2018)
27. Matthews-Brownell, M., Hall, J.A.: Utilization of 3-D printing in treatment of veterans with impaired hand function: applications for custom orthotics. J. Hand Ther. **31**, 156–157 (2018)
28. Yang, L.: Application of 3D printing in the surgical planning of Trimalleolar fracture and doctor-patient communication. BioMed Res. Int. (2016). https://doi.org/10.1155/2016/2482086
29. Moreta-Matrinez, R., Calva-Haro, J.A., Perez-Mananes, R., Garcia-Sevilla, M., Mediavilla-Santos, L., Pascau, J.: Desktop 3D printing: KEY for surgical navigation in acral tumors? Appl. Sci. **10**, 8984 (2020). https://doi.org/10.3390/app10248984
30. Wong, K.C., Kumta, S.M., Geel, N.V., Demol, J.: One-step reconstruction with a 3D-printed, biomechanically evaluated custom implant after complex pelvic tumor resection. Comput. Aided Surg. **20**(1), 14–23 (2015)
31. Hale, L., Linley, E., Kalaskar, D.M.: A digital workflow for design and fabrication of bespoke orthoses using 3D scanning and 3D printing, a patient-based case study. Sci. Rep. **10**, 7028 (2020). https://doi.org/10.1038/s41598-020-639337-1
32. Li, B., et al.: Clinical value of 3D printing guide plate in core decompression plus porous bioceramics rod placement for the treatment of early osteonecrosis of the femoral head. J. Orthop. Surg. Res. **13**, 130 (2018). https://doi.org/10.1186/s13018-018-0812-3

33. Ferras-Tarrago, J., Jorda-Gomez, P., Catala-de-Las-Marinas, J., Antequera-Cano, J.M., Barres-Carsi, M.: A new universal 3D-printable device to prevent excessive drilling in orthopedic surgery. Eur. J. Trauma Emerg. Surg. (2020). https://doi.org/10.1007/s00068-020-01465-3
34. Ozturk, A.M., Ozer, M.A., Suer, O., Derin, O., Govsa, F., Aktuglu, K.: Evaluation of the effects of using 3D - patient specific models of displaced intra - articular calcaneal fractures in surgery. Injury (2020). https://doi.org/10.1016/j.injury.2020.04.057
35. Zheng, W., et al.: Comparison of the conventional surgery and the surgery assisted by 3D printing technology in the treatment of calcaneal fractures. J. Invest. Surg. **31**(6), 557–567 (2017)
36. Chen, C., Cai, L., Zheng, W., Wang, J., Guo, X., Chen, H.: The efficacy of using 3D printing models in the treatment of fractures: a randomised clinical trial. BMC Musculskeletal Disord. **20**, 65 (2019). https://doi.org/10.1186/s12891-019-2448-9
37. Zheng, W., et al.: Application of 3D-printing technology in the treatment of humeral intercondylar fractures. Orthop. Traumatol. Surg. Res. **104**, 83–88 (2018)
38. You, W., et al.: Application of 3D printing technology on the treatment of complex proximal humeral fractures (Neer3-part and 4-part) in old people. Orthop. Traumatol. Surg. Res. **102**, 897–903 (2016)
39. Wong, T.T., Lynch, T.S., Popkin, C.A., Kazam, J.K.: Preoperative use of a 3D printed model for femoroacetabular impingement surgery and its effect on planned osteoplasty. AJR **211**, W116–W121 (2018)
40. Wan, L., et al.: Clinical feasibility and application value of computer virtual reduction combined with 3d printing technique in complex acetabular fractures. Exp. Ther. Med. **17**, 3630–3636 (2018)
41. Yang, L., Grottkau, B., He, Z., Ye, C.: Three dimensional printing technology and materials for treatment of elbow fractures. Int. Orthop. **41**, 1–7 (2017). https://doi.org/10.1007/s00264-017-3627-7
42. Kong, L., et al.: Surgical treatment of intra-articular distal radius fractures with the assistance of three-dimensional printing technique. Medicine **99**(8), e19259 (2020). https://doi.org/10.1097/MD.0000000000019259
43. National Institutes of Health: Study Quality Assessment Tools (2020). https://www.nhlbi.nih.gov/health-topics/study-quality-assessment-tools. Accessed 16 Nov 2021
44. Auricchio, F., Marconi, S.: 3D printing: clinical applications in orthopaedics and traumatology. Effort Open Rev. **1**, 112–127 (2016)
45. Bizzotto, N., et al.: 3D Printed replica of articular fractures for surgical planning and patient consent: a two years multi-centric experience. 3D Printing Med. **2**(1), 1–6 (2016). https://doi.org/10.1186/s41205-016-0006-8
46. Landi, A., Mancarekka, C., Gregori, F., Delfini, R.: Spinal neuronavigation and 3D-printed tubular guide for pedicle screw placement: a really new tool to improve safety and accuracy of the suigcal technique? J. Spine **4**(5), e118 (2015), doi:https://doi.org/10.4172/21657939.1000e118
47. Axial3D: How Axial3D leverages AWS to help surgeons save lives with 3D medical models (2020). https://www.axial3d.com/blog/how-axial3d-leverages-aws-to-help-surgeons-save-lives-with-3d-medical-models/. Accessed 16 Nov 2021

Assessing Methodological Uncertainty of *In-Vitro* Digital Volume Correlation Bone Strain Measurements in Total Shoulder Arthroplasties

Jakub Targosinski[1], Jonathan Kusins[2], Nicole Martensson[3], Andrew Nelson[4], Nikolas Knowles[5], and Louis Ferreira[1,2](✉)

[1] School of Biomedical Engineering, Western University, London, ON, Canada
lferreir@uwo.ca

[2] Department of Mechanical and Materials Engineering, Western University, London, ON, Canada

[3] Department of Electrical and Computer Engineering, Western University, London, ON, Canada

[4] Department of Anthropology, Western University, London, ON, Canada

[5] Department of Radiology, University of Calgary, Calgary, AB, Canada

Abstract. Digital volume correlation (DVC) performed on micro-computed tomography (CT) imagery provides a measurement technique which can measure full-field deformations of loaded osseous tissues. This experimental approach is of interest in the investigation of the failure mechanisms of glenoid implants in total shoulder arthroplasties, as it allows for direct experimental measurement of strains at the bone-cement-implant interface. It is therefore important to understand the methodological limitations of the bone strain measurements made and the inherent uncertainty present in this approach. Micro-CT scans of two cadaveric scapulae from donors who had been treated with shoulder arthroplasty while living were acquired with differing numbers of CT projections under loaded and unloaded conditions. DVC strain measurements were quantified from the unloaded and loaded volumetric images with five distinct sub-volume sizes. The mean absolute error and measurement sensitivity were quantified in the DVC strains as a function of projection count and sub-volume size, establishing relationships between measurement spatial resolution, image quality, and strain measurement error. Observations revealed that with careful selection of DVC spatial resolution and CT projection count, CT acquisition times can be halved with little impact on DVC strain accuracy. Using 3141 CT projection images as the basis of loaded DVC yielded a maximum mean relative error of 587 $\mu\varepsilon$, which increased to 689 $\mu\varepsilon$ using 1570 CT projection imagery. Thus, with careful consideration of the imaging parameters used, DVC can be a useful preclinical evaluation tool to quantify the internal strain within bone-implant constructs.

Keywords: Digital volume correlation · Micro-computed tomography · Glenoid implants · Osseous strains

J. M. R. S. Tavares et al. (Eds.): CMBBE 2021, LNCVB 38, pp. 183–193, 2023.
https://doi.org/10.1007/978-3-031-10015-4_16

1 Introduction

In combination with micro-computed tomography (micro-CT) imaging, digital volume correlation (DVC) has the potential to enable full-field strain measurements within bone [1–4]. The DVC method offers the ability to assess internal bone strain, not possible with conventional surface measurement techniques such as strain gauges and digital image correlation (DIC), which are further limited to discrete locations (strain gauges) or sub-regions on the surface of the specimen (DIC) [5].

Glenoid implants, which serve as the replacement for the articular surface of the scapula in patients who undergo total shoulder arthroplasty procedures, are a class of implant which are of particular interest to the practical application of DVC-based strain measurement methods. The long-term outcomes of glenoid implants remain relatively poor compared to other joint arthroplasties, with symptomatic glenoid failure occurring at a mean annualized rate of 1.2% after primary surgery [6]. The most common failure mode of these implants is the loosening of the glenoid implant from its fixation within bone. It is currently believed that the cause of glenoid loosening failure is mechanical: the rocking-horse effect, where cyclic eccentric loading of the glenoid rim induces excessive mechanical stresses causing progressive damage to the cement mantle-bone interface [7, 8]. However, the direct measurement of the mechanical behavior at this interface is difficult, and currently the only experimental measurement technique that can probe this interface directly involves the combination of high resolution volumetric imaging with DVC [9, 10]. Finite element models have also investigated the behavior of the implant-bone-cement interface, but rely on bone density-material property relationships that can have a large impact on the predicted outcomes of these models [11–15].

Strain measurements derived from DVC face issues with validation, as no other measurement tool can provide full-field strain measurements of the same caliber; thus, it is difficult to comparatively assess the accuracy and precision of DVC strains made against previously validated measurements acquired through other metrological approaches. It is therefore of importance to assess the strain measurement error associated with methodologies that aim to employ DVC to quantify strain in the glenoid implant-cement-bone interface and in other arthroplasties. Various methods exist in the literature for assessing DVC measurement error. Some have used a zero-strain assumption approach in which they comparatively assess the strain between two unloaded volumetric images and use the disagreement between the DVC-measured strain and the idealized zero-strain condition as a measure of the methodological error [3]. Others have used digital affine transformations of volumetric images to impose a known virtual strain and then quantified the error as a function of the disagreement between the known virtual strain and the DVC-measured strain [4]. *In-vitro* DVC strain measurement error in the glenoid implant-bone-cement interface has been recently analyzed, but the relationship between image acquisition time and DVC sub-volume size was not reported [16]. The goal of the current study was to quantify the experimental uncertainties associated with the use of DVC to measure full-field strains in underlying scapular bone following a shoulder arthroplasty procedure. Specifically, this study quantified (1) the influence of imaging acquisition time and (2) the global DVC sub-volume size on the accuracy and precision of the measured osseous strains in cadaveric shoulders under unloaded and loaded conditions.

2 Methods

Two independent variables were established representing factors of image quality and spatial resolution of the strain measurement: 1) CT projection count was varied to investigate the effect of image quality; and 2) the global DVC sub-volume size was varied to investigate the effect of varying strain measurement resolution. Improved CT image quality can alter the experimental uncertainty of associated DVC strain measurements by providing bone structural information unconfounded by CT noise. Similarly, increasing the sub-volume size also improves the DVC strain uncertainty, but both changes can have a significant impact on the time it takes to complete any single analysis [3]. Therefore, it is of practical concern to quantify the error response of both parameters in order to optimize the time needed to complete a DVC measurement with an acceptable level of error.

2.1 Specimen Preparation and Experimental Protocol

Two cadaveric scapulae, from donors previously treated with total shoulder arthroplasty (79-year old male and 83-year old male), were recovered and denuded of all soft tissue, in accordance with institutional ethics (HSREB#113023). One scapula contained a pegged glenoid implant and the other contained a keeled glenoid implant; two implant designs with significant clinical relevance due to their overwhelming prevalence. The medial borders of the scapulae were cut parallel to the glenoid cavity at a distance of 6 cm from the articular surface and subsequently potted using polymethyl methacrylate (PMMA) bone cement. A CT-compatible loading hexapod robot was used to apply external loads to the cadaveric specimens directly within a cone-beam micro-CT scanner (Nikon XTH 225ST, Nikon Metrology, Brighton, MI, USA) [14]. The hexapod robot was augmented with carbon-fiber struts for radiolucency, and a vise clamp fixed within the robot was used to locate and clamp the specimen.

For both cadaveric scapulae, the experimental loading protocol was as follows: prior to loading the scapula, two sets of micro-CT images were captured in an unloaded state (33.5 μm isotropic voxel size, 95 kVp, 64 μA, 1000 ms exposure). A 10 N load was applied to stabilize the specimen and ensure no bulk relative motion between the two captured volumetric images for the unloaded state. This load was measured through a 6-axis load cell (ATI mini45, ATI Industrial Automation, Apex, NC, USA) instrumented within the hexapod. To investigate the influence of acquisition time and image quality on the experimental uncertainties, two sets of unloaded scans were acquired with a varied number of projections (3141, 1571, 785, 393, 196) corresponding to various scan acquisition times (52, 26, 13, 6 and 3 min, respectively). The scapula was then subjected to a 750 N compressive axial load (ASTM F2028-17, 2002), and after allowing for viscoelastic relaxation of the bone to a steady state, additional micro-CT images were captured of the loaded scapulae at the same CT projection counts and imaging parameters (Fig. 1).

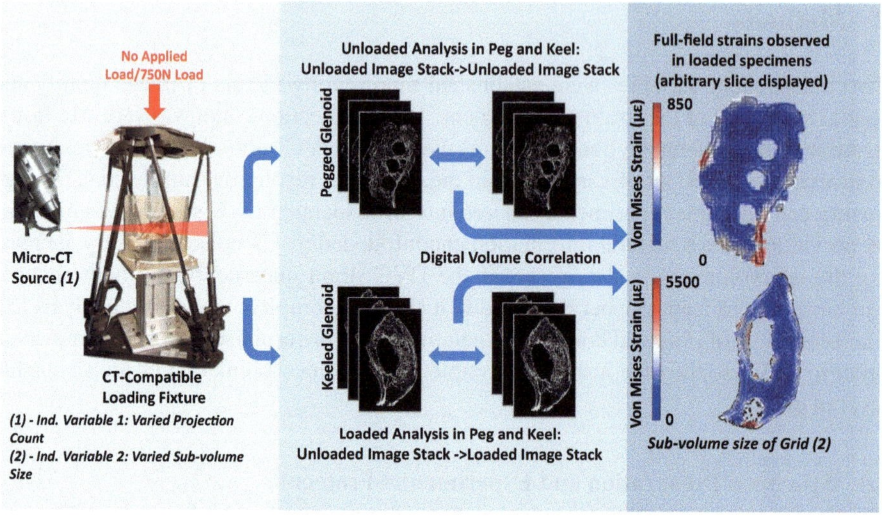

Fig. 1. Simplified workflow of the unloaded and loaded DVC analyses performed. For the unloaded analysis, every combination of the two independent variables (projection count, sub-volume size) was investigated. For the loaded analysis, the spatial resolution of the DVC measurement was fixed based on the findings of the unloaded analysis at a value of 1072 μm (32 voxels).

2.2 Image Processing for DVC

Prior to the DVC registrations, the micro-CT images were post-processed using a consistent workflow to isolate the trabecular structure of the glenoid. The images were first cropped to a region of interest centered on the glenoid. The region of interest was chosen based on the 3141-projection volumetric images for both the pegged and keeled implants, and subsequently used for the remainder of the volumes. A specimen-specific thresholding operation was applied to each image to isolate the trabecular structure from the background image with threshold limits decided subjectively by an experienced user based on the 3141-projection volumes, and afterwards applied to all subsequent CT projection-varied volumes. This was followed by a region growing operation (Materialise Mimics 20.0.0, Materialise, Leuven, Belgium) with a seed point selected within the trabecular structure isolated using the thresholding operation. The images were then converted to 8-bit grey scale, and manually registered to the unloaded images using an interactive program (MeVis Lab 3.1.1, MeVis Medical Solutions, Bremen, Germany) prior to running the DVC analyses.

A commercially available DVC implementation (LaVision DaVis 10.1.1, LaVision, Gottingen, Germany) was used to perform DVC at varied sub-volume sizes (8, 16, 32, 64, 128 voxels) on the two sets of unloaded images. An FFT pre-shift window size of 128, and a valid voxel requirement of 50% was used. A 0% sub-volume overlap was used for each sub-volume size. DaVis uses a hybrid DVC approach, which combines an initial FFT-DVC step to find large particle shifts and to create an initial predictor field, and subsequent direct DVC passes to find a solution for each shifted sub-volume. The local displacements calculated by DVC between the two images are then differentiated using a center finite difference (CFD) scheme to calculate strain.

2.3 Unloaded Uncertainty Analysis

An unloaded uncertainty analysis was performed in order to determine the accuracy and precision of the DVC strain measurements taken of the glenoid bone [1, 3]. For each set of DVC measurements, a Matlab script (Matlab R2020a, MathWorks, Natick, MA, USA) extracted the strain tensors at each node, and two measures were defined to represent the accuracy and precision of the DVC approach employed. The mean absolute error (MAE) (Eq. 1) was defined as the mean of the absolute values of the strain tensor elements from each voxel, and the standard deviation of error (SDE) (Eq. 2) was defined as the standard deviation of the absolute values of the strain tensor elements from each voxel [4]:

$$MAE = \frac{\sum_{(i,j,k)} \frac{\left(|\varepsilon_{xx}|+|\varepsilon_{yy}|+|\varepsilon_{zz}|+|\gamma_{xy}|+|\gamma_{xz}|+|\gamma_{yz}|\right)}{6}_{i,j,k}}{N} \tag{1}$$

$$SDE = \sqrt{\frac{\sum_{i,j,k} \left| \frac{\left(|\varepsilon_{xx}|+|\varepsilon_{yy}|+|\varepsilon_{zz}|+|\gamma_{xy}|+|\gamma_{xz}|+|\gamma_{yz}|\right)_{i,j,k}}{6} - MAE \right|^2}{N}} \tag{2}$$

where $\varepsilon_{m,n}$ and $\gamma_{m,n}$ are the components of the strain tensor at the sub-volume located at position i, j, k, and N is the total number of sub-volumes in the DVC analysis. Power law fits were constructed for each set of MAE and SDE measurements of error by first linearizing the data and then fitting a first-order polynomial. Power law coefficients were then extracted from the first-order polynomial coefficients. The linearization was performed to achieve a uniform regression response across the data set.

2.4 Loaded Uncertainty Analysis

Based on the accuracy (*i.e.*, MAE) and precision (*i.e.*, SDE) determined by the sensitivity analysis of the unloaded scans, a DVC sub-volume size of 1072 μm (32 voxels) was chosen as the best compromise between spatial resolution and the apparent strain measurement error. DVC bone strain measurements were taken from the loaded scans of the glenoids at each number of CT projections (3141, 1571, 785, 393, 196). For each set of DVC measurements, the difference in strain between the DVC measurement and the corresponding 3141-projection measurement was calculated for each DVC sub-volume. The mean difference in sub-volume strain and standard deviation of the strain differences between the sub-volumes served as analogous measures to the MAE and SDE used in the unloaded analysis. These were used as a measure of the overall methodological error and repeatability inherent in our DVC strain measurement process.

3 Results

Improving image quality by increasing the number of projections used to capture the volumetric images decreased the MAE and SDE. Likewise, decreasing the DVC measurement spatial resolution decreased the MAE and SDE for both the pegged and keeled glenoid implant designs.

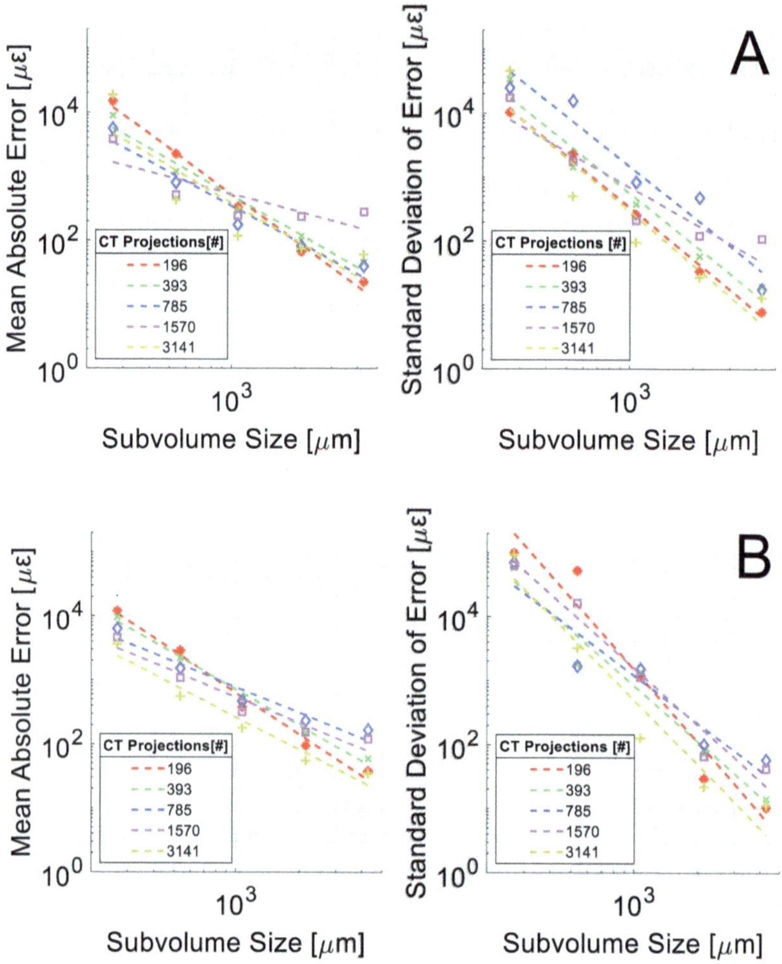

Fig. 2. Mean absolute error (MAE) and standard deviation of error (SDE) in the unloaded scans of the keeled implant (A) and pegged implant (B) displayed with power law regressions corresponding to CT projection counts.

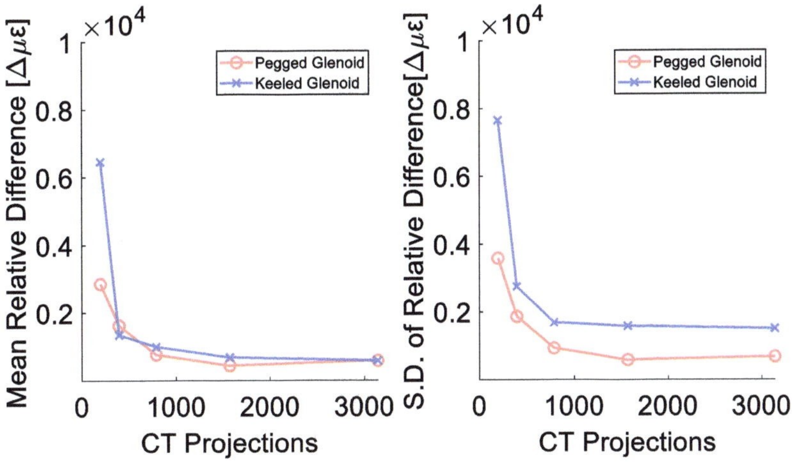

Fig. 3. Performance of DVC as a function of CT projections and implant fixation type. Mean relative and 1 standard deviation of the strain difference are shown for projection counts of 3141, 1570, 785, 393, and 196 as a fraction of the 3141 reference DVC performance metrics. All measurements were taken at a sub-volume size of 32 voxels (1072 μm).

For a spatial resolution of 1072 μm (32 voxel sub-volume size), the accuracy and precision (MAE and SDE) of the strain measurements with the keeled implant ranged between 118–330 με and 96–833 με, respectively (Fig. 2). With the pegged implant, the values were similar as determined through the unloaded analysis, with accuracy and precision ranging between 180–518 με and 128–1206 με, respectively.

At the selected strain measurement resolution of 32 voxels (1072 μm) in the loaded analysis, both the mean relative difference between the sub-volume strains, and the standard deviation of the relative difference, grew as a function of decreasing CT projection count for both implant designs (Fig. 3). For CT projection counts decreasing from 3141 to 196, the corresponding mean relative differences ranged 588–6473 με for the keeled implant, and 583–2871 με for the pegged implant. Similarly, one standard deviation ranged 1508–7664 με and 675–3587 με for the keeled and pegged implants, respectively. A CT projection count of 1570 produced similar error to the 3141 projection measurements across both implants.

Locations of high strains were in the cortical shell throughout the glenoid, though the strains were particularly high underneath the glenoid implant pegs and keel in areas where the glenoid narrowed as it joined with the subscapular fossa. The trabeculae in these areas also experienced high strains, as did those adjacent to the cement mantle of the implants (Fig. 4). The capability to make these observations decreased as the number of projections decreased as the strain field was overcome by apparent error.

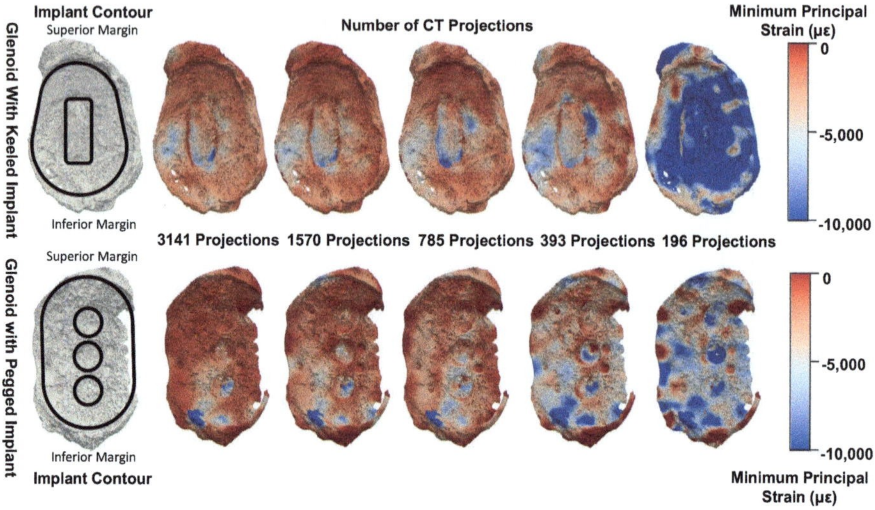

Fig. 4. Minimum principal strains measured through DVC in the glenoid under both implants at 750N from DVC analyses performed on 3141, 1570, 785, 393 and 196 CT projection images. As the number of CT projections falls, the strain measurements rise erroneously.

4 Discussion

This study provides guidance to optimize DVC bone strain measurement error while minimizing scan time using commercially available local DVC software. This investigation quantified the methodological uncertainties associated with measuring full-field strains within glenoid implant fixation in cadaveric scapular specimens from donors who had been treated with total shoulder arthroplasty in life. Based on the accuracy and precision determined by a sensitivity analysis in unloaded scans with a zero-strain assumption, a DVC sub-volume size of 1072 μm (32 voxels) was chosen as the best compromise between spatial resolution and the apparent strain measurement error. Comparatively, the mean relative difference and the standard deviation of the relative difference, which can be thought of as the error associated with performing the whole DVC-based methodology on loaded specimens, was approximately a half order of magnitude higher across each measured level. This could be due to methodological errors such as not allowing for enough relaxation time to reach a steady state in the osseous tissue. Other studies have reported similar magnitudes of MAE and SDE for in-vitro DVC within the glenoid peri-implant space [16].

Image quality as affected by the number of CT projections, had a strong effect on the accuracy and precision of loaded strain measurements, reaching a plateau beyond 785 CT projections (Fig. 3). From our results, there was no benefit beyond 1570 CT projections with the scanner used, which translates to a decrease in acquisition time of 50% relative to the 3141 CT projection scans. This can compound to substantial savings in stepwise loading study protocols that often require several loading levels multiplied by the number of implants being compared, along with any other factors being tested. A limitation of this study is that, in the loaded condition, the two 3141-projection count scans (one for reference) showed a mean relative difference of approximately 600 $\mu\varepsilon$ (Fig. 3). Ideally, this would have been near zero. This encapsulates the repeatability of the entire DVC+micro-CT strain measurement methodology used, which would be influenced by factors such as relaxation of the cadaveric tissue, thermal expansion of the cadaveric tissue as it reached an equilibrium temperature within the micro-CT and relaxation of the loading fixture over time. Notably, relaxation in the fixture and relaxation of the cadaveric tissue would likely be more impactful at higher loads, which was consistent with the observation that the relative error between loaded DVC analyses was higher than the absolute errors observed in the unloaded DVC analyses.

A compressive yield strain of ~7,000–40,000 $\mu\varepsilon$ at the apparent level has been reported for trabecular bone [18–21]. It is important to contextualize that in the loaded measurements of this study, the mean relative error did not exceed 9.8% of the lower bound of this value using the 1570 CT projection imagery. Moreover, the standard deviation of the relative error in the 1570 CT projection measurements was 22.7% of the lower bound of this value. This means that the DVC strain measurements captured through the methodology employed in this paper are physiologically relevant in the trabecular network of the glenoid. It has been observed that local strains in individual trabeculae can be substantially higher than those observed at the apparent level, which may mean the threshold for the error of DVC measurements at lower sub-volume sizes could also be significantly higher [22]. It should also be noted that mechanical properties of trabecular bone are thought to be sensitive to anatomical site, and DVC strain measurement error considerations should be made with respect to any specific application [18]. Nonetheless, these error metrics indicate DVC strain measurements at the bone-cement-implant interface can be reliable if care is taken to select appropriate spatial resolutions for the measurements and due consideration is given to imaging quality.

5 Conclusion

The combined use of micro-CT imaging and DVC strain measurements has the potential to provide clinically relevant bone strain measurements at the bone-cement-implant interface of glenoid implants. Careful selection of CT projection count and DVC spatial resolution can halve the time of CT acquisition, with no negative impact on DVC strain measurements. This can ameliorate the time constraints of future studies with stepwise load protocols. This improvement will facilitate the use of DVC to improve implant designs, and in turn the long-term outcomes of shoulder arthroplasty procedures.

References

1. Liu, L., Morgan, E.F.: Accuracy and precision of digital volume correlation in quantifying displacements and strains in trabecular bone. J. Biomech. **40**, 3516–3520 (2007). https://doi.org/10.1016/j.jbiomech.2007.04.019
2. Palanca, M., Tozzi, G., Cristofolini, L., Viceconti, M., Dall'Ara, E.: Three-dimensional local measurements of bone strain and displacement: comparison of three digital volume correlation approaches. J. Biomech. Eng. **137** (2015). https://doi.org/10.1115/1.4030174

3. Dall'Ara, E., Peña-Fernández, M., Palanca, M., Giorgi, M., Cristofolini, L., Tozzi, G.: Precision of digital volume correlation approaches for strain analysis in bone imaged with micro-computed tomography at different dimensional levels. Front. Mater. **4**, 31 (2017). https://doi.org/10.3389/fmats.2017.00031
4. Comini, F., Palanca, M., Cristofolini, L., Dall'Ara, E.: Uncertainties of synchrotron microCT-based digital volume correlation bone strain measurements under simulated deformation. J. Biomech. **86**, 232–237 (2019). https://doi.org/10.1016/j.jbiomech.2019.01.041
5. Grassi, L., Isaksson, H.: Extracting accurate strain measurements in bone mechanics: a critical review of current methods (2015). https://doi.org/10.1016/j.jmbbm.2015.06.006
6. Papadonikolakis, A., Neradilek, M.B., Matsen, F.A.: Failure of the glenoid component in anatomic total shoulder arthroplasty: a systematic review of the English-language literature between 2006 and 2012. J. Bone Jt. Surg. Ser. A. **95**, 2205–2212 (2013). https://doi.org/10.2106/JBJS.L.00552
7. Gregory, T., et al.: Glenoid loosening after total shoulder arthroplasty: an in vitro CT-scan study. J. Orthop. Res. **27**, 1589–1595 (2009). https://doi.org/10.1002/jor.20912
8. Jackins, S.E., Barrett, W.P., Matsen, F.A.: Glenoid loosening in total shoulder arthroplasty: association with rotator cuff deficiency. J. Arthroplasty. **3**, 39–46 (1988). https://doi.org/10.1016/S0883-5403(88)80051-2
9. Zhou, Y., Gong, C., Lewis, G.S., Armstrong, A.D., Du, J.: 3D full-field biomechanical testing of a glenoid before and after implant placement. Extrem. Mech. Lett. **35**, 100614 (2020). https://doi.org/10.1016/j.eml.2019.100614
10. Sukjamsri, C., et al.: Digital volume correlation and micro-CT: an in-vitro technique for measuring full-field interface micromotion around polyethylene implants. J. Biomech. **48**, 3447–3454 (2015). https://doi.org/10.1016/j.jbiomech.2015.05.024
11. Cristofolini, L., Schileo, E., Juszczyk, M., Taddei, F., Martelli, S., Viceconti, M.: Mechanical testing of bones: the positive synergy of finite-element models and in vitro experiments. Philos. Trans. R. Soc. A Math. Phys. Eng. Sci. **368**, 2725–2763 (2010). https://doi.org/10.1098/rsta.2010.0046
12. Knowles, N.K., Kusins, J., Faieghi, M., Ryan, M., Dall'Ara, E., Ferreira, L.M.: Material mapping of QCT-derived scapular models: a comparison with micro-CT loaded specimens using digital volume correlation. Ann. Biomed. Eng. **47**, 1–11 (2019). https://doi.org/10.1007/s10439-019-02312-2
13. Kluess, D., et al.: A round-robin finite element analysis of human femur mechanics between seven participating laboratories with experimental validation. Comput. Methods Biomech. Biomed. Engin. **22**, 1020–1031 (2019). https://doi.org/10.1080/10255842.2019.1615481
14. Kusins, J., Knowles, N., Ryan, M., Dall'Ara, E., Ferreira, L.: Performance of QCT-derived scapula finite element models in predicting local displacements using digital volume correlation. J. Mech. Behav. Biomed. Mater. **97**, 339–345 (2019). https://doi.org/10.1016/j.jmbbm.2019.05.021
15. Knowles, N.K., Langohr, G.D.G., Faieghi, M., Nelson, A.J., Ferreira, L.M.: A comparison of density–modulus relationships used in finite element modeling of the shoulder. Med. Eng. Phys. **66**, 40–46 (2019). https://doi.org/10.1016/j.medengphy.2019.02.005
16. Boulanaache, Y., Becce, F., Farron, A., Pioletti, D.P., Terrier, A.: Glenoid bone strain after anatomical total shoulder arthroplasty: in vitro measurements with micro-CT and digital volume correlation. Med. Eng. Phys. **85**, 48–54 (2020). https://doi.org/10.1016/j.medengphy.2020.09.009
17. ASTM: Astm F2028-17, pp. 1–15 (2002). https://doi.org/10.1520/F2028-17.Copyright
18. Morgan, E.F., Keaveny, T.M.: Dependence of yield strain of human trabecular bone on anatomic site. J. Biomech. **34**, 569–577 (2001). https://doi.org/10.1016/S0021-9290(01)00011-2

19. Odgaard, A., Hvid, I., Linde, F.: Compressive axial strain distributions in cancellous bone specimens. J. Biomech. **22**, 829–835 (1989). https://doi.org/10.1016/0021-9290(89)90066-3
20. Bayraktar, H.H., Morgan, E.F., Niebur, G.L., Morris, G.E., Wong, E.K., Keaveny, T.M.: Comparison of the elastic and yield properties of human femoral trabecular and cortical bone tissue. J. Biomech. **37**, 27–35 (2004). https://doi.org/10.1016/S0021-9290(03)00257-4
21. Nagaraja, S., Couse, T.L., Guldberg, R.E.: Trabecular bone microdamage and microstructural stresses under uniaxial compression. J. Biomech. **38**, 707–716 (2005). https://doi.org/10.1016/J.JBIOMECH.2004.05.013
22. Turunen, M.J., et al.: Sub-trabecular strain evolution in human trabecular bone. Sci. Rep. **101**(10), 1–14 (2020). https://doi.org/10.1038/s41598-020-69850-x

Evidence for the Applicability of Musculoskeletal Human Models to Improve Outcomes of Total Hip Arthroplasty

David Scherb[1(✉)], Christopher Fleischmann[2], Stefan Sesselmann[2], Jörg Miehling[1], and Sandro Wartzack[1]

[1] Engineering Design, Friedrich-Alexander-Universität Erlangen-Nürnberg (FAU), Martensstraße 9, 91058 Erlangen, Germany
{scherb,miehling,wartzack}@mfk.fau.de

[2] Institut für Medizintechnik, Technical University of Applied Sciences Amberg-Weiden, Hetzenrichter Weg 15, 92637 Weiden, Germany
{c.fleischmann,s.sesselmann}@oth-aw.de

Abstract. During the preoperative planning of total hip replacements (THR) the patient's biomechanical condition is widely neglected. The result can be a suboptimal implant selection and positioning, which may cause muscular dysfunctions, especially for revision THR. Thus, this study provides evidence in the use of musculoskeletal human models (MHM) during preoperative planning to improve postoperative outcomes. Therefore, a patient-specific MHM is placed in the pose of single leg stance. First, gluteus medius muscle activation is simulated in three adapted situations: healthy, after primary THR and after revision THR. In a second step, a parameter study with adjustable implant parameters (neck-shaft angle, femoral offset and antetorsion angle) is executed to investigate the effects on the gluteus medius activation. The gluteus medius activation of all three situations shows a remarkable increase to the status after revision THR, with a high risk of muscular impairments occurring due to muscle weakening. The parameter study demonstrates a considerable influence of the adjustable parameters not only on gluteus medius activation but also on hip joint contact force. Concluding, MHM not only allow the analysis of a patient's individual biomechanical condition, but also improve biomechanical outcomes by enabling the simulation of different geometrical THR implant parameters during surgery planning to support the surgeon in identifying the optimal implant parameters and positioning from both the muscle function and a kinematical (bony) perspective.

Keywords: Musculoskeletal human models · Patient-specific modelling · Revision hip arthroplasty · Biomechanics · Fatty degeneration · Computer aided surgery

1 Introduction

Total Hip replacement (THR) is one of the most executed surgical interventions in Germany [1] and all around the world [2, 3]. Analyzing the implantations performed in

J. M. R. S. Tavares et al. (Eds.): CMBBE 2021, LNCVB 38, pp. 194–207, 2023.
https://doi.org/10.1007/978-3-031-10015-4_17

the recent years, a continuous rise can be observed until today, which is mainly caused by the demographic change in industrialized countries. It is projected that the number of implantations will continue to increase at the same rate or even at a higher rate in the future [4, 5].

Prior to the operation, the surgeon has to decide preoperatively which implant to select for a specific patient. The decision depends on a preoperative planning process based on imaging, normally radiographs. This offers the surgeon the oppurtunity to choose the one implant best fitting to the patient´s individual anatomical parameters (e.g. neck-shaft angle, offset) out of a variety of different endoprostheses [6]. Indeed, there are some issues related to preoperative planning, especially based on radiographs like magnification or incorrect recording that can be solved by using 3D imaging such as MRI or CT [7]. However, a pitfall during the planning process may be the focus lying mainly on the patient's bony structures (kinematical condition) and less on the the patient's abilities determined by muscles (kinetic condition). This can lead to muscular dysfunctions, observable as a pathological gait like trendelenburg gait [8], caused by a not optimally selected or positioned implant entailing elongated rehabilitation time or even persisting disability.

For THR, two occasions have to be distinguished. The first one is the so-called primary total hip replacement (primary THR). In this case, parts of the bones that form the hip, more specifically the femoral neck and head and the acetabulum, are replaced by an endoprothesis. The main reason for primary THR is osteoarthritis [9]. The second type of THR is revision total hip replacement (revision THR). In this context, a revision THR describes the complete replacement of the implanted endoprosthesis from the primary THR by a new endoprothesis. There can be many possible reasons for the necessity of a new implant, e.g. aseptic loosening, fracture of the implant or infections [10]. These can occur shortly after the executed operation or as late as 15–20 years after the implantation, when the endoprosthesis exceeds its regular life span [5].

Depending on the surgical approach chosen for THR (lateral, posterior…), muscular damages occur to the hip muscles, more specifically to the gluteal muscles (gluteus maximus, medius, minimus), leading to a weakening of the muscles. The causes are scars, nerve damage [11] or a process called fatty degeneration. Fatty degeneration describes on the one hand the infiltration and accumulation of fat tissue in the muscle and on the other hand the transformation of muscle tissue to fat, which both prove to be irreversible [12]. The result is a decrease in the fraction of the muscle fiber volume according the total muscle volume. The muscle fibers are the main contributors to the muscle's force-generating capacity. Therefore, with a loss of muscle fiber volume, the force that a muscle can exert decreases and the muscle gets weaker accordingly. The negative influence of fatty infiltration on the functional outcome of a muscle was first demonstrated in cuff muscles [13], but has now also been proven for gluteal muscles [14]. Fatty degeneration of muscles is primarily a factor of aging [12, 15], but can be considerably enforced under specific conditions like muscle injury [12]. These two factors apply to revision THR, in which the muscles are damaged for at least the second time due to the surgical procedure and the age of the patients is higher on average compared to the patients undergoing primary THR.

Hence, a prediction of the biomechanical postoperative outcomes of the chosen implant (parameters) already during the preoperative planning would be extremely beneficial for the patient, especially for revision THR. This would allow the surgeon to adjust the parameter selections, if necessary, to prevent any muscular dysfunctions and consequently increase the quality of operative outcome and patient satisfaction [16]. The usage of musculoskeletal human models (MHMs) [17] during preoperative planning of THR could provide these benefits. Thus, the patient can be represented virtually and the planned operation could be executed on the model. Such a model allows simulations of different movements during daily activities and at the end, the surgeon would receive biomechanical feedback on the planned implant configuration.

Therefore, our aim in this study was to provide evidence that it is possible to use MHMs in the preoperative planning process to evaluate and improve the outcome of THR, especially in the case of revision THR. To achieve this, MHMs must be able to provide a hint for muscular issues occurring after revision THR. Furthermore, the transfer of the planned implant configuration of the surgeon to the model has to be enabled and additionally, the effect of the varied geometrical hip condition on the patient's biomechanics needs to be recognisable.

2 Material and Methods

2.1 Patient-Specific Model

In order to obtain valid simulation results by the MHM for the outcomes of the THR of a specific patient, it is obvious that not any model can be used. The use of a so-called patient-specific model is required. The principle creation process from segmentation to adjustment of the passive musculoskeletal system (bones) to modeling the patient's muscle paths and muscle parameters was described in detail in previous published work [18–20]. In this study, we used a generic whole body model [21] in OpenSim 4.1 [17] to scale it to the anthropometry of a patient. The available patient data were his body size and an MRI image from his pelvis and femora. The remaining body dimensions were estimated using a population database of the german society. The patient's pelvis and femora bones were segmented from the MRI image. During the integration of the segmented bones in the patient model, the coordinate system of each segmented bone was defined as that of the corresponding bone replaced from the generic model. The hip joint rotation centers were adopted from the segmented MRI image of the patient and the center of mass and inertia of each bone were adopted from bones in the scaled generic model. Muscle modeling focused on the gluteal muscles, because they are the ones that are mainly affected by a THR. The muscle origin and insertion points were derived from the segmented patient data. The optimal fiber length L_m^{opt} and pennation angle θ of each muscle compartment was taken from literature [17, 22]. Each gluteal muscle was divided into three single compartments (1-anterior, 2-medial and 3-posterior) to enable the muscles to reproduce their real working ratio on different joint coordinates, as it was implemented by Delp et al. [17]. The maximal isometric force ($F_{max,iso}$) of every muscle compartment was calculated based on the determined muscle volume (Vol) from the segmented patient data using the following Eq. (1):

$$F_{max,iso} = \frac{Vol * \cos\theta}{L_m^{opt}} * \sigma \quad (1)$$

The specific muscle tension σ was assumed to be 60 N/mm^2 as it has already been done in literature [22]. The finally created patient-specific model with only the adjusted gluteal muscles is shown in Fig. 1a.

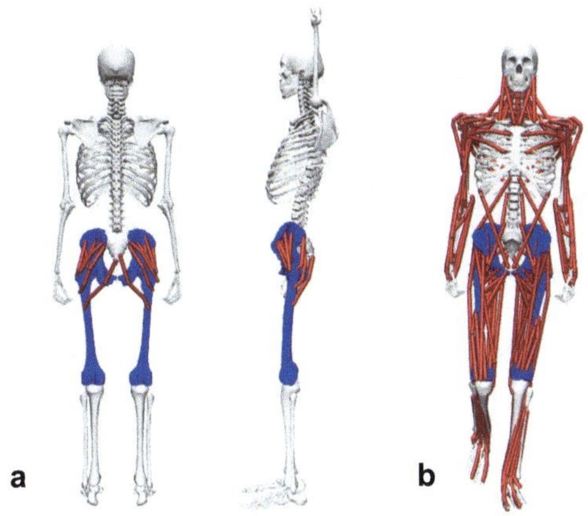

Fig. 1. a) Patient-specific model with inserted segmented bone data of pelvis and femora and adjusted muscle paths and parameters of the gluteal muscles; b) Simulated Position of single leg stance to compute the load on the left hip abductors.

2.2 Movement and Simulation

To simulate the activation and force applied by the patient-specific muscles, a suitable movement has to be performed by the patient-specific model. The gluteal muscles are mainly associated to the hip abductors (gluteus medius, gluteus minimus and part of the gluteus maximus). The main function of the hip abductors is pelvic stabilization during the stance phase of gait [23]. If the moment of stance phase during the gait cycle is extracted, the situation looks very similar to a single leg stance. Thus, we positioned our patient model in a single leg stance and let it perform a static optimization (SO) to calculate the required activation and force of the hip abductors. As it can be seen in Fig. 1b, we chose to position our model on the left foot, resulting in a load on the left hip abductors. The ground reaction forces were estimated by point and torque actuators in all three spatial directions in the calcaneus and the toes and were afterwards defined as external forces in the SO.

2.3 Weakening of the Muscles

A major remaining problem is that we have the data of a young, healthy human who has no diagnosed muscular impairments. Hence, the expected results of the simulation would be that the model has no problems executing the movement. Thus, we decided to apply the previously described issue of weakened gluteal muscles and associated decreased ability to generate force after THR to our patient-specific model, in order to represent a condition, in which the biomechanics is more similar to a damaged constitution as it occurs after THR.

Because of the little influence of the gluteus maximus to hip abduction, the investigation of the effects of fatty degeneration of this muscle will be neglected. When analyzing the relevance of fatty degeneration in the gluteus medius and minimus, many studies make the same conclusions. First, an anteroposterior gradient is existent in both muscles and second, the amount of fat in the gluteus minimus is higher than in the gluteus medius [24–26], which results in a greater weakening rate of the gluteus minimus. However, the gluteus medius is the stronger muscle compared to the gluteus minimus and is also considered as the main hip abductor. Furthermore, a weakened gluteus medius is directly related to the aforementioned trendelenburg gait, underlining the importance for investigating this muscle for THR. Most studies analyze the fat in the gluteus medius with a classification system that assigns a grade to the amount of fat in the muscle or the muscle compartments [13]. Von Roth et al. [27] were the only ones, to our knowledge, that computed the average percentage of remaining muscle fibers compared to the total muscle volume of multiple patients. Additionally, they explicitly distinguished between primary and revision THR. The results are shown in Table 1.

Table 1. Remaining fraction of muscle fiber volume compared to the total muscle volume of gluteus medius determined after primary and revision THR from von Roth et al. [27]

Muscle fiber volume [%]	Healthy	Primary THR	Revision THR
Glut med 1	100	84.7	48.5
Glut med 2	100	90.1	67.4
Glut med 3	100	93.5	72.7

By mutiplying the remaining muscle fiber fraction to the equivalent segmented muscle compartment volume of our patient, a new maximal isometric force can be computed with Eq. (1). In this way, two further muscular conditions (after primary THR and after revision THR) were generated in addition to the healthy status of the patient. With all three muscular conditions (represented as three MHMs with different gluteus medius condition each), simulation of the single leg stance was performed and the activation of the gluteus medius and the gluteus medius compartments was compared.

2.4 Parameter Study

During the preoperative planning of THR, the surgeon can modify the geometrical parameters of the hip by choosing an implant with the desired configuration. To investigate the effect of varied parameters for the biomechanics of the patient, we performed a parameter study with some typical hip parameters (see Fig. 2). The first considered parameter was the neck-shaft-angle (NSA). We defined the NSA as the angle between the femoral neck axis originating from the hip joint center and the femoral shaft axis in the frontal plane. The second parameter was the femoral offset. This is defined as the perpendicular distance from the femoral shaft axis to the hip joint center. The third parameter was the antetorsion angle (AT). The AT is defined as the angle between the plane at the femoral condyles and the femoral neck axis. All these parameters have an initial value given by the patient. First, these parameters were determined. The patient value of the NSA was 130°, AT was at 20° and femoral offset was measured with 3.5 cm (see Table 2). In the second step, these parameters were varied in a certain range with a certain step size for each parameter (Table 2). For the variation of the parameters, the femur was shifted in all spatial directions to obtain the desired parameter combination, which conversely simulates a possible parameter change in the context of a THR. The leg length difference was assumed to be zero for every generated configuration. Then, the single leg stance simulation was performed with each parameter combination and the activation of the gluteus medius was computed and compared for all configurations.

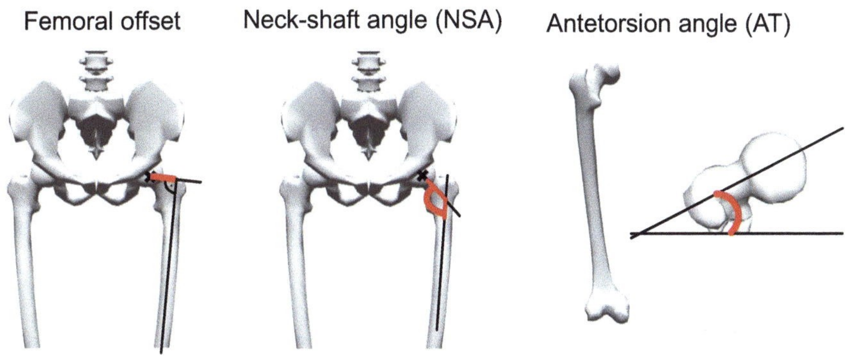

Fig. 2. Definition of some typical geometrical hip parameters the surgeon can adjust from an implant to match the anatomical condition of the patient during preoperative planning

Table 2. Variation of parameter values and parameter step size used for the parameter study

	Femoral offset	NSA	AT
Patient initial value	3.5 cm	130°	20°
Variation	1–5 cm	110–150°	10–40°
Step size	0.5 cm	5°	5°

3 Results

The comparison of required gluteus medius activity of all three muscular conditions to keep the model in the single leg stance is shown in Fig. 3. It can be seen that an

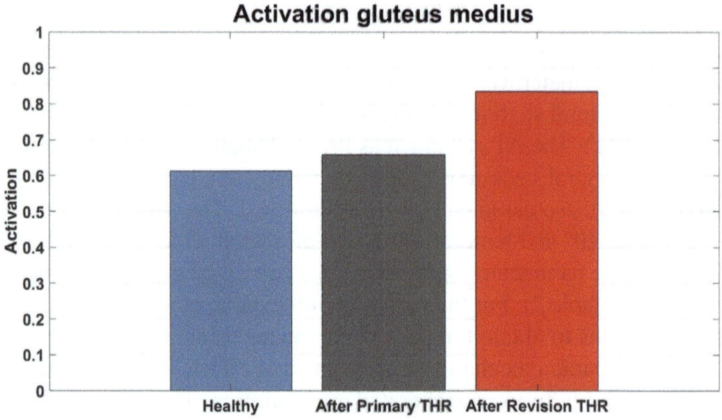

Fig. 3. Activation of total gluteus medius muscle for the three defined conditions: healthy, after primary THR and after revision THR

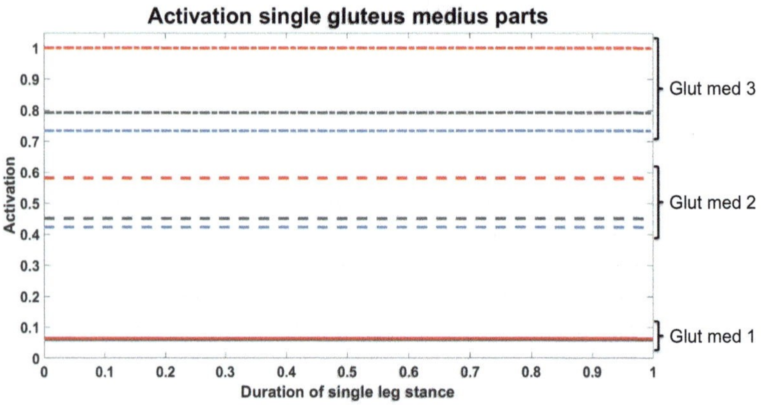

Fig. 4. Activation of the single gluteus medius compartments for the three defined conditions: healthy, after primary THR and after revision THR

activation of 60% is necessary for the healthy status. This activation rises quite slightly for the patient condition after primary THR to around 66%. A remarkably big increase is then noticed for the patient status after revision THR, where an activation of the gluteus medius of 83% is required. Splitting the activation of the entire gluteus medius muscle to its single parts for all conditions reveals some major differences of the muscle weakening (see Fig. 4). While the gluteus medius 1 activation is almost unchanged for all three conditions, the gluteus medius 2 activation is already considerably increased from the healthy status to the condition after revision THR. The highest rise of activation is current for the gluteus medius 3 activation, where activation increases to nearly 100% after revision THR. The muscle activity for the status after primary THR shows a recognizable increase for gluteus medius 2 and 3, but is way less compared to the status after revision THR and therefore less stressful for the patient. Thus, the parameter study was just

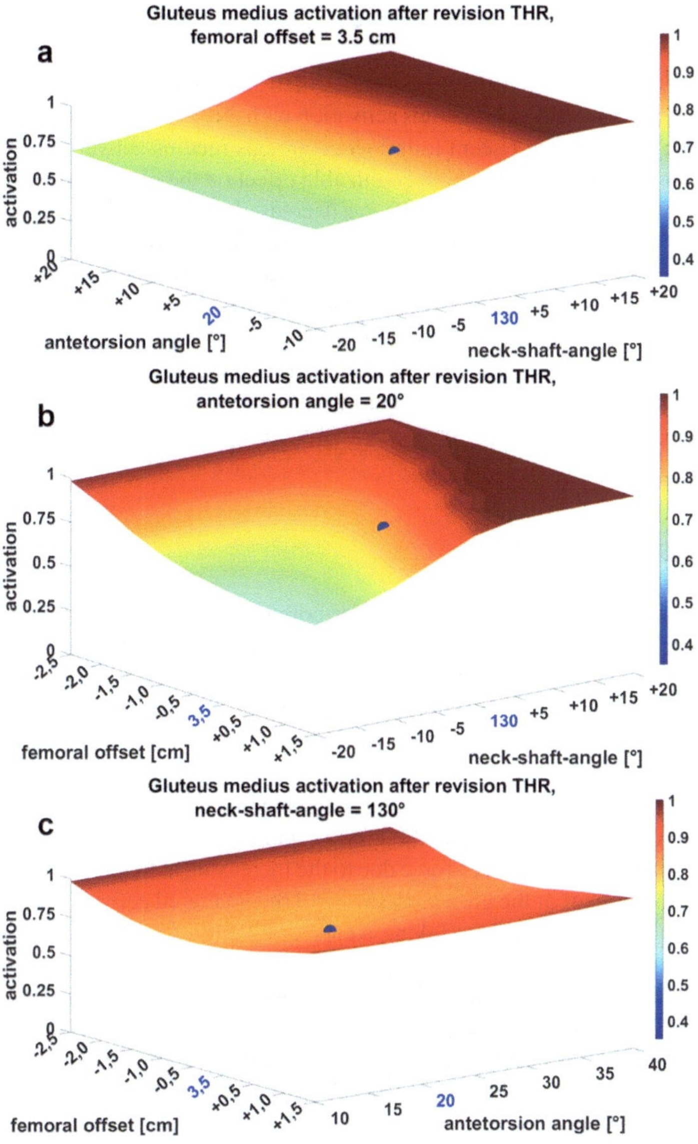

Fig. 5. Dependency of gluteus medius activation from the varied geometrical hip parameters; Blue marks and points identify the patient's anatomical condition and respectively the gluteus medius activation at the patient's anatomical condition

performed for the more severe case, the status after revision THR. The three plots in Fig. 5 show the influence of a variation of respectively two parameters on the gluteus medius activation, while the third one remained at its initial value. The blue marks and points represent the patient's initial anatomical condition and the patient's gluteus medius activation at the initial anatomical condition, respectively (same activation as in Fig. 3).

The results show decreased gluteus medius activation with decreasing NSA and correspondingly increased gluteus medius activation with increasing NSA (Fig. 5a). Changing the value of AT doesn't seem to influence gluteus medius activation strongly. The femoral offset also seems to have a recognizable effect on the gluteus medius activation as the NSA. With increasing femoral offset, a decreased gluteus medius activation results and accordingly by decreasing the femoral offset the activation increases (Fig. 5b). In combination with a variation of the NSA, activation can even be decreased to a minimal value. Figure 5c, indeed, suggests another effect on the gluteus medius activation with altering femoral offset at a constant NSA value. As femoral offset increases or decreases, the activation is now increased. The minimal value can be identified at an average femoral offset value close to the initial anatomical condition. The results for AT show the same behaviour as previous, that there is no influence on gluteus medius activation.

4 Discussion

Weakening of the gluteus medius muscle for both conditions - after primary THR and after revision THR - results in an increased muscle activation required to keep the position of the single leg stance, especially after revision THR. If a muscle weakens, having the consequence of a decreased maximal isometric force, the muscle has to be activated stronger to apply the same necessary force, which is unaffected due to the unchanged anatomical situation. Hence, the increased activation seems to be a reasonable result. Considering the activation of the single gluteus medius compartments of the different situations, the results indicate that the less weakened muscle compartments are used preferred to generate the reqiured force due to the increasing activation of the medial and posterior compartment of the gluteus medius and the unaltered activation of the anterior compartment of gluteus medius. Taking a closer look at the posterior compartment of the gluteus medius, the activation of nearly 100% after revision THR implies a very high load for the muscle. Transferred to the real situation, this finding indicates that the muscle would have to use almost its complete available muscle force in the stance phase of the left leg during the gait cycle. Thus, a damage to the muscle is highly probable, resulting in a higher risk of muscular issues for the patient, which can become apparent by e.g. a trendelenburg gait.

Furthermore, the results of the parameter study identify that the NSA and the femoral offset have a impact on the activity of the gluteus medius. The increasing activation with decrasing NSA und increasing femoral offset and, conversely, the decreasing activation with increasing NSA und decreasing femoral offset is in good agreement with the findings of Lenaerts et al. [28]. However, the influence of the femoral offset is linked to a varying NSA. At constant NSA, an increase of the femoral offset causes an enormous stretch of the muscle, which brings it close to its physiological limits. This means the muscle is stretched to the point that it is technically unable to generate forces, resulting in the high activation of the gluteus medius. Thus, the NSA is the main contributor for the activity of the gluteus medius.

Based on the results of the parameter study, the suggestions might be made to simply implant an endoprosthesis with a low NSA and additionally a high femoral offset to a patient with risk of muscular issues after THR, which is correct for the muscular condition in the hip. The low NSA effects a high lever arm for the muscles that requires a lower generated force by the muscles and thus unloads them. On the negative side, tensile stresses are occurring on the lateral side of the femoral neck, leading to a high fracture risk of the bone. The bones prefer the situation characterized by a high NSA. In this situation, only desirable compressive stresses occur in the femoral neck. However, this situation represents the worst case for the muscles due to the low lever arm and the high muscle force required. [29] Therefore, the existing NSAs in the hip joints of the human body can be interpreted as a compromise between both extremes. Following this procedure, the selection of the optimal implant parameter and position has to be done by considering both, the muscular and the skeletal status, in relation to their respective conditions. Hence, for the previous suggestion, the effect on the bones has to be included just as the current preoperative planning process neglects the effect on the muscles of the chosen implant so far. A suitable parameter to investigate the effect on the joint is the so-called hip joint reaction force (F_R). This force appears as a consequence of the weight force and the muscle force and can be ultilized to simulate the effects of the implant configuration on the bone and the implant [30]. MHMs therefore have the additional advantage of being able to examine the magnitude and direction of the patient's F_R and its influence on the intraoperative adjustable parameters, as it can be seen in Fig. 6. Thus, the resulting F_R of a specific implant configuration can be computed with the MHM and inserted in a parallel process (e.g. FE analysis) to ensure the chosen configuration also from the perspective of the bones (kinematical perspective).

There are also some limitations in our simulation study. The first one is that we did not execute a comparison between our model and the real patient. To ensure valid results of the model, a comparison of the patient's muscle activation recorded via EMG with the simulated muscle activation of the patient-specific model has to be conducted. This allows the validation of the model creation process and, especially, the values used for muscle (compartment) parameters, like optimal fiber length, to be checked and adjusted, if necessary. The second limitation is the weakened gluteus medius muscle according to literature values. Due to the lack of suitable data, we chose this approach to get a first impression of the effect of atrophied gluteal muscles after THR. The effect has to be further evaluated with real patient data. The findings of van Roth et al. [27], in general consistent with other results [24, 26], show a wide range in the results of the fatty infiltration for each gluteus medius compartment. For example, the range of computed fat fraction after revision THR in the posterior gluteus medius compartment differs from 0.6 to 98.4 percent, implying strong patient specificity for fatty degeneration of the muscles. Furthermore, the study of von Roth included only patients having undergone a direct lateral approach for THR. Several studies, however, demonstrated that other approaches reach better functional outcomes and thereby better results regarding fatty infiltration [8, 31, 32]. Nevertheless, fatty degeneration can never be eliminated completely and the surgeon has to decide which approach is even possible to be executed for the individual patient. In revision THR, the choice of applicable approaches is anyway already reduced compared to primary THR and, in particular, some approaches classified as tissue conservative are among these, as the minimally invasive anterolateral approach. In addition, fatty degeneration is a process of aging, as mentioned above. Therefore, the utility of biomechanical evaluation of THR and of the effects of THR may be given for every patient.

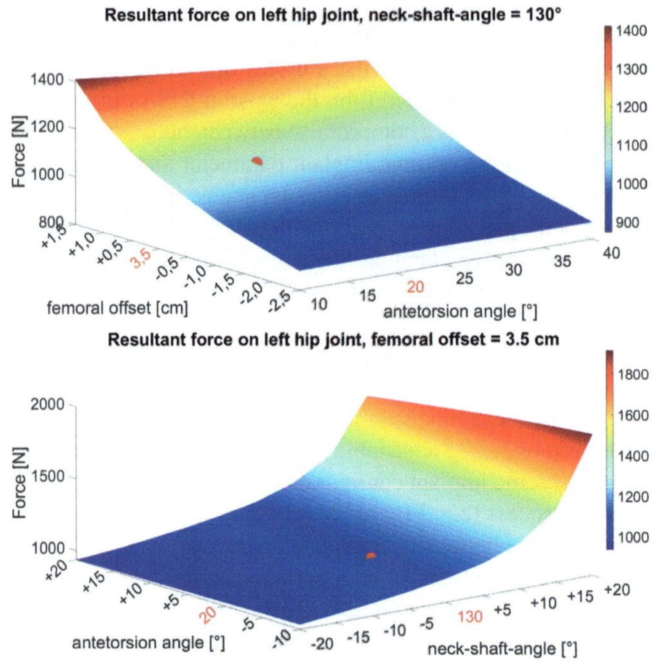

Fig. 6. Dependency of the hip joint reaction force from the varied geometrical hip parameters; Red marks and points identify the patient's anatomical condition and respectively the hip joint reaction force at the patient's anatomical condition

Another limitation in our performed study is that we simulated a "static movement", meaning that we did not really conduct a real dynamic movement like gait. However, the load on the gluteal muscles in the considered movement is given to a comparable extent as in the single stance phase of gait. Therefore, the results obtained should be reasonable, at least for a first investigation. Future work has to include the recording and analysis of different movements with a specific loading on the gluteal muscles. With these movements, the existing potential of remaining muscle force could be estimated in a better way and the effect of varying intraoperative adjustable parameters during THR could be better characterized.

5 Conclusion

As a conclusion, we have shown that it is possible to analyse the patient's individual biomechanical condition with MHMs. Here, the simulation with MHMs provides a good hint for possibly occurring muscular issues as they are associated to appear at higher risk following revision THR. Thus, revision THR is a suitable use case for the application of MHMs to improve the overall surgical outcome. By inclusion of MHMs in the preoperative planning process, the effect of different intraoperative adjustable parameters can be simulated and evaluated. Consequently, the surgeon is enabled to have a better opportunity to identify the optimal implant parameter and implant position from both a biomechanical and kinematical view.

References

1. Grimberg, A., Jansson, V., Melsheimer, O.: Jahresbericht 2019. EPRD Deutsche Endoprothesenregister, Berlin (2019)
2. Wengler, A., Nimptsch, U., Mansky, T.: Hip and knee replacement in Germany and the USA: analysis of individual inpatient data from German and US hospitals for the years 2005 to 2011. Deutsches Arzteblatt international (2014). https://doi.org/10.3238/arztebl.2014.0407
3. Graves, S.E., et al.: The Australian orthopaedic association national joint replacement registry. Med. J. Aust. (2004). https://doi.org/10.5694/j.1326-5377.2004.tb05911.x
4. Kurtz, S., Ong, K., Lau, E., Mowat, F., Halpern, M.: Projections of primary and revision hip and knee arthroplasty in the United States from 2005 to 2030. J. Bone Joint Surg. Am. Vol. (2007). https://doi.org/10.2106/JBJS.F.00222
5. Rupp, M., Lau, E., Kurtz, S.M., Alt, V.: Projections of primary TKA and THA in Germany from 2016 through 2040. Clin. Orthop. Relat. Res. (2020). https://doi.org/10.1097/CORR.0000000000001214
6. Eggli, S., Pisan, M., Müller, M.E.: The value of preoperative planning for total hip arthroplasty. J. Bone Joint Surg. Br. Vol. (1998). https://doi.org/10.1302/0301-620X.80B3.7764
7. Huppertz, A., Radmer, S., Wagner, M., Roessler, T., Hamm, B., Sparmann, M.: Computed tomography for preoperative planning in total hip arthroplasty: what radiologists need to know. Skeletal Radiol. **43**(8), 1041–1051 (2014). https://doi.org/10.1007/s00256-014-1853-2
8. Madsen, M.S., et al.: The effect of total hip arthroplasty surgical approach on gait. J. Orthop. Res. (2004). https://doi.org/10.1016/S0736-0266(03)00151-7
9. Chang, R.W.: A cost-effectiveness analysis of total hip arthroplasty for osteoarthritis of the Hip. JAMA (1996). https://doi.org/10.1001/jama.1996.03530350040032
10. Ulrich, S.D., et al.: Total hip arthroplasties: what are the reasons for revision? Int. Orthop. (2008). https://doi.org/10.1007/s00264-007-0364-3
11. Abitbol, J.J., Gendron, D., Laurin, C.A., Beaulieu, M.A.: Gluteal nerve damage following total hip arthroplasty. J. Arthroplasty (1990). https://doi.org/10.1016/S0883-5403(08)80090-3
12. Hamrick, M.W., McGee-Lawrence, M.E., Frechette, D.M.: Fatty infiltration of skeletal muscle: mechanisms and comparisons with bone marrow adiposity. Front. Endocrinol. (2016). https://doi.org/10.3389/fendo.2016.00069
13. Goutallier, D., Postel, J.-M., Gleyze, P., Leguilloux, P., van Driessche, S.: Influence of cuff muscle fatty degeneration on anatomic and functional outcomes after simple suture of full-thickness tears. J. Shoulder Elbow Surg. (2003). https://doi.org/10.1016/S1058-2746(03)00211-8
14. Thaunat, M., et al.: Influence of muscle fatty degeneration on functional outcomes after endoscopic gluteus medius repair. Arthroscopy J. Arthroscopic Related Surg. Official Publ. Arthroscopy Assoc. North Am. Int. Arthroscopy Assoc. (2018). https://doi.org/10.1016/j.arthro.2018.01.005

15. Correa-de-Araujo, R., et al.: Myosteatosis in the context of skeletal muscle function deficit: an interdisciplinary workshop at the national institute on aging. Front. Physiol. (2020). https://doi.org/10.3389/fphys.2020.00963
16. Patil, S., Garbuz, D.S., Greidanus, N.V., Masri, B.A., Duncan, C.P.: Quality of life outcomes in revision vs primary total hip arthroplasty: a prospective cohort study. J. Arthroplasty (2008). https://doi.org/10.1016/j.arth.2007.04.035
17. Delp, S.L., et al.: OpenSim: open-source software to create and analyze dynamic simulations of movement. I.E.E.E. Trans. Biomed. Eng. (2007). https://doi.org/10.1109/TBME.2007.901024
18. Fleischmann, C., et al.: Segmentation of musculotendinous structures of the hip from 3D imaging for patient-specific individualization of biomechanical simulations. In: Ahram, T., Taiar, R., Langlois, K., Choplin, A. (eds.) IHIET 2020. AISC, vol. 1253, pp. 331–337. Springer, Cham (2021). https://doi.org/10.1007/978-3-030-55307-4_50
19. Scherb, D., Kurz, M., Fleischmann, C., Wolf, A., Sesselmann, S., Miehling, J.: Patientenspezifische Modellierung des passiven Bewegungsapparates als Grundlage für die präoperative Abschätzung postoperativer Folgeerscheinungen des endoprothetischen Hüftgelenkersatzes. In: Proceedings of the 31st Symposium Design for X (DFX2020). Proceedings of the 31st Symposium Design for X, 16–17 December 2020, pp. 1–10. The Design Society (2020). https://doi.org/10.35199/dfx2020.1
20. Leher, I., et al.: Patient-specific modelling for preoperative estimation of hip mechanics for improved planning of total hip endoprosthesis using multibody simulations. In: Ahram, T., Taiar, R. (eds.) IHIET 2021. LNNS, vol. 319, pp. 1088–1096. Springer, Cham (2022). https://doi.org/10.1007/978-3-030-85540-6_140
21. Miehling, J.: Musculoskeletal modeling of user groups for virtual product and process development. Comput. Methods Biomech. Biomed. Engin. (2019). https://doi.org/10.1080/10255842.2019.1651296
22. Rajagopal, A., Dembia, C.L., DeMers, M.S., Delp, D.D., Hicks, J.L., Delp, S.L.: Full-body musculoskeletal model for muscle-driven simulation of human gait. I.E.E.E. Trans. Biomed. Eng. (2016). https://doi.org/10.1109/TBME.2016.2586891
23. Flack, N.A.M.S., Nicholson, H.D., Woodley, S.J.: A review of the anatomy of the hip abductor muscles, gluteus medius, gluteus minimus, and tensor fascia lata. Clin. Anatomy (New York, N.Y.) (2012). https://doi.org/10.1002/ca.22004
24. Daguet, E., et al.: Fat content of hip muscles: an anteroposterior gradient. J. Bone Joint Surg. Am. Vol. (2011). https://doi.org/10.2106/JBJS.J.00509
25. Kovalak, E., Özdemir, H., Ermutlu, C., Obut, A.: Assessment of hip abductors by MRI after total hip arthroplasty and effect of fatty atrophy on functional outcome. Acta Orthop. Traumatol. Turc. (2018). https://doi.org/10.1016/j.aott.2017.10.005
26. Pfirrmann, C.W.A., Notzli, H.P., Dora, C., Hodler, J., Zanetti, M.: Abductor tendons and muscles assessed at MR imaging after total hip arthroplasty in asymptomatic and symptomatic patients. Radiology (2005). https://doi.org/10.1148/radiol.2353040403
27. von Roth, P., et al.: Significant muscle damage after multiple revision total hip replacements through the direct lateral approach. Bone Joint J. (2014). https://doi.org/10.1302/0301-620X.96B12.34256
28. Lenaerts, G., et al.: Subject-specific hip geometry and hip joint centre location affects calculated contact forces at the hip during gait. J. Biomech. (2009). https://doi.org/10.1016/j.jbiomech.2009.03.037
29. Pauwels, F.: Atlas zur Biomechanik der gesunden und kranken Hüfte. Springer, Heidelberg (1973)

30. Stolk, J., Verdonschot, N., Cristofolini, L., Toni, A., Huiskes, R.: Finite element and experimental models of cemented hip joint reconstructions can produce similar bone and cement strains in pre-clinical tests. J. Biomech. (2002). https://doi.org/10.1016/S0021-9290(01)00213-5
31. Müller, M., Tohtz, S., Dewey, M., Springer, I., Perka, C.: Evidence of reduced muscle trauma through a minimally invasive anterolateral approach by means of MRI. Clin. Orthop. Relat. Res. (2010). https://doi.org/10.1007/s11999-010-1378-5
32. Bremer, A.K., Kalberer, F., Pfirrmann, C.W.A., Dora, C.: Soft-tissue changes in hip abductor muscles and tendons after total hip replacement: comparison between the direct anterior and the transgluteal approaches. J. Bone Joint Surg. Br Vol. (2011). https://doi.org/10.1302/0301-620X.93B7.25058

Development of a Silent Speech Interface for Augmented Reality Applications

Christine Walck(✉), Tania Rivas, Riley Flanagan, and Michael Fornito

Embry Riddle Aeronautical University, Daytona Beach, FL 32114, USA
Christine.Walck@erau.edu

Abstract. Silent speech interfaces using non-invasive electromyography (EMG) sensors have been utilized to control internet-of-things devices [1] and provide communications in acoustically challenging environments [2]. However, they have yet to be implemented into Augmented Reality displays, an area they can potentially revolutionize as a human-machine interface by offering low-profile and fluid input. This study overviews the development of a silent speech interface that receives and decodes input from subvocalizations recorded by skin surface EMG sensors, to be used to control a heads-up-display built on a Microsoft HoloLens. Measured muscle activation of the anterior cervical region while a subject subvocalized words from a predetermined library were collected. Trials consisting of subvocalized words were parsed for individual subvocalizations to build a dataset for training of a speech recognition model. The speech recognition model based on a one dimensional convolutional neural network employed to classify subvocalized words was built with the Keras application programming interface in Python, using the TensorFlow library. Preliminary results demonstrate effectiveness in classifying commands, with classification accuracies for ten trained models showing promise. Successful classification was achieved with models showing accuracy in the range of 66.6% to 100%. An average word classification accuracy of 82.5% between all models is observed. While all models were trained and tested on the same datasets, the stochastic nature of the model has significant effects on output, with the dropout layer adding artificial noise to training, and the gradient-descent based optimization algorithm adding random variance to the completed model effectiveness.

Keywords: Electromyography · Convolutional neural network · Silent speech interface

1 Background

Speech is considered silent when no intelligible acoustic signal is produced or received. Silent speech interfaces are assistive devices which aid in reestablishing communication when acoustic signals are unattainable or undesired. General research efforts into this technology were driven by the need for speech assistive devices. About 40 million U.S. citizens have communication disorders, as reported by the American Speech Language Hearing Association, and annually costs about $154–$186 billion [3]. Interest peaked

J. M. R. S. Tavares et al. (Eds.): CMBBE 2021, LNCVB 38, pp. 208–214, 2023.
https://doi.org/10.1007/978-3-031-10015-4_18

in the early 2000s prompted by the Defense Advanced Research Projects Agency's Advanced Speech Encoding program. Since then, seven central data collection methods have risen which will be discussed in their respective subsection below. As to date, all methods have either been used for speech to text, translating speech biosignals into text, or direct speech synthesis, modeling biosignals into acoustic waveforms [4].

1.1 Electromagnetic Articulography Sensors

Because the vocal tract plays an important role in speech production, creating an approach based on its different positions is instinctive. Currently, there are two methods commonly used. One involves the use of electromagnetic articulography (EMA) sensors, which track the movement of fixed points within the vocal tract. Multiple researchers utilize implanted coils connected electrically to external equipment and electromagnetically paired with excitation coils. This systems tracks the precise Cartesian coordinates of the implanted coils. However, measuring the defined Cartesian positions of the defined points is needed to be achieved in a significant, convenient manner which has still not been accomplished [5].

The other method involves nonlinear mapping between the vocal tract position and its produced sound. This system places permanent magnets on the vocal tract linked to magnetic sensors placed on the subject's head. This overrides an electric connection and provides placement flexibility [6].

1.2 Ultrasound Plus Optical Imaging

Another position centered method is using ultrasound imagery plus optical cameras to gather vocal tract configuration data. An ultrasound transducer is placed underneath the chin to obtain a partial view of the tongue's surface and a video camera is placed in front of the subject's mouth [7]. Together, non-acoustic features are collected and utilized to create a speech synthesizer. Lip profiles and tongue contours are taken from the videos and mapped into a "standard vocal tract configurations" representative space. From there, a visuo-phonetic decoder interprets the data and outputs speech through a vocoder. However, this approach only predicts spectral features limiting the speech synthesis to linear predictive coding without finding the excitation signal needed. This leads to only a 60% accuracy rate when classifying phones making this approach unable to output intelligible synthesis.

1.3 Nonaudible Murmur Microphone

Nonaudible microphones (NAM) are coined by low amplitude sounds made by laryngeal airflow and vocal tract resonance. NAM sounds are not obvious to others nearby but can be identified by highly sensitive contact microphones. The main design goal is to detect tissue vibrations during speech while ignoring environmental noise. What began as a simple stethoscope placed on the underside of the chin is now a small microphone encased in silicone material to efficiently transmit the skin vibrations but ignore external ones. The NAM microphone is traditionally placed on a large neck muscle below the mastoid [8].

1.4 Electromagnetic Sensors Analyzing Glottal Activity

There have been multiple studies investigating glottal waveforms produced during speech. Glottal waveforms travel through tissue and bone but not through air. This categoeirzes them as non-acoustic. Currently researchers use either vibration or EMA sensors to collect glottal waveform data. Vibration techniques use microphones such as throat (a double budded microphone placed on each side of the Adam's apple to obtain speech signals), bone (a vibration sensor built into helmets as a way to collect bone propagated speech signals on points of the cheekbones, forehead and crown), in-ear (a microphone placed into the ear canal and covered by an earphone to block external noise), and physiological (a microphone built into a neckband with a piezoelectric sensor) microphone [9, 10]. Whereas, EMA sensors use coils and excitation signals to act like transformers to measure the distances between them and calculate defined positions. There are currently three types of EMA sensors: electroglottograph (EEG), General Electromagnetic Motion System (GEMS) and Tuned Electromagnetic Resonating Collar (TERC). The EGG sensors measure the electrical impedance changes across the throat during speech production. It is made of two gold plated electrodes placed on either side of the larynx with an applied potential. The GEMS uses an antenna attached to the throat on the laryngeal notch and can detect the electromagnetic frequency changes [11]. The TERC is comprised of copper electrodes tuned at a specific resonance to detect glottal vibration signals by measuring the changes of the glottal intrinsic electrical capacitance [12]. A drawback to using any kind of EMA sensor is the sensitivity to the exact positioning of the electrodes, and antennae. Any movements to these devices cause large deviations.

1.5 Electroencephalographic Sensors

Electroencephalography (EEG) has been utilized to control brain computer interfaces (BCI) to take thoughts and intentions to direct computers and prostheses. Recently, researchers will have a subject imagine saying a word or imagine individual vowels as the EEG sensors placed are positioned around the head, record the data. EEG recordings have been seen to reflect isolated words [13]. However, the process for subjects to learn to manipulate their brain activity takes months which makes this method unfavorable [14].

1.6 Speech-Motor Cortex Implants

Attempts to help restore speech for those who have been paralyzed has been researched using an invasive method involving intracortical microelectrode technology [15]. Applications involve controlling computer cursors and virtual keyboards. Currently there are two main electrode choices: the Utah array and Neurotropic Electrode. The Utah array is a single silicon wafer with many electrodes whereas the Neurotrophic Electrode user significantly fewer low impedance wires covered by a glass cone [16]. Both are implanted completely in the cerebral cortex and have been operational for several years at a time. The most common way to decode these modalities are through movement kinematics – relating movements to its neural activity. This has been applied to speech motor control by looking at speech production as a motor task rather than something abstract.

1.7 Surface Electromyography Sensors

Surface EMG sensors record muscles' electrical activity in a non-invasive manner as they record the ion flows travelling through the central nervous system from the surface of the skin which creates a potential difference in the body. The voltages collected can be entered into a variety of electronic devices to be processed further. In addition, EMG sensors can recognize the intention of movement even if there is no explicit movement such as isometric contractions. This makes sEMG sensors prime detectors of subvocalizations [17]. A benefit to using EMG sensors is the freedom of placement on the subject's head, neck, or facial regions allowing subvocalized or mouthed words and phonemes to be collected.

2 Methodology

A healthy 22 year old Caucasian male with no speech impairments in a controlled lab environment was prompted to subvocalize command words from a library of six words: Up, Down, Left, Right, Yes, and No. To ensure high accuracy rates while reducing potential error from differences within individual subvocalizations, the subject was prompted to subvocalize commands forty times. Preliminary subvocalization data was collected using EMG sensors (Delsys, Natick, MA) applied to the Digastric, Sternohyoid, Stylohyoid, and Cricothyroid, as seen in Fig. 1. This placement selection was determined through an amplitude qualities and characteristic waveform comparison during the speech production process. Sensors with poor signal to noise ratios were excluded as they were considered low quality.

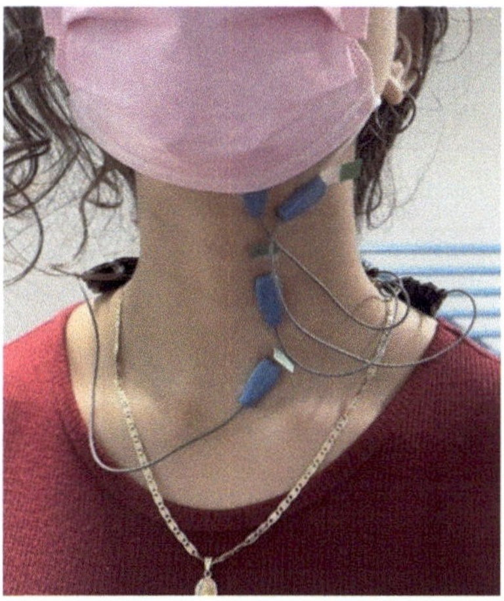

Fig. 1. EMG sensor head setup parallel to the Digastric, Sternohyoid, Stylohyoid and Cricothyroid muscle fibers.

Data was filtered to remove environmental noise that could potentially interfere with the recognition model. According to the sensor manufacturer, the EMG electrodes have a bandwidth of 10–450 Hz establishing the value of a band-pass filter to remove high and low frequency noise from the signal at a window of 10–400 Hz. Applying the filter allowed the band of frequencies describing the subvocal activity to be isolated. Individual command words were parsed from the EMG signal data and placed into individual training datasets. A one dimensional, seven layered Convolutional Neural Network (CNN) was created with the Keras library in Python. Important characteristics of this model are the two dropout layers integrated to reduce overfit by randomly compensating for small training datasets and reduce the model's dependency on the current training data while simultaneously improving the accuracy of unseen data.

3 Results

The current CNN model was tested using 12 training datasets, resulting in a mean recognition accuracy of 82.5% with a 10.172 variance, a maximum of 100% accuracy and minimum of 66.66%. The effect of using the same datasets to both train and test the classifier is reflected in the high variance value (Table 1).

Table 1. Summarized training accuracies for models 1–10.

Trained model	Accuracy
1	66.7%
2	83.3%
3	91.7%
4	83.3%
5	91.7%
6	83.3%
7	83.3%
8	100.0%
9	66.7%
10	75.0%
Mean	82.5%

4 Conclusion

Though these values are promising, there are probably still large effects of overfit due to the small amounts of data inputted despite the addition of dropout layers. A large source of error stems from overfit coming from the classifier's tendency to fitting too closely to the training sets. Trials 2, 6, and 7 resulted in the same values of accuracy, yet another potential consequence of overfit. Furthermore, the CNN has a naturally occurring random probability distribution leading to significant differences in output despite being trained with the same datasets. To combat this, a larger and more diverse set of training datasets will be needed. This will not only improve the CNN's accuracy but maximize the applicability of the interface to multiple users as well. Currently, the interface is a proof-of-concept of being able to classify previously acquired words in an ideal lab environment after being processed.

5 Discussion

Current AR displays use explicit input such as hand gestures or buttons making them susceptible to privacy issues, movement restrictions, and unintentional input. This causes the integration of AR interfaces in the aerospace and defense fields to be inhibited by conspicuous and cumbersome inputs. Therefore, the application of subvocalizations, allowing a convolutional neural network to control the AR display within a multitude of environments, is twofold. This study posits subvocal input as the answer as a silent speech interface can be used where speech is masked from background noise or military operations requiring silence. Moreover, this type of input can also be beneficial to persons with speech disabilities and/or movement disorders. Today's world is built for the abled-bodied making the 7.5 million Americans who have speech disabilities and 42 million Americans with movement disorders constantly left behind [18]. Augmented reality bridges the gap to allow those with disabilities to learn new skills, recover movements and enjoy entertainment that were not possible before. With a silent speech interface controlling AR devices, those with both speech and movement disorders will be able to immerse themselves in a minimally constrained life. This includes technical job training where bulky equipment was previously used or stimulating parts of the brain relevant to speech production even with no sound being generated.

With the need of this technology, future studies should include a multitude of data to complete the training datasets making the classifier for transitional to industry. To reduce the prevalent effect of overfit being seen and the large amount of error it comes with, the data should also be normalized with respect to the individual EMG sensor and shuffled more thoroughly before being fed into the neural network.

Acknowledgements. Special thanks to Embry-Riddle Aeronautical University's biomedical laboratory facilities and management, which is supported by the Mechanical Engineering Department.

Declaration Statements. N/A
Funding: Ignite
Ethical Approval: IRB 21-117

References

1. Wadkins, E.: A continuous silent speech recognition system for AlterEgo, a silent speech interface, 24 May 2019. https://dspace.mit.edu/bitstream/handle/1721.1/123121/1128187233-MIT.pdf?sequence=1&isAllowed=y

2. Meltzner, G.S., Heaton, J.T., Deng, Y.: The MUTE silent speech recognition system. In: INTERSPEECH, Burlington (2013)
3. ASHA American Speech - Language - Hearing Association, "Quick Facts,". Accessed 12 Sept 2021
4. Gonzalez-Lopez, J.A.. Gomez-Alanis, A., Martin-Donas, J.M., Perez-Cordoba, J.L., Gomez, A.M.: Silent speech interfaces for speech restoration: a review, Granada (2020)
5. Hummel, J., et al.: Evaluation of a new electromagnetic tracking system using a standardized assessment protocol. Phys. Med. Biol. **51**(10), 27 (2006)
6. Fagan, M.J., Ell, S.R., Gilbert, J.M., Sarrazin, E., Chapman, P.M.: Development of a (silent) Speech Recognition System for Patients Following Laryngectomy. Med. Eng. Phys. **30**, 419–425 (2008)
7. Hueber, T., Chollet, G., Denby, B., Stone, M., Zouari, L.: Ouisper: corpus based synthesis driven by articulatory data. In: 16th International Congress of Phonetic Sciences (2007)
8. Nakajima, Y.: Development and evaluation of soft silicone NAM. In: IEICE, pp. 7–12 (2005)
9. Bos, J., Tack, D.: Speech input hardware investigation for future dismounted soldier computer systems (2005)
10. Hansen, J.H., Patil, S.A.: The physiological microphone (PMIC): a competitive alternative for speaker assessment in stress detection and speaker verification. Speech Commun. **52**(4), 327–340 (2010)
11. Titze, I.R., Story, B.H., Burnett, G.C., Holzrichter, J.F., Ng, L.C., Lea, W.A.: Comparison between electroglottography and electromagnetic glottography. J. Acoust. Soc. Am. **107**(1), 581–588 (2000)
12. Spinlab: Tuned Electromagnetic Resonator Collor Sensor (2004). Accessed 2021
13. Tamm, M.-O., Muhammad, Y., Muhammad, N.: Classification of vowels from imagined speech with convolutional neural networks, University of Tartu: Institute of Computer Science (2020)
14. Neuper, C., Müller, G.R., Kübler, A., Birbaumer, N., Pfurtscheller, G.: Clinical Application of an EEG-based brain computer interface: a case study in a patient. Clin. Nueropychol. **114**, 399–409 (2003)
15. Bartels, J., et al.: Neurotrophic electrode: method of assembly and implantation. J. Neurosci. Methods **174**(2), 168–176 (2008)
16. Hochberg, L.R., et al.: Neuronal ensemble control of prosthetic devices by a human with tetraplegia. Nature **442**(7099), 164–171 (2006)
17. Huang, Y., Low, K., Lim, H.: Initial analysis of EMG signals of hand functions associated to rehabilitation tasks. In: International Conference on Robotics and Biomimetics, Singapore (2009)
18. Regents of the University of Michigan: Neurosciences: Movement Disorders (2021). https://www.uofmhealth.org/conditions-treatments/brain-neurological-conditions/movement-disorders

Hand Gesture Recognition for Sign Languages Using 3DCNN for Efficient Detection

Taranya Elangovan(✉), R. Arockia Xavier Annie, Keerthana Sundaresan, and J. D. Pradhakshya

College of Engineering, Guindy, Anna University, Chennai 600025, India
taranya@hotmail.co.uk

Abstract. Sign Language Recognition aims at providing an efficient and accurate mechanism for the recognition of hand gestures made in sign languages and converting them into text and speech. Sign language is a means of communication using bodily movements, especially using the hands and arms. With sign language recognition methods, dialog communication between the deaf and hearing society can become a reality. In this project, we carry out sign language recognition by building 3D convolutional neural network (3DCNN) models that can perform multi-class prediction on input videos containing hand gestures. On detection of the input, both text and speech are generated and presented as output to the user. In addition to this, we also implement real-time video recognition and continuous sign language recognition for multi-word videos. We present a method for recognition of words in three languages – Tamil Sign Language (TSL), Indian Sign Language (ISL), and American Sign Language (ASL), and outperform state-of-the-art alternatives in with accuracies of 97.5%, 99.75% and 98% respectively.

Keywords: 3DCNN · Sign language recognition · Video Classification

1 Introduction

With the widespread use of technology in our day-to-day lives, traditional Human-Computer Interaction (HCI) is becoming increasingly more important in real life applications. One of the domains in which HCI has made breakthroughs is in facilitating ease of communication. Sign Language Recognition (SLR) falls into this domain. Sign language is a means of communication using bodily movements, especially using the hands and arms. It is used when spoken communication is not possible or desirable. The goal of SLR is to provide a convenient and accurate mechanism to convert sign language into text or speech so that dialog communication between the deaf and hearing society can become a reality.

Wherever vocal communication is not an option i.e., when one or more participant communicators cannot speak or hear, sign language can be used to bridge the gap. Using sign language, one can communicate letters, words, or even sentences of spoken languages using different hand signs and hand gestures. Sign languages vary from country to country. These languages have their own vocabulary and grammar, slightly different

J. M. R. S. Tavares et al. (Eds.): CMBBE 2021, LNCVB 38, pp. 215–233, 2023.
https://doi.org/10.1007/978-3-031-10015-4_19

from the corresponding spoken language. Furthermore, there may be multiple different forms of sign language for a single spoken language. Each non-speaking community can develop its own sign language. Even within a country, sign language can vary from region to region, just like spoken languages. However, some sign languages are not completely independent of each other. Sign languages such as the British Sign Language (BSL), Auslan and New Zealand Sign Language (NZSL) are very similar.

2 Literature Survey

In the SLR system presented by Huang et al. [1], an attention-based 3D Convolutional Neural Network (3DCNN) was presented for SLR. 3DCNN learns the spatiotemporal features from the raw input video, and spatial attention is used to focus on areas of interest. Instead of feeding video directly into the network, viewpoint selection is performed to integrate spatial attention mechanism into the model. Attention is used after feature extraction to identify the most significant motions for classification. This proposed method was evaluated on two large scale sign language datasets – a Chinese Sign Language (CSL) dataset collected by Huang et al. themselves, with 500 categories, and the ChaLearn14 benchmark. While performing the experiment on the CSL dataset, accuracy was used as an evaluation metric. They compared their proposed approach to other methods such as Gaussian Mixture Model-Hidden Markov Model (GMM-HMM) with an accuracy of 78.1%, and found that their novel proposed method had the highest accuracy of 88.7%. While performing the experiment on the ChaLearn14 dataset, Jaccard Index was used to evaluate performance. Here, the proposed method was compared to other deep neural networks like CNN which has an accuracy of 78.9%, and was found to outperform other methods with an accuracy of 95.3%. Hence, it can be seen that the method proposed by Huang et al. performs better on both datasets than several current day algorithms. This work was focused on isolated SLR. Huang et al. have defined the future scope of this work to be with attention to continuous SLR – translating a sign language video into a semantic sentence.

In the SLR system using multi-view hand skeleton presented by Rastgoo et al. [2], a deep learning-based pipeline consisting of Single Shot Detector (SSD), 2DCNN, 3DCNN, and Long Short-Term Memory (LSTM) was proposed for efficient automatic hand sign language recognition from RGB videos. A CNN based model was used to estimate 3D hand key points from the 2D input frames, and these 3D key points were then connected to form a hand skeleton. The pixel map from the input frames, the projections of the 3D hand skeleton, and a heatmap image of detected key points were used in a stacked fashion. 3DCNNs were applied to these stacked inputs, and the outputs of the 3DCNNs combined and fed to an LSTM. The fully connected layer was finally where the classification was carried out. They performed evaluation on three datasets, NYU, First-Person, and a new dataset presented by Rastgoo et al. themselves, RKS-PERSIANSIGN. Average error and accuracy were used to evaluate their model against the different datasets. While evaluating the model on the NYU dataset, they achieved an average error of only 4.46%, as compared to models like REN proposed by Guo et al. (2017) which had an error rate of 12.7%. While evaluating the model on the First-Person dataset, they achieved an accuracy of 91.12%, higher than other methods like the

First-Person model proposed by Garcia-Hernando et al. (2018), which had an accuracy of 85.6%. Hence, the novel method proposed by Rastgoo et al. outperformed all the state-of-the-art algorithms that it was evaluated against, showing good improvement in sign language detection. Their method showed significant improvement when compared to contemporaries for hand sign language recognition, hand pose estimation, and hand action recognition.

The SLR system using 3DCNN presented by Al-Hammadi et al. [3] focused on hand gesture recognition by deep convolutional neural networks, with a transfer learning-based approach. A special focus was given to achieving normalization of the spatial dimensions that were contributed by facial position, facial length and human body part ratios, so that the signer would not need to be in a fixed position in frame or fixed distance from the camera. Two proposed approaches for spatiotemporal feature learning were implemented; the first was a single 3DCNN structure, and the second was a fusion of parallel 3DCNN structure with multilayer perceptron (MLP), LSTM and stacked autoencoder. The model was trained using a sequence of RGB frames captured by a basic camera, removing the need for accessories such as hand gloves. Classification was enabled with a SoftMax activated layer. The experiments conducted included two modes, namely a signer-dependent and signer-independent mode. The model performance was evaluated by making use of three gesture datasets from colour videos. Based on the accuracy achieved in all methods, the parallel 3DCNN fusion-based architecture was shown to have superior performance. The single 3DCNN mode showed 72.32% accuracy, while the parallel 3DCNN with LSTM showed 82.19% accuracy and MLP showed 84.38% accuracy. Confusion matrices were also plotted to further understand the results and make inferences. Al-Hammadi et al. have defined the future scope of this work to be with attention on receiving a live video feed, and optimizing all the hyperparameters.

Kishore et al. [4] proposed a system for recognizing ISL sentences by performing optical flow tracking and hand shape feature extraction. Horn Schunck optical flow (HSOF) was used to extract tracking features and Active Contours (AC) was used to extract shape features. Continuous sign videos were used, with a feature matrix developed to characterize the signs in these videos. A neural network object with backpropagation training algorithm was used to classify the signs into various word sequences in digital format. Digital word sequences were translated into text by matching, after which the text was voice translated using Windows API (Win-API). The dataset consisted of 580 sentences in total, with hand gestures from ten signers, each gesture sequence consisting of sentences having 30 words long. This was used to test the performance of the algorithm by computing word matching score (WMS). The average WMS was found to be 89.48% for the classification, on par with other researchers for American Sign Language and Chinese Sign Language. By employing multiple features in the continuous SLR process, this method showed improved word classification rates as compared to single feature SLR models.

A deep neural network with iterative training was proposed by Cui et al. [5] for continuous SLR. The proposed system was based on recurrent convolutional neural networks, trained on multi-modal data of RGB frames and optical flow images. The entire architecture consisted of a feature extraction module composed of deep CNN followed by temporal fusion layers, and a sequence learning module using Recurrent Neural Networks (RNNs) with bidirectional LSTM (Bi-LSTM). The RNNs were employed as the sequence learning module in an effort to increase the learning capacity in comparison to state-of-the-art continuous SLR systems which used Hidden Markov Models (HMM). The iterative training was focused on to highlight learning with a limited dataset. A multimodal fusion model was used to integrate appearance and motion cues from sign language videos to present better spatiotemporal representations for the gestures. The experiments were carried out on two publicly available datasets, the RWTH-PHOENIX-Weather multi-signer 2014 database and the SIGNUM signer-dependent set. The model was evaluated on the SLR benchmarks. The experimental performance was quantitatively evaluated using word error rate (WER), which would quantify the dissimilarity of the predicted sequence of labels and the ground truth transcription. For the RWTH-PHOENIX-Weather dataset, the WER was shown to improve to 24.43% in comparison to 26.80% using state-of-the-art methods. On the SIGNUM benchmark, the WER was shown to be 2.80%, compared to state-of-the-art 4.80%. Cui et al. have highlighted several directions for future work, including exploration on integration of multiple modalities and prior knowledge of sign language, to help the model learn on limited data.

A deep architecture to learn spatiotemporal features for gesture recognition was proposed by Zhang et al. [6]. The architecture employed 3DCNN and bidirectional convolutional LSTM (ConvLSTM) to learn 2D spatiotemporal feature maps. The 2D feature maps had the capacity to simultaneously encode global temporal information and local spatial information. 2DCNN was then further utilized for higher level learning for the final gesture recognition. Experiments were conducted on the ChaLearn LAP large-scale isolated gesture dataset (IsoGD) and Sheffield Kinect Gesture (SKIG) dataset. Temporal pooling and SoftMax layers were used in the classification module, along with fusion methods to improve prediction accuracy on multi-modal data. The model was evaluated by using the maximum and average fusion for score fusion, as well as maximum and average pooling for temporal pooling. Evaluation on the IsoGD test subset showed accuracy of 67.26%, and on the SKIG dataset, 99.52% accuracy. Improvements on the existing model were suggested for future scope in order to obtain higher recognition accuracy.

In previous ventures, SLR for a single language had been implemented. This does not provide universality between users of different sign languages. In order to overcome this barrier, we aim to implement a system that can recognize three different sign languages and transform them into text and speech format. It has also been observed that many pre-existing systems provided detection from pre-recorded videos. In this project, we intend to enhance dynamicity by providing real-time recognition. We also aim to implement continuous SLR i.e., recognition of multiple words one after another to form sentences. In addition to this, we aim to use a diverse dataset in order to make the recognition system robust and accepting of a wide range of user inputs.

3 Architecture

The input to the SLR system is in the form of a video given by the user. This video may either be a pre-recorded video that the user can select as input, or a real-time video recorded at the time of giving input. The video is split into a number of frames. These frames are converted to grayscale in order to reduce noise and complexity while training the model. Finally, the set of frames for each video is converted to a NumPy array, which is fed as input to the trained model, which then predicts the sign in the input video. The text form of the sign word is displayed to the user, and also played as an audio. In the event of multi-word prediction, the output after recognition will be in the form of multiple words in a list, which then undergoes sentence construction. The output of sentence construction is finally presented to the user in text and speech forms. A detailed architecture diagram is shown in Fig. 1. as follows:

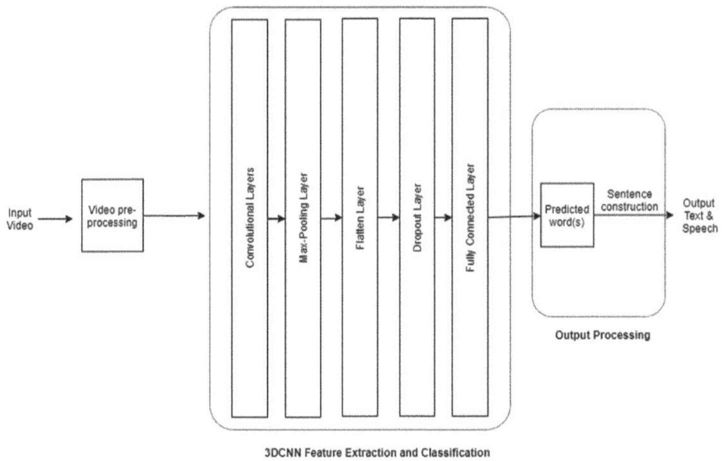

Fig. 1. Overall architecture

4 Detailed Module Description

With the rapid development of Human-Computer Interaction (HCI), SLR has become a complex and interesting problem to tackle. It provides a user-friendly way of an efficient and socially relevant mode of communication between humans, aided by technology. For this application, we train a ML model using the 3DCNN algorithm which is better on a spatiotemporal scale in comparison to other existing algorithms. The user can give their input in two formats i.e., a recorded video or a real-time video fed by the system webcam. The video is then pre-processed accordingly before being classified by the 3DCNN model to generate an appropriate output. Our application is designed for isolated and continuous SLR. In case of continuous SLR, the isolated words are further processed to generate meaningful phrases or sentences. The final output is provided to the end user in the form of both text and audio.

4.1 Input Gesture

The input video can be of two types: recorded video, which is a pre-recorded video separate from the dataset is fed to the classifier for recognition, and real-time video, where the input gesture is signed by the user to a webcam and fed to the classifier for recognition.

4.2 Video Pre-processing

The application deals with recognition of both individual words and sentences from hand gestures. The first step under this module is identifying the number of words signed in the given input video. The length of the video is divided by 2 to retrieve the number of words. Based on this number, the input video is split into smaller videos, and subfolders are created for further pre-processing. The input video gesture is split into several frames and stored in the folders. Key-frame extraction is carried out by retaining every fifth frame from the set of frames generated to satisfy a count of 20 frames for each input video. These frames are then converted to grayscale in order to reduce noise and complexity while training the model. Finally, the set of frames for each video is converted to a NumPy array that is used in forming the train, validation and test datasets for training the model (Fig. 2).

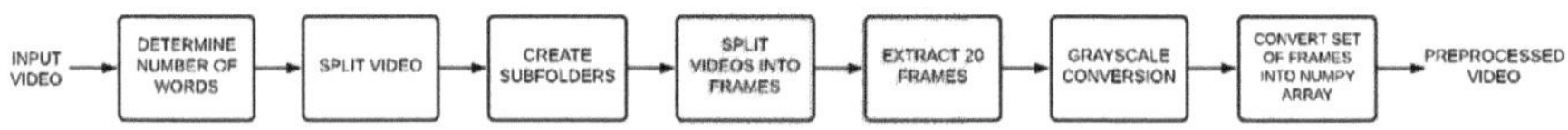

Fig. 2. Video pre-processing

Algorithm 1: Determining number of words in the input video

Input: input video

```
video_length=input_video.duration
numwords=video_length/2
if numwords is 1
   time_slots=[[0,2]]
if numwords is 2
   time_slots=[[0,2],[2,4]]
if numwords is 3
   time_slots=[[0,2],[2,4],[4,6]]
```

Algorithm 2: Creating subfolders for further pre-processing

Function: video_folder_create
Input: input_video_path, video_path, pre_path, time_slots

```
for i in np.arange(len(time_slots)):
        vid_name="vid%d.mp4" % i
        output_video_path=video_path+"/"+vid_name

      with VideoFileClip(input_video_path) as video:
         new = video.subclip(time_slots[i][0], time_slots[i][1])
new.write_videofile(output_video_path,audio_codec='aac')
return True
```

Algorithm 3: Splitting video into frames & converting frames into grayscale

```
Function: video_capturing_function
Input: dataset, folder_name

for each i = 0 to length(dataset)
video_name=dataset[i]
video_path=dataset_path+video_name
cap = videoCapture object to capture from video_path
make video_frames_directory for video frames
count = 0
while cap is open
frame_id = cap.getFrame()
frame = cap.read()
if(frame_id % 5 == 0)
filename = "frame"+count+".jpg"
count++
convert frame to grayscale
        write frame to video_frames_directory
```

Algorithm 4: Loading frames into numpy array

```
Function: load_frames
Input: directory
Output: NumPy array of frames is returned

for each video in dataset
vid_name = video_name[i]
vid_path = video_path_name[i]
 for l=0 to 20
 frames_to_select.append(frame%l.jpg)
 for each frame in frames_to_select
 resize frame to size 250*250
numpyarray conversion
return np.array(frames)
```

4.3 3DCNN Training and Classification

The model trained and used for prediction contains the following layers: 6 3D Convolutional layers, 4 Max-Pooling layers, 3 fully connected Dense layers, 2 Dropout layers, and an output layer.

4.4 Sentence Construction

The words predicted by the model are combined together to form a meaningful sentence.

Algorithm 5: Constructing sentence from list of words

```
Input: List of predicted words
Output: sentence

valid -> list of lists of valid sentence structures
def checktype(predicted_words):
  marker=[] #list of word types
  For each word in predicted_words:
      check grammar of word, whether noun, verb, etc.
      marker.append(word_type)
while marker not in valid:
      predicted_words.append(words and punctuation)
      checktype(predicted_words)
return predicted_words
```

4.5 Output

Output is returned to the user in the form of text and speech. The output obtained as text is converted into audio using the Google Text-To-Speech (gTTS) Python library.

5 Implementation

5.1 Dataset

Indian Sign Language. For ISL, we use the dataset "Hand gestures for emergency situations: A video dataset based on words from Indian sign language collected" by V. Adithya and R. Rajesh, Department of Computer Science, Central University of Kerala [7]. The dataset consists of videos for eight ISL words – accident, call, doctor, help, hot, lose, pain, thief, with 52 videos for each word. Videos were collected from 26 individuals with 2 samples per individual, shot in indoor lighting. We use 20 videos for each word in implementation of the 3DCNN model (for training, testing, and validation), hence a total of 160 videos. For each word, 15 videos are taken for training and 5 for testing. Of the 15 training videos, 3 videos (20%) are used for validation.

Tamil Sign Language. For TSL, we collected a dataset of videos for 9 TSL words – friend, good morning, he, hello, how, my, who, work, you. The dataset consists of 18 videos for each word from 12 different signers. The dataset consists of a total of 162 videos. For each word, 14 videos are taken for training and 4 for testing. Of the 14 training videos, 3 videos (20%) are used for validation.

American Sign Language. For ASL, we collected a dataset modelled based on the Purdue RVL-SLLL American Sign Language Database [8]. This dataset consists of videos for 39 words, with 28 videos per word contributed by 14 signers. Based on this dataset, we recorded a dataset of 6 ASL words – day, drop, forget, my, someone, throw. Our dataset consists of 18 videos per word, hence a total of 108 videos. For each word, 14 videos are used for training and 4 for testing. Of the 14 videos used for training, 3 videos (20%) are used for validation.

5.2 Video Pre-processing

The input video is split into sub-videos depending on the number of words, which is determined by the length. Prior to training, each video in the dataset is pre-processed and converted into frames. Each video is split into 20 frames, which are converted into grayscale. The 20 frames for each video are then loaded, resized into 250 × 250 pixels each, and converted into a NumPy array. The final train set after pre-processing consists of NumPy arrays for each video, and this dataset can be used for model training.

5.3 Model Training

A 3DCNN Keras Sequential model is trained on the dataset. In this project, three trained models are obtained – one for each language trained on the respective dataset for that language. The model has 6 3D Convolutional layers, 4 Max-Pooling layers, 3 fully connected Dense layers, 2 Dropout layers, and an output layer.

ISL Training. The model was trained on the ISL dataset for 20 epochs with a batch size of 5. An Adam optimiser with learning rate 0.00001 was used. A final train accuracy of 99.75% and validation accuracy of 87.5% is obtained after training (Fig. 3).

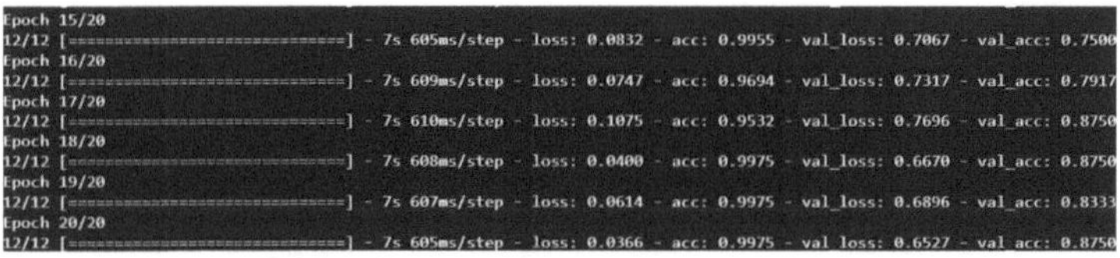
```
Epoch 15/20
12/12 [==============================] - 7s 605ms/step - loss: 0.0832 - acc: 0.9955 - val_loss: 0.7067 - val_acc: 0.7500
Epoch 16/20
12/12 [==============================] - 7s 609ms/step - loss: 0.0747 - acc: 0.9694 - val_loss: 0.7317 - val_acc: 0.7917
Epoch 17/20
12/12 [==============================] - 7s 610ms/step - loss: 0.1075 - acc: 0.9532 - val_loss: 0.7696 - val_acc: 0.8750
Epoch 18/20
12/12 [==============================] - 7s 608ms/step - loss: 0.0400 - acc: 0.9975 - val_loss: 0.6670 - val_acc: 0.8750
Epoch 19/20
12/12 [==============================] - 7s 607ms/step - loss: 0.0614 - acc: 0.9975 - val_loss: 0.6896 - val_acc: 0.8333
Epoch 20/20
12/12 [==============================] - 7s 605ms/step - loss: 0.0366 - acc: 0.9975 - val_loss: 0.6527 - val_acc: 0.8750
```

Fig. 3. Training model on ISL dataset for 20 epochs

TSL Training. The model was trained on the TSL dataset for 25 epochs with a batch size of 8. An Adam optimiser with learning rate 0.00001 was used. A final train accuracy of 98% and validation accuracy of 85% is obtained after training.

ASL Training. The model was trained on the ASL dataset for 20 epochs with a batch size of 8. An Adam optimiser with learning rate 0.00001 was used. A final train accuracy of 97.5% and validation accuracy of 61.1% is obtained after training.

5.4 Model Prediction

The trained models are used to predict the signs as captured in the video fed to the system - a pre-recorded video or real-time video input recorded on the webcam. The predicted sign is represented in the form of text and speech. The steps include selection of a sign language, selecting the input video, pre-processing and prediction of the video using the trained model, and finally displaying the output (Figs. 4, 5 and 6).

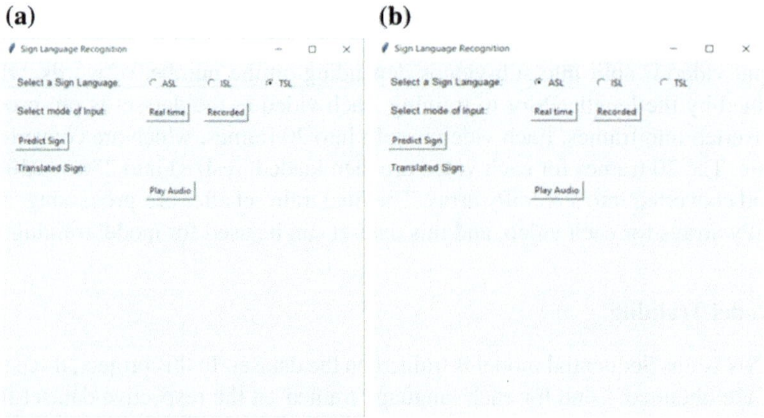

Fig. 4. Selecting sign languages (a) TSL (b) ASL

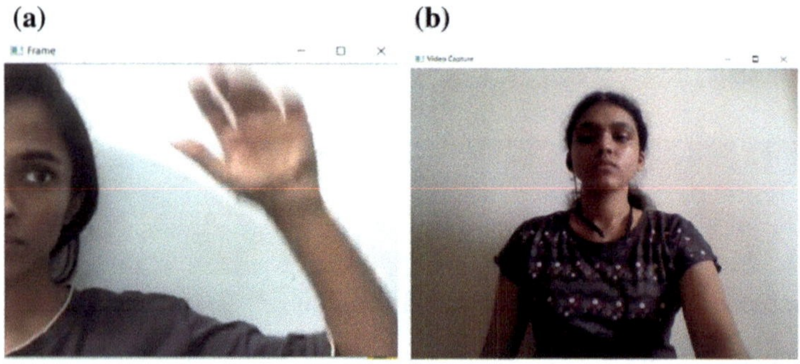

Fig. 5. Selecting input as (a) pre-recorded video using "Recorded" button (TSL: "hello" "good morning") (b) Real-time webcam input using "Real time" button (ASL: "day")

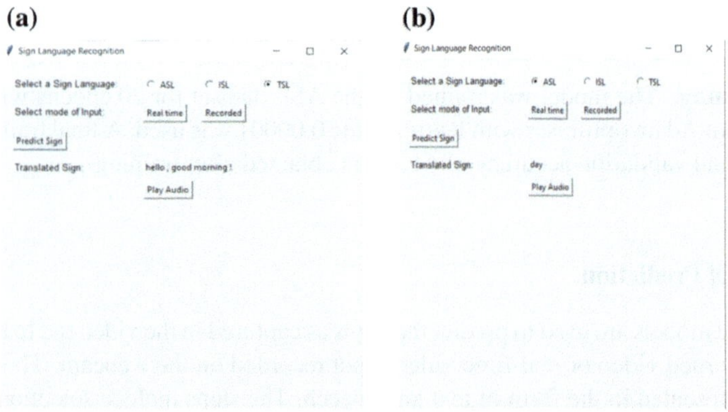

Fig. 6. After "Predict Sign", output in the form of audio ("Play Audio") and text ((a) sentence (b) word)

6 Results

6.1 Metrics for Evaluation

Definitions

True Positives (TP). Number of classifications where the actual label is positive and class is correctly predicted by the model to be positive.

False Positives (FP). Number of classifications where the actual label is negative and class is incorrectly predicted by the model to be positive.

True Negatives (TN). Number of classifications where the actual label is negative and class is correctly predicted by the model to be negative.

False Negatives (FN). Number of classifications where the actual label is positive and class is incorrectly predicted to be negative.

Accuracy. Accuracy is the fraction of predictions the model has correctly classified. Formally, accuracy can be defined as follows:

$$\text{Accuracy} = (\text{TP} + \text{TN}) / (\text{TP} + \text{TN} + \text{FP} + \text{FN}) \tag{1}$$

Loss. Loss is a number indicating how the model's prediction was on a single example. The loss function used in this model is the categorical cross-entropy loss function, as the model performs multi-class classification.

6.2 Model Training History

The training accuracy, training loss, validation accuracy, and validation loss per epoch in the process of training the 3DCNN models were plotted on graphs as shown in Fig. 7, 8,

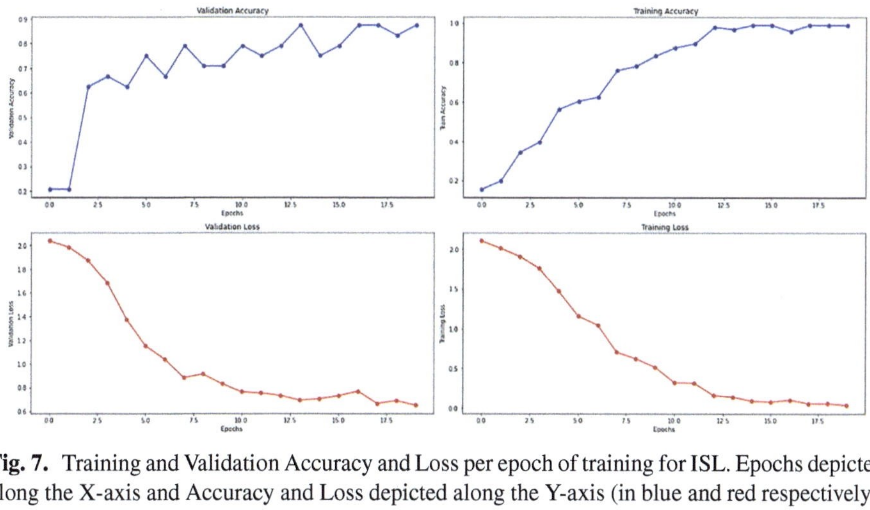

Fig. 7. Training and Validation Accuracy and Loss per epoch of training for ISL. Epochs depicted along the X-axis and Accuracy and Loss depicted along the Y-axis (in blue and red respectively).

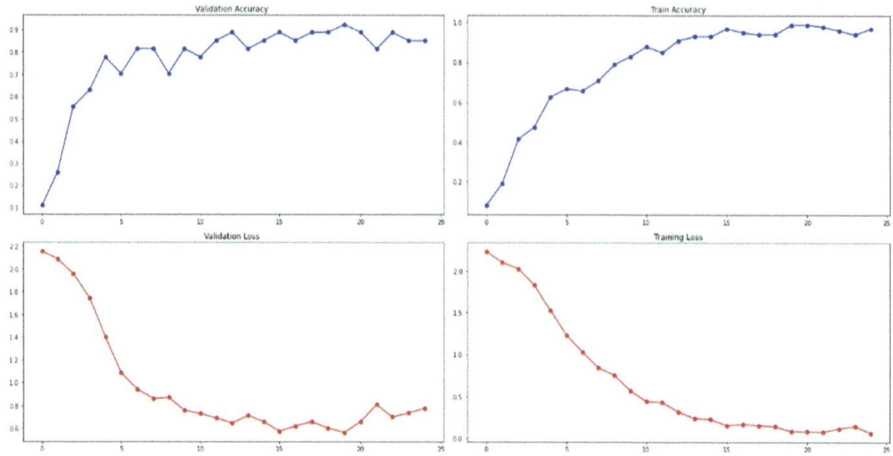

Fig. 8. Training and Validation Accuracy and Loss per epoch of training for TSL. Epochs depicted along the X-axis and Accuracy and Loss depicted along the Y-axis (in blue and red respectively).

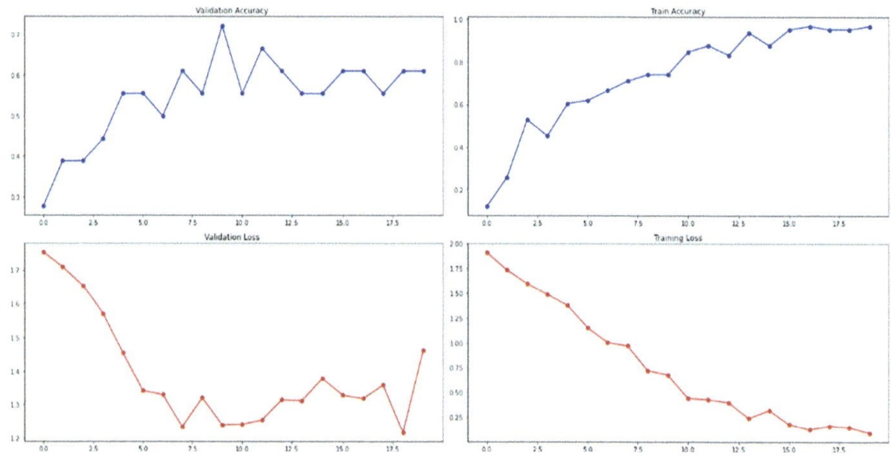

Fig. 9. Training and Validation Accuracy and Loss per epoch of training for ASL. Epochs depicted along the X-axis and Accuracy and Loss depicted along the Y-axis (in blue and red respectively).

and 9 for ISL, TSL and ASL respectively. The accuracies increase as the epoch number increases, i.e., the accuracy of the model increases as training progresses. Similarly, the losses decrease as the model progresses through training. The increase in accuracies and decrease in losses over the course of training show that the model is being trained to perform classification correctly.

6.3 Model Testing

Figure 10 shows the confusion matrix for the ISL test dataset. This describes the performance of the trained 3DCNN ISL model. The ISL model was evaluated against 5 test videos for each of the 8 words. A total of 34 test videos out of 40 predicted correctly. The words "accident", "doctor", "hot", and "thief" showed perfect accuracy on the test set with all 5 videos in each of these categories being correctly predicted by the trained model.

Figure 11 shows the confusion matrix for the TSL test dataset. This describes the performance of the trained 3DCNN TSL model. The TSL model was evaluated against 4 test videos for each of the 9 words. A total of 26 test videos out of 36 predicted correctly. The words "friend", "how", and "work" showed perfect accuracy on the test set with all 4 videos in each of these categories being correctly predicted by the trained model.

Figure 12 shows the confusion matrix for the ASL test dataset. This describes the performance of the trained 3DCNN ASL model. The ASL model was evaluated against 4 test videos for each of the 6 words. A total of 17 test videos out of 24 predicted correctly. The words "forget" and "someone" showed perfect accuracy on the test set with all 4 videos in each of these categories being correctly predicted by the trained model.

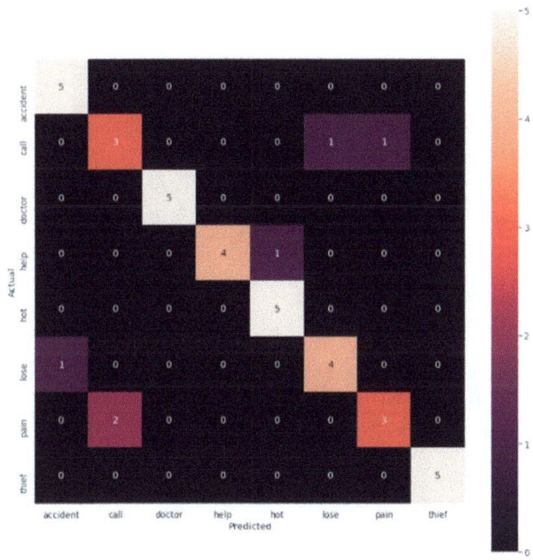

Fig. 10. Confusion matrix for ISL test data

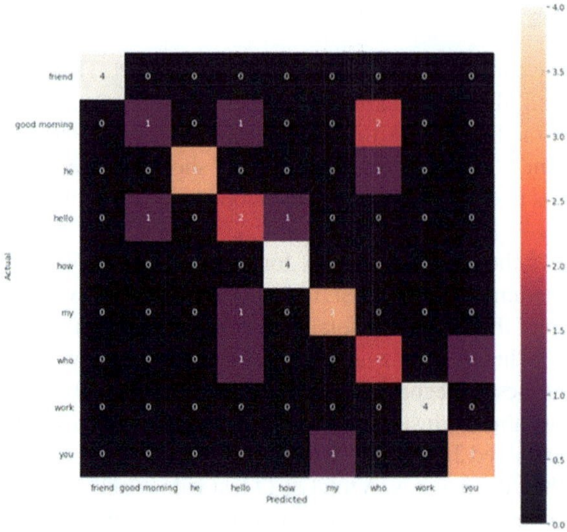

Fig. 11. Confusion matrix for TSL test data

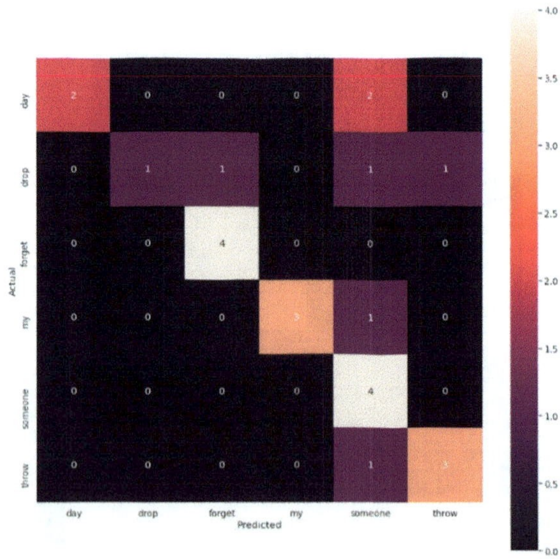

Fig. 12. Confusion matrix for ASL test data

6.4 Test Cases

Correct Classifications. With a plain background and good lighting on hand, the model predicts the signs accurately as "how" and "good morning" respectively as shown in Fig. 13.

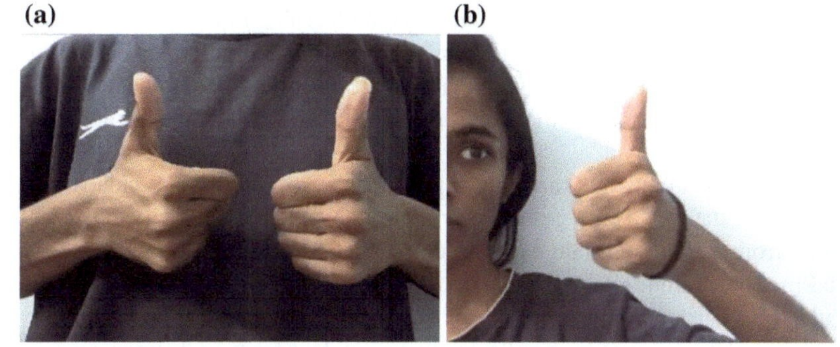

Fig. 13. Frame of TSL sign for (a) "how" (b) "good morning"

Incorrect Classifications. If the lighting in the input video is poor, as in the sign frame depicted in the image, the model fails to predict the correct output as shown in the image. As shown in Fig. 14. (a), the sign for "drop" in ASL is incorrectly predicted as "day" due to poor lighting. With increased background noise, the trained model fails to generate the correct output as shown in the image. As shown in Fig. 14. (b), the sign for "throw" in ASL is incorrectly predicted as "someone" due to background noise.

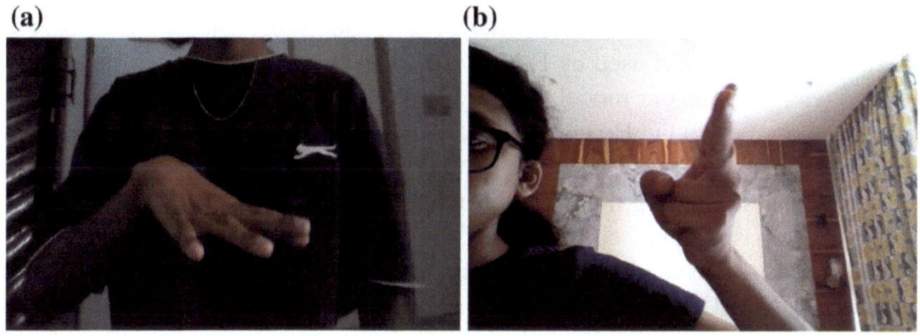

Fig. 14. Frame of ASL sign for (a) "drop" (b) "throw"

6.5 Comparative Analysis

This project comprises three identical 3DCNN models trained on three different datasets – ISL, TSL and ASL. The ISL dataset consists of 160 videos encompassing 8 words. The ISL model was trained for 20 epochs with a batch size of 8 and using an Adam optimiser with default learning rate 0.00001, and achieved a final accuracy of 99.75%, with validation accuracy 87.5%. The TSL dataset consists of 162 videos encompassing 9 words. The TSL model was trained for 25 epochs with a batch size of 8 and using an Adam optimiser with default learning rate 0.00001 and achieved a final accuracy of 98%, with validation accuracy 85%. The ASL dataset consists of 108 videos encompassing 6 words. The ASL model was trained for 20 epochs with a batch size of 8 and using an Adam optimiser with default learning rate 0.00001, and achieved a final accuracy of 97.5%, with validation accuracy 61.1%. Different batch sizes and learning rates were used for the three datasets, but the structure and layers of the three 3DCNNs are identical over all three languages. ISL showed the highest accuracy with almost 99.75%.

The ISL model likely had the highest performance of three as the videos were collected from the dataset "Hand gestures for emergency situations: A video dataset based on words from Indian sign language collected" [7]. In this dataset, videos were contributed by 26 different signers. All videos were shot in the same lighting and with a plain black background, and hence all focus was on the hand gesture movements. The TSL and ASL datasets were both collected by us – videos from 12 different signers were incorporated into the TSL dataset, and videos from 3 different signers into the ASL dataset. The inclusion of more signers in the TSL dataset may have been the reason for achieving higher accuracy than the ASL model. The diverse nature of the TSL dataset could have contributed to the robustness of the trained model.

In the SLR system proposed by Huang et al. [1], attention-based 3DCNNs were trained on two different datasets – a Chinese Sign Language (CSL) dataset collected by Huang et al., with 500 categories, and the ChaLearn14 benchmark. They obtained an accuracy of 88.7% after training on the CSL dataset, and an accuracy of 95.3% after training on the ChaLearn14 dataset. All three of our models outperformed these models. The CSL dataset used for training consisted of 180 videos per category, whereas the datasets used in this project consists of only 14 or 15 videos per category. It is also worth mentioning that the model used by Huang et al. also contained an LSTM layer and an attention layer, whereas our model did not have these layers. We were able to achieve comparable accuracy to the Huang et al. CSL model with a much smaller dataset, and less model layers.

In the SLR system using multi-view hand skeleton presented by Rastgoo et al. [2], a deep learning-based model consisting of 3DCNN and LSTM was proposed for SLR. They used the First-Person dataset for training and evaluation. This dataset contained more than 100,000 RGB + D frames of 45 hand action categories for 26 different objects in several hand configurations. We only used the RGB sequences for model evaluation. On training this model on the First-Person dataset, they achieved an accuracy of 91.1%. The accuracies obtained in our project were higher than the accuracy obtained by Rastgoo et al.

In the SLR system using 3DCNN presented by Al-Hammadi et al. [3], three different CNN architectures were used - a single 3DCNN structure, a parallel fusion of 3DCNN with LSTM, and a parallel fusion of 3DCNN and multilayer perceptron (MLP). The single 3DCNN mode showed 72.3% accuracy, the 3DCNN + LSTM showed 82.2% accuracy, and the 3DCNN + MLP showed 84.4% accuracy. Our models outperformed all three of these models.

Kishore et al. [4] proposed a system for recognising ISL sentences by performing optical flow tracking and hand shape feature extraction. Horn Schunck optical flow (HSOF) was used to extract tracking features and Active Contours (AC) was used to extract shape features. A neural network object (ANN) with backpropagation training algorithm was used to classify the signs into various word sequences. The dataset used for training was contributed by 5 test subjects, each performing the same set of sentences for Indian Sign Language. The average WMS was found to be 89.5% for the classification, which is a lower accuracy than we obtained.

The architecture proposed for gesture recognition proposed by Zhang et al. [6] employed 3DCNN and bidirectional convolutional LSTM (ConvLSTM). They performed training on two datasets - the ChaLearn LAP large-scale isolated gesture dataset (IsoGD) and Sheffield Kinect Gesture (SKIG) dataset. The IsoGD dataset contained 47,933 RGB + D gesture videos divided into 249 categories of gestures performed by 21 individuals. SKIG contained 1,080 RGB + D videos of 10 kinds of gestures, performed by 6 individuals with 3 kinds of hand postures under 2 illumination conditions and 3 backgrounds. Training on the IsoGD dataset gave an accuracy of 67.2%, and on the SKIG dataset, 99.5%. Comparing this to our project, all three of our models outperformed the IsoGD-trained model proposed by Zhang et al., and our ISL model outperformed the SKIG-trained model.

For the training of our ISL model, we used the dataset of hand gestures presented by Aditya et al. [7]. This video dataset comprises different hand gestures of Indian sign language (ISL) words used in emergency situations. The dataset encompassed 8 different words – "accident", "call", "doctor", "help", "hot", "lose", "pain" and "thief", with 52 videos contributed by 26 signers for each word. They analysed this dataset using two different classification techniques – Support Vector Machine (SVM), and Deep Learning. In the Deep Learning approach, they used a pre-trained CNN, GoogleNet, combined with an LSTM network for classification. They obtained an accuracy of 90% using SVM-based classification and an accuracy of 96.25% using Deep Learning. On

Table 1. Comparative analysis

Model	Dataset	Accuracy (%)
Attention-based 3DCNN [1]	CSL	88.7
	ChaLearn14	95.3
3DCNN + LSTM [2]	First-Person	91.1
3DCNN [3]	ArSL, KSU-SSL, RVL-SLLL-ASL	72.3
3DCNN + LSTM [3]		82.2
3DCNN + MLP [3]		84.4
HSOF + AC + ANN [4]	ISL [4]	89.5
3DCNN + ConvLSTM [6]	IsoGD	67.2
	SKIG	99.5
SVM [7]	ISL [7]	90
CNN + LSTM [7]	ISL [7]	96.25
3DCNN (proposed)	**ISL** [7]	**99.75**
	TSL	**98**
	ASL	**97.5**

using the same dataset in the training of our ISL model, we achieved a much higher accuracy of 99.75%.

On evaluating our project against previous ventures, all three of our models performed on par with or better than many previous models. Our ISL model outperformed all other ventures, with our ASL and TSL models outperforming all but one. We were able to achieve this accuracy with a much smaller training set than used in other ventures (Table 1).

7 Conclusions

In this project, we have implemented a Hand Gesture Recognition system for Sign Languages using 3DCNN for efficient detection. This system comprises three 3DCNN models, one for each of three sign languages - Indian Sign Language, Tamil Sign Language and American Sign Language. 3DCNN models are used due to their spatiotemporal feature learning capability. Each of these models is trained on a video dataset of more than 100 isolated sign words for each respective language. The models are then used in detection of signed video inputs that can be given in two formats - pre-recorded and real-time. The models show good accuracies across the different languages, and predict reliably for many different user inputs. The system can predict both single and multi-word user inputs. In case of multi-word user inputs, the input video is split into sub-videos of each word, and prediction is performed on each of these sub-videos. After prediction, the list of multiple words obtained undergoes sentence construction to form a grammatically correct sentence. In this way, this project implements continuous SLR (recognition of multiple words signed one after another) as well as isolated SLR (single word recognition), hence overcoming a key fallback of other similar projects, where continuous SLR is not satisfied and only isolated prediction is carried out.

7.1 Future Work

In future ventures, focus should be given to collecting and using larger and more diverse datasets for training in order to achieve higher accuracies and more robust models. More attention also needs to be paid to real-time recognition, where the accuracy of prediction is not as high as pre-recorded recognition due to factors like background noise. This can potentially be implemented with the use of feature extraction algorithms to isolate salient objects and disregard background noise.

References

1. Huang, J., Zhou, W., Li, H., Li, W.: Attention-based 3D-CNNs for large-vocabulary sign language recognition. IEEE Trans. Circuits Syst. Video Technol. **29**(9), 2822–2832 (2018)
2. Rastgoo, R., Kiani, K., Escalera, S.: Hand sign language recognition using multi-view hand skeleton. Exp. Syst. Appl. **150**, 113336 (2020)
3. Al-Hammadi, M., Muhammad, G., Abdul, W., Alsulaiman, M., Bencherif, M.A., Mekhtiche, M.A.: Hand gesture recognition for sign language using 3DCNN. IEEE Access **8**, 79491–79509 (2020)

4. Kishore, P.V.V., Prasad, M.V.D., Kumar, D.A., Sastry, A.S.C.S.: Optical flow hand tracking and active contour hand shape features for continuous sign language recognition with artificial neural networks. In: 2016 IEEE 6th International Conference on Advanced Computing (IACC), pp. 346–351 (2016)
5. Cui, R., Liu, H., Zhang, C.: A deep neural framework for continuous sign language recognition by iterative training. IEEE Trans. Multimedia **21**(7), 1880–1891 (2019)
6. Zhang, L., Zhu, G., Shen, P., Song, J., Afaq Shah, S., Bennamoun, M.: Learning spatiotemporal features using 3DCNN and convolutional LSTM for gesture recognition. In: IEEE International Conference on Computer Vision Workshops (ICCVW), pp. 3120–3128 (2017)
7. Adithya, V., Rajesh, R.: Hand gestures for emergency situations: a video dataset based on words from Indian sign language. Data Brief **31**, 106016 (2020)
8. Martinez, A.M., Wilbur, R.B., Shay, R., Kak, A.C.: Purdue RVL-SLLL ASL database for automatic recognition of American Sign Language. In: Proceedings. Fourth IEEE International Conference on Multimodal Interfaces, pp. 167–172 (2002)

Simulation Study to Investigate the Accuracy of in Vivo Motor-Unit Twitch Force Measurements

Harnoor Saini(✉) and Oliver Röhrle

Institute of Modelling and Simulation of Biomechanical Systems, University of Stuttgart, Pfaffenwaldring 4f, 70569 Stuttgart, Germany
harnoor.saini@imsb.uni-stuttgart.de

Abstract. Motor-unit behaviour is central to understanding muscle contractile function. Commonly, in vivo experimental measurements of motor-unit twitches are taken at points biomechanically linked to the muscle. Here, we investigate whether these measurements accurately represent muscle twitch force behaviour by using a 3D, continuum-mechanical masticatory system model. Selected motor-units within the masseters were individually stimulated, and twitch force measured at the occlusal surface (typical for experiments) and at the muscle origin (representing true twitch force). Occlusal measurements underpredicted twitch force across all motor-units. This was due to force (re)distribution caused by motor-unit location and muscle architecture. Additionally, we investigated the assumption of twitch force summation by comparing twitch forces from co-contracting motor-units with linearly summed counterparts. Shear stress comparisons within the muscle revealed a lower lateral force dispersion for the co-contracting case, partly explaining the underpredicted magnitudes of the linearly summed forces. These findings have implications for twitch force characterisation and motor-unit based muscle modelling.

Keywords: Twitch force · Motor-units · Finite-element · Continuum-mechanics · Skeletal-muscle

1 Introduction

The basic functional element of force production in the neuromuscular system is the motor-unit, which consists of all muscle fibres innervated by a single alpha motor-neuron [19]. The force exerted by a muscle depends on the number of motor-units recruited, together with the rate of their stimulation [1]. Recruitment typically occurs in order of increasing motor-unit twitch force, referred to as the size-principle; so-called because motor-units that produce a larger force tend to consist of more fibres [6]. Therefore, identification of individual motor-unit behaviour is paramount to understanding healthy and pathological neuromuscular system function.

J. M. R. S. Tavares et al. (Eds.): CMBBE 2021, LNCVB 38, pp. 234–243, 2023.
https://doi.org/10.1007/978-3-031-10015-4_20

Motor-units are typically characterised by their so-called twitch response. This refers to the temporal force behaviour of a motor-unit in response to a single action-potential delivered by its parent alpha motor-neuron. Common properties of motor-unit twitch are peak force, the time taken to attain it, and the (more gradual) force drop-off. Spike-triggered-averaging is a commonly used technique to measure these properties, e.g., [3,12,20], which involves simultaneously recording intramuscular electrical activity and joint force and using motor-unit discharge times or "spikes", as a trigger to average the force signal. While this technique provides a convenient way to measure in vivo twitch properties, it is not without problems. Several of these shortcomings have been investigated by experimental-, simulation-, and theoretical-methods, e.g., [2,8,21,22]. For example, animal experiments revealed non-linearity in twitch force summation [22] and theoretical analyses revealed how motor-unit synchronisation impact the averaged twitch response [8]. These, and other limitations, can lead to mismatches even under ideal conditions [2,8]. Therefore, alternative methods have been proposed to measure in vivo twitch properties. For example, Negro et al. [9] use a deconvolution approach on subpopulations of motor-units to determine the average twitch response.

Nevertheless, regardless of the method to characterise in vivo twitch forces, force measurements cannot be taken directly at the muscle due to ethical reasons. Instead, measurements occur at a point that is biomechanically linked to the muscle. For example, in masticatory muscles, bite force is typically used as an indirect measurement of masseter twitch properties, e.g., [7,23]. Obviously, there is a series of mechanical linkages between the occlusal surface and the muscle, e.g., enamel, dentin, periodontal ligament, cortical and cancellous bones, and tendons. While moment arms may be used to account for this; such analytical methods typically oversimplify the biomechanical linkage. For example, the geometry is typically treated as two-dimensional, or the anatomy is treated as rigid. Furthermore, motor-unit location within the muscle is seldom taken into account. Therefore, the accuracy of such indirect twitch force measurements to capture true muscle twitch behaviour remains unclear.

In this paper, we aim to investigate whether twitch forces measured at locations other than at the muscle accurately represent muscle behaviour. To do this, we use an anatomically realistic, 3D, continuum-mechanical computer model of the masticatory system. This gives us access to measurement locations not ethically or technically feasible in experiments. Individual motor-units are stimulated within the masseter, and the twitch forces measured at the masseter origin—taken as the true twitch force—are compared to those measured at the molar occlusal surface.

Twitch forces are commonly related to each other via linear summation. That is, total muscle force is assumed to be an algebraic sum of its constituent motor-unit twitches. However, such an assumption neglects the interaction of motor-units, for example via lateral force transmission [11]. To investigate this we co-contract individual motor-units within the masseters to compute a so-called compound twitch. We then compare compound twitches to linearly summed counterparts to both determine the accuracy of linear summation, and to investigate lateral force transmission within the muscle.

2 Methods

First, we briefly describe the material modelling of the musculotendon-complex and the means by which motor-unit information is integrated (Sect. 2.1). Second, the masticatory model is detailed, including the geometry and masseter motor-unit physiology (Sect. 2.2). Third, we describe the simulations of individual and compound twitch forces (Sect. 2.3).

2.1 Constitutive Relation of the Musculotendon-Complex

Details of the musculotendon-complex (MTC) constitutive relation, together with material properties, can be found in [13,15,18]. In short, the MTC is treated as hyperelastic, transversely isotropic, and nearly incompressible. A tissue parameter $\gamma_{\mathrm{m}}(\boldsymbol{X}) \in [0,1]$ varies the volume fraction of muscle and tendon at any point $\boldsymbol{X}$ within the geometric model of the MTC, where $\gamma_{\mathrm{m}} = 1$ for pure muscle and $\gamma_{\mathrm{m}} = 0$ for pure tendon.

Passive muscle behaviour, differentiated into contributions from the bulk ($\boldsymbol{S}_{\mathrm{iso}}$) and along the fibres ($\boldsymbol{S}_{\mathrm{ansio}}$), is additively split from the active behaviour ($\boldsymbol{S}_{\mathrm{act}}$). The MTC (second Piola-Kirchhoff) stress tensor is then given by

$$\boldsymbol{S}^{\mathrm{MTC}} = \boldsymbol{S}_{\mathrm{iso}}(\boldsymbol{C}, \gamma_{\mathrm{m}}) + \boldsymbol{S}_{\mathrm{aniso}}(\boldsymbol{C}, \boldsymbol{M}, \gamma_{\mathrm{m}}) + \gamma_{\mathrm{m}}\, \boldsymbol{S}_{\mathrm{act}}(\boldsymbol{C}, \boldsymbol{M}, \alpha), \tag{1}$$

where $\boldsymbol{C}$ is the right Cauchy-Green deformation tensor and $\boldsymbol{M}$ is a structural tensor that acts to direct the anisotropic stresses along the fibre direction. Note that for the passive behaviour γ_{m} linearly interpolates material parameters between pure tendon and muscle, whereas γ_{m} scales the active behaviour directly.

Lastly, the muscle activation parameter $\alpha \in [0,1]$ dictates the degree of muscle activation and is computed via a weighted sum of individual motor-unit activity $\alpha_i(t)$, $i = 1 \ldots N_{\mathrm{MU}}$, where N_{MU} are the total number of motor-units in the muscle. The weighting factor is taken as the volumetric distribution of the motor-unit within the muscle $\kappa_i(\boldsymbol{X}) \in [0,1]$. In other words, $\kappa_i(\boldsymbol{X})$ acts to spatially weight the temporal activity of the motor-unit $\alpha_i(t)$. Total muscle activity is then given at any point and time by summation over i, i.e., $\alpha(t, \boldsymbol{X}) = \sum_i^{N_{\mathrm{MU}}} \alpha_i(t) \kappa_i(\boldsymbol{X})$ [17].

2.2 Masticatory Model

Details of the meshed geometries of the masticatory model, together with material properties, can be found in [14,16,17]. Briefly, the model consists of the mandible (cortical bone), molar (comprising dentin, enamel, and pulp), periodontal ligament (PDL), left and right articular discs, maxillary fossae, and the masseters (Fig. 1). The masseters are divided into deep- and superficial-heads and have fibres oriented posterio-cranially and anterio-cranially, respectively. Furthermore, tendinous regions ($\gamma_{\mathrm{m}} = 0$) are defined at masseter insertion and origin. The middle of the masseters are pure muscle ($\gamma_{\mathrm{m}} = 1$), and the tissue parameter is gradually reduced to zero ($\gamma_{\mathrm{m}} = 1 \rightarrow 0$) between the pure muscle and tendon regions.

The masticatory model is fixed in space via (zero displacement) Dirichlet boundary conditions, applied at: the cranial surface of the maxillary fossae, the occlusal surface of the molar, and the origin of the masseters (Fig. 1). At their insertion, the masseters are affixed to the mandible by "tying" the attachment surfaces of the muscle and bone together. Therefore, when the masseters generate contractile stresses, they are transmitted to both their origin and to the mandible via the attachment area. From there, the stresses are transmitted to the molar surface (via the PDL) and the maxillary fossae (via the articular discs), attaining static equilibrium.

The masseters were populated with 50 motor-units each, with their sizes following an exponential distribution, cf. [4]. The motor-units were distributed within the masseter according to the method described in [17] and produced a distribution that has limited overlap between neighbouring motor-units. Smaller (type-S) motor-units were located anterio-medially in the masseter, i.e., anteriorly in the deep-head. Larger (type-FF, or -FR) units were located posterio-superficially, i.e., posteriorly in the superficial-head (Fig. 1).

Motor-unit activities were precomputed using models of calcium- and cross-bridge dynamics, see [5,17] and references therein. Essentially, motor-neuron stimuli serve as inputs to produce corresponding motor-unit (twitch) activities $\alpha_i(t)$. Rate coefficients were adjusted such that smaller motor-units attained a low force, slowly (slow-twitch, type-S) compared to larger motor-units (fast-twitch, type-FR or -FF). Furthermore, instead of distinct motor-unit types, rate coefficients were varied continuously between the "slowest" (smallest) and "fastest" (largest) motor-units. Note that motor-unit activities are normalised, since the size of the motor-unit is captured by the weighting factor $\kappa_i(\boldsymbol{X})$. That is, if a fully activated motor-unit occupies a small region, it automatically produces a low twitch force.

2.3 Simulating Motor-Unit Twitches

Four motor-units in the left masseter were stimulated individually. We chose motor-units spread across the motor-unit pool to cover a range of sizes and locations. Specifically, motor-units 10, 20, 30, and 40 were chosen (Fig. 1b).

To investigate the influence of twitch force measurement location, we recorded twitch forces at both the origin of the masseter (measurement point $\mathrm{M_o}$, taken as the true twitch force), and also at the occlusal surface of the masseter (measurement point $\mathrm{M_m}$, representing a typical location for experiments) (Fig. 1a). Forces (all in N) were recorded at a frequency of 2 kHz in the left–right, posterior–anterior and caudal–cranial directions, denoted: $F_j^{\mathrm{LR}}(t)$, $F_j^{\mathrm{PA}}(t)$ and $F_j^{\mathrm{CC}}(t)$, respectively. The subscript $j = \mathrm{o}$ for forces measured at the masseter

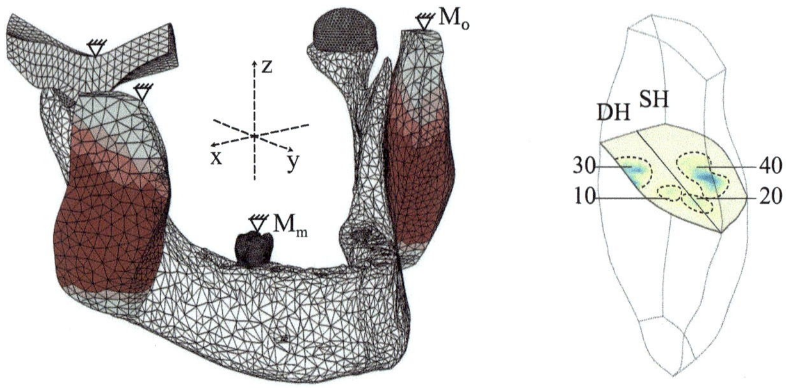

(a) Finite-element model. (b) Motor-unit territories.

Fig. 1. Details of the masticatory model. (a) The finite-element masticatory model (left) with a total of approximately 70,000 tetrahedral elements and 35,000 nodes. The open triangles represent zero displacement boundary conditions. The measurement locations at the masseter origin and molar surface are denoted M_o and M_m, respectively. The x-, y-, and z- directions correspond to the lateral–medial, posterior–anterior, and cranial–caudal axes, respectively. (b) Motor-unit territory volume fractions $\kappa_{\{10,20,30,40\}}(\boldsymbol{X})$ are shown over the left masseter cross-section. Approximate boundaries are sketched and motor-units are labelled. The approximate division between the deep-head (DH) and superior-head (SH) is also indicated.

origin and j = m for measurements at the molar. Force magnitude was computed via $F_j^{\mathrm{MAG}}(t) = \sqrt{(F_j^{\mathrm{LR}})^2 + (F_j^{\mathrm{PA}})^2 + (F_j^{\mathrm{CC}})^2}$. Additionally, peak twitch force was measured along with the time at which it occurs, i.e., the time-to-peak.

To compare twitch force summation models, we stimulated different combinations of motor-units simultaneously to produce so-called compound-twitches. Specifically, motor-unit combinations: 10 + 20, 10 + 20 + 30, and 10 + 20 + 30 + 40 were co-stimulated. These compound-twitches, measured at the molar M_m, were taken as the true twitch forces. In the linear summation model, the individual twitch forces were algebraically summed, e.g., $F_{\mathrm{m},10}^{\mathrm{MAG}}(t) + F_{\mathrm{m},20}^{\mathrm{MAG}}(t)$, and compared to the corresponding compound twitch force, i.e., $F_{\mathrm{m},10+20}^{\mathrm{MAG}}(t)$.

We additionally investigated the impact of compound-twitch 10 + 20 + 30 + 40 on the lateral force transmission within the masseter. For the linearly summed case, we recorded the shear stresses at each point of the muscle $\tau(\boldsymbol{x})$. This was done at the time-to-peak (TTP), and for each motor-unit 10, 20, 30, and 40, i.e., $\tau_{10}(\boldsymbol{x}, \mathrm{TTP})$ for motor-unit 10. These shear stresses were then linearly summed to give the linearly summed shear stress distribution, i.e.,

$$\tau_{\mathrm{lin}}(\boldsymbol{x}, \mathrm{TTP}) = \tau_{10} + \tau_{20} + \tau_{30} + \tau_{40}. \tag{2}$$

The shear stresses for the co-contracting motor-units were also recorded $\tau_{\mathrm{comp}}(\boldsymbol{x}, \mathrm{TTP})$, i.e., when motor-units 10, 20, 30, and 40 were simultaneously contracting. Finally, the shear stress difference at each point within the masseter was computed by $\tau_{\text{diff}}(\boldsymbol{x}, \text{TTP}) = \tau_{\text{lin}}(\boldsymbol{x}, \text{TTP}) - \tau_{\text{comp}}(\boldsymbol{x}, \text{TTP})$.

3 Results

The first pair of charts (Fig. 2) compare twitch force measurement locations. Twitch force magnitudes for each motor-unit, measured at the masseter origin ($F_{\text{o}}^{\text{MAG}}(t)$) and the molar occlusal surface ($F_{\text{m}}^{\text{MAG}}(t)$), are shown in Fig. 2a. The occlusal twitch forces underpredicted the muscle twitches by 11%, 23%, 5%, and 40% for motor-units 10, 20, 30, and 40, respectively. The same data is presented in Fig. 2b, but here the peak twitch force is plotted against the time-to-peak.

The second pair of charts (Fig. 3) compare the twitch force summation models. Specifically, the linearly summed and compound twitch forces for the different combinations of co-contracting motor-units are given in Fig. 3a. The linearly summed twitch forces underpredicted the compound twitches by 19%, 21%, and 23% for motor-unit combinations 10 + 20, 10 + 20 + 30, and 10 + 20 + 30 + 40, respectively. And, lastly, the distribution of shear stress differences $\tau_{\text{diff}}(\boldsymbol{x}, \text{TTP})$ in the masseter for the linearly summed and compound-twitches is shown is Fig. 3b. The right-skewness of the shear stress difference histogram reveals that the linearly summed shear stresses were higher than that for the compound-twitch.

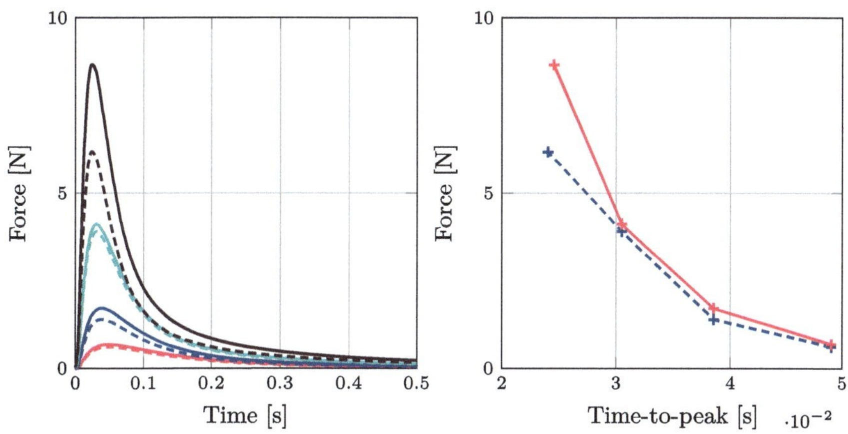

(a) Individual twitches.

(b) Maximum twitch force.

Fig. 2. Comparison of measurement location. (a) Twitch magnitude measured at the origin (solid) and molars (dashed), with motor-units 10, 20, 30, and 40 in red, blue, green, and black, respectively. (b) Peak twitch force and TTP measured at the molar in dashed blue, and at the muscle origin in solid red.

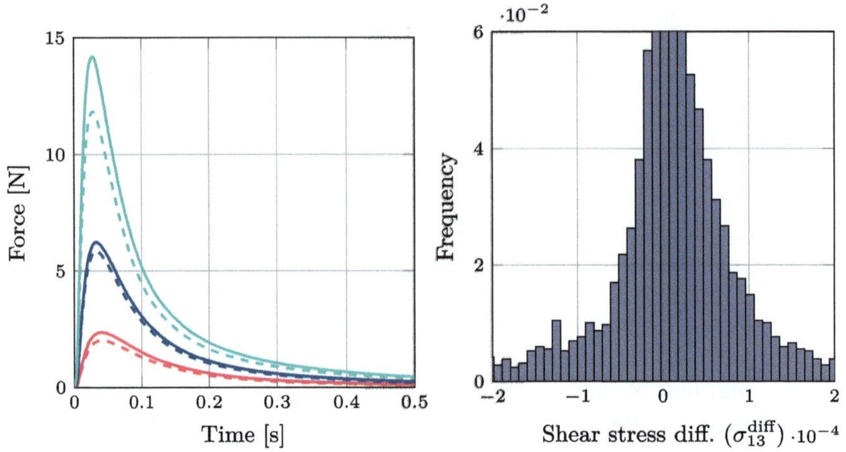

(a) Summation model comparison. (b) Shear stress differences.

Fig. 3. Comparison of twitch force summation models (all measurements at the molar M_O). (a) The dotted lines represent a linear summation, and solid lines the compound-twitch, with: red, blue and green representing co-contracting motor-units 10 + 20, 10 + 20 + 30, and 10 + 20 + 30 + 40, respectively. (b) Distribution of shear stress differences at the point of peak twitch force for the case 10 + 20 + 30 + 40.

4 Discussion

In this paper, we used a simulation study to investigate, first, the accuracy of twitch forces measured at some point removed from the muscle, and second, the accuracy of linearly summing individual twitches to represent co-contracting motor-units.

Concerning twitch force measurement location, the simulation study shows that twitch forces measured at the molar underestimated the twitch force measured at the muscle origin by an average of 20%. This cannot be explained due to energy dissipation, e.g., via viscous effects, or via frictional heat, since a purely mechanical analysis was carried out with perfectly elastic materials. Instead, we posit force distribution as the primary cause of this mismatch. First, due to the direction of the twitch force produced by the motor-unit, and second, due to the location of this force in relation to the attachment area. The location of the motor-unit, via the local fibre-orientation of the muscle, determines the direction of the contractile force. For example, motor-unit 10 is located in the deep-head, meaning that the force will be oriented posterio-cranio-medially. On the other hand, motor-unit 40 is located in the superficial-head, producing an anterio-cranio-laterally oriented force. This means that the force produced by motor-unit 40 is directed towards the masseter origin and fossae (or temporo-mandibular joint) to a greater degree than for motor-unit 10, explaining the lower force appearing at the molar. Additionally, motor-unit location influences the position at which the force is transmitted across the attachment area, which alters the moment arms in relation to the molar. However, this most likely plays a small role since the (much stiffer) tendons distribute twitch force evenly over the attachment area.

Concerning the summation of twitch forces, the simulation study shows that co-contracting units produced a higher force than linearly summing individual twitches, resulting in a 127% force on average. This is inline with experimental results, for example, Troiani et al. [22] found "more than linear summation" (of co-contracting units), where compound tetanic contractions resulted (on average) in 124% force of the linearly summed counterparts. Often, the hypothesis of lateral force dispersion is posited as an explanation of such observations, e.g., [10,22]. To investigate this hypothesis, we analysed the shear stress deformation within the masseters. Indeed, linearly summing shear stresses resulted in higher values when compared to the compound twitch case. This confirms the hypothesis that when a larger number of motor-units co-contract, there is less relative (shearing) movement within the muscle. As a result, a greater amount of force is directed along the muscle's fibres, resulting in a higher twitch force (at the molars).

5 Conclusions

Twitch forces measured at the masseter underpredicted the true twitch force behaviour, measured at the masseter origin. These discrepancies were related to the position of the motor-unit within the muscle, altering twitch force distribution between the molars, muscle origin and temporomandibular joint. Compound-twitches elicited a higher twitch force than their linearly summed counterparts. This was partly due to more efficient transfer of forces along the fibres towards the tendons. Therefore, models which linearly sum twitch force magnitudes may underpredict muscle forces. These results imply that the interplay of these factors makes it difficult to predict absolute twitch force from measurements carried out at some point biomechanically linked to the muscle. While relative differences in twitch behaviour seems to hold, discrepancies in absolute behaviour are more apparent for many co-contracting motor-units.

References

1. Adrian, E.D., Bronk, D.W.: The discharge of impulses in motor nerve fibres: Part II. The frequency of discharge in reflex and voluntary contractions. J. Physiol. **67**(2), 9–151 (1929). https://doi.org/10.1113/jphysiol.1929.sp002557
2. Dideriksen, J.L., Negro, F.: Spike-triggered averaging provides inaccurate estimates of motor unit twitch properties under optimal conditions. J. Electromyogr. Kinesiol. **43**, 104–110 (2018). https://doi.org/10.1016/j.jelekin.2018.09.008
3. Clark, R.W., Luschei, E.S., Hoffman, D.S.: Recruitment order, contractile characteristics, and firing patterns of motor units in the temporalis muscle of monkeys. Exp. Neurol. **61**(1), 31–52 (1978). https://doi.org/10.1016/0014-4886(78)90179-6

4. Fuglevand, A.J., Winter, D.A., Patla, A.E.: Models of recruitment and rate coding organization in motor-unit pools. J. Neurophysiol. **70**(6), 2470–2488 (1993). https://doi.org/10.1152/jn.1993.70.6.2470
5. Heidlauf, T., Klotz, T., Rode, C., Siebert, T., Röhrle, O.: A continuum-mechanical skeletal muscle model including actin-titin interaction predicts stable contractions on the descending limb of the force-length relation. PLOS Comput. Biol. **13**(10), e1005773 (2017). https://doi.org/10.1371/journal.pcbi.1005773
6. Henneman, E.: Relation between size of neurons and their susceptibility to discharge. Science **126**(3287), 1345–1347 (1957)
7. McMillan, A.S., Sasaki, K., Hannam, A.G.: The estimation of motor unit twitch tensions in the human masseter muscle by spike-triggered averaging. Muscle Nerve **13**(8), 697–703 (1990). https://doi.org/10.1002/mus.880130806
8. Negro, F., Yavuz, U.Ş., Farina, D.: Limitations of the spike-triggered averaging for estimating motor unit twitch force: a theoretical analysis. PLOS ONE **9**(3), e92390 (2014). https://doi.org/10.1371/journal.pone.0092390
9. Negro, F., Orizio, C.: Robust estimation of average twitch contraction forces of populations of motor units in humans. J. Electromyogr. Kinesiol. **37**, 132–140 (2017). https://doi.org/10.1016/j.jelekin.2017.10.005
10. Powers, R.K., Binder, M.D.: Summation of motor unit tensions in the tibialis posterior muscle of the cat under isometric and nonisometric conditions. J. Neurophysiol. **66**(6), 1838–1846 (1991). https://doi.org/10.1152/jn.1991.66.6.1838
11. Purslow, P.P.: Muscle fascia and force transmission. J. Bodywork Mov. Ther. **14**(4), 411–417 (2010). https://doi.org/10.1016/j.jbmt.2010.01.005
12. Roatta, S., Arendt-Nielsen, L., Farina, D.: Sympathetic-induced changes in discharge rate and spike-triggered average twitch torque of low-threshold motor units in humans. J. Physiol. **586**(22), 5561–5574 (2008). https://doi.org/10.1113/jphysiol.2008.160770
13. Röhrle, O., Sprenger, M., Schmitt, S.: A two-muscle, continuum-mechanical forward simulation of the upper limb. Biomech. Modeling Mechanobiol. **16**(3), 743–762 (2016). https://doi.org/10.1007/s10237-016-0850-x
14. Röhrle, O., Saini, H., Lee, P.V.S., Ackland, D.C.: A novel computational method to determine subject-specific bite force and occlusal loading during mastication. Comput. Methods Biomech. Biomed. Eng. **21**(6), 453–460 (2018). https://doi.org/10.1080/10255842.2018.1479744
15. Röhrle, O., Yavuz, U.Ş., Klotz, T., Negro, F., Heidlauf, T.: Multiscale modeling of the neuromuscular system: coupling neurophysiology and skeletal muscle mechanics. Wiley Interdisc. Rev. Syst. Biol. Med. **11**(6), e1457 (2019). https://doi.org/10.1002/wsbm.1457
16. Saini, H., Ackland, D.C., Gong, L., Cheng, Röhrle, O.: Occlusal load modelling significantly impacts the predicted tooth stress response during biting: a simulation study. Comput. Methods Biomech. Biomed. Eng. **23**(7), 261–270. https://doi.org/10.1080/10255842.2020.1711886
17. Saini, H., Klotz, T., Röhrle, O.: Enriching macroscopic skeletal muscle models on a motor-unit level reveals differences in joint force: a simulation study (in Preparation)
18. Saini, H., Röhrle, O.: A biophysically guided constitutive law of the musculotendon-complex: modelling and numerical implementation in Abaqus (2021). https://papers.ssrn.com/sol3/papers.cfm?abstract_id=4102677
19. Sherrington, C.S.: Ferrier lecture.-some functional problems attaching to convergence. Proc. Roy. Soc. London B Biolog. Sci. **105**(737), 332–362 (1929). https://doi.org/10.1098/rspb.1929.0047

20. Stein, R.B., French, A.S., Mannard, A., Yemm, R.: New methods for analysing motor function in man and animals. Brain Res. **40**(1), 187–192 (1972). https://doi.org/10.1016/0006-8993(72)90126--6
21. Taylor, A.M., Steege, J.W., Enoka, R.M.: Motor-unit synchronization alters spike-triggered average force in simulated contractions. J. Neurophysiol. **88**(1), 265–276 (2002). https://doi.org/10.1152/jn.2002.88.1.265
22. Troiani, D., Filippi, G.M., Bassi, F.A.: Nonlinear tension summation of different combinations of motor units in the anesthetized cat peroneus longus muscle. J. Neurophysiol. **81**(2), 771–780 (1999). https://doi.org/10.1152/jn.1999.81.2.771
23. Yemm, R.: The orderly recruitment of motor units of the masseter and temporal muscles during voluntary isometric contraction in man. J. Physiol. **265**(1), 163–174 (1977). https://doi.org/10.1113/jphysiol.1977.sp011710

On the Use of Mesh-Based Joint Contact Models Within Simulations Using Automatic Differentiation

Gil Serrancolí[1(✉)], Jordi Torner[1], Simone Perelli[2,3], and Joan Carles Monllau[2,3]

[1] Multimedia Applications Lab, Universitat Politècnica de Catalunya, Barcelona, Spain
gil.serrancoli@upc.edu

[2] Institut Català de Traumatologia i Medicina de l'Esport (ICATME), Hosp. Dexeus, UAB, Barcelona, Spain

[3] Department of Orthopaedic Surgery, Hospital del Mar, UAB, Barcelona, Spain

Abstract. Biomechanical simulations integrating computations of joint contact pressures within optimizations typically use mesh-based contact models. However, these models are usually not continuous, which could make convergence difficult. Moreover, they can be difficult to tackle when using expression graphs to calculate derivatives (e.g., when using automatic differentiation) during the resolution of optimization problems. This is due to the computational need to use branches when dealing with conditionals. This study presents a mesh-based contact model adapted to deal with gradient-based optimizations and automatic differentiation. A tracking example of a knee prosthesis contact, integrated into a full-body muscle-driven model, is presented. Kinematics and dynamics data were tracked during an overground gait trial. The results show that the accuracy of knee contact force estimations is comparable to other studies (RMSE medial contact force = 50.8 N, RMSE lateral = 81.6 N), and the computational time (3 h and 20 min) is acceptable to be run in a conventional PC. Future work is oriented to formulate predictive simulations of muscle-driven full-body models to predict kinematics and dynamics data of subjects post-surgery simultaneously.

Keywords: Mesh-based contact · Muscle-driven model · Automatic differentiation · Gradient-based optimization

1 Introduction

Musculoskeletal simulations can be used to estimate or predict biomechanical variables in silico. Recent advances integrating automatic differentiation (AD) tools into OpenSim software [7] improve the efficiency of those simulations, usually obtained by solving an optimal control problem. This is because AD allows calculating the derivatives much faster than using traditional methods (up to 20 times for complex musculoskeletal methods), such as finite differences (FD). AD tools create an expression graph of the mathematical expressions and compute the derivatives. This optimal control framework has

J. M. R. S. Tavares et al. (Eds.): CMBBE 2021, LNCVB 38, pp. 244–249, 2023.
https://doi.org/10.1007/978-3-031-10015-4_21

been used to predict human gait in terms of musculoskeletal movement and forces [1], to predict subject-exoskeleton contact forces and their collaborative movement [2], to track sprint movements and their forces [3], or to estimate muscle forces in animals [4].

One of the main drawbacks is that collision-detection algorithms are not easy to handle when using gradient-based optimization and AD tools, since they are discontinuous and require many branches of the expression graph. A point contact force model was developed by smoothing the transition between contact and no-contact events in a previous work [2]. Although it is computationally efficient and avoids discontinuities (without using conditional statements), it only computes the resultant contact wrench (punctual contact force) and it does not allow the estimation of contact pressures. To compute those pressures, the geometry is typically discretized in a mesh and the computation of contact detection between two or more meshes is required. In this study, we present a mesh-based elastic foundation contact model, which can be used within optimal control problems and solved using gradient-based methods and AD tools.

2 Methods

2.1 Mesh-Based Contact Model

The contact model estimates the pressures at each face of a body mesh in contact with another. Instead of detecting the collisions at each evaluation, the pressures are calculated at each potential pair using a continuous function. Therefore, before the use of the contact model (before the optimization), all potential pairs between both bodies need to be identified. Then, the process of calculating the pressures at each optimization iteration is as follows:

1. Calculate distances of all potential pairs.
2. Identify minimum distance d (max penetration δ) at each face of one body with respect to all faces of the other body using a continuous function:

$$\min d = -\frac{\log\left(\sum_{i}^{nfaces} e^{kd_i}\right)}{k} \tag{1}$$

3. Calculate contact pressures using those penetration values, similar to Bei and Fregly [5], but using a continuous function (similar to [2]) to avoid negative values and obtain smooth transitions:

$$p = \frac{(1-\nu)E}{(1+\nu)(1-2\nu)} \ln\left(1 - \frac{\delta}{h}\right) \tag{2}$$

$$p_{smooth} = p(1 + \tanh(k\delta))/2 \tag{3}$$

4. Calculate contact forces based on the pressure and the face area.
5. Calculate the resultant wrench (3 forces and 3 moments) at the center of the body.

2.2 Optimal Control Problem and Experimental Data

This contact model was applied to a knee prosthesis contact (tibial tray and femoral component). First, the meshes were simplified (reducing the number of faces) by keeping the interesting regions and shapes, and the tibial tray was split into medial and lateral compartments.

Figure 1A shows an example of the distance between one face of the tibial tray with respect to all potential contact faces of the femoral component during a gait trial, with the ones in contact highlighted in green, yellow and blue (black line is the minimum value throughout the cycle). Negative values of these distances indicate positive indentation, therefore the contact pressures produced with those positive indentation values will be significant. When indentation is negative, the pressures will be non-zero (according to Eq. (3)), but the values will be negligible. Figure 1B shows as example the potential contact faces between one face of the tibial tray and the femoral component. The residual contact forces at the knee (between tibial tray and femoral component) were obtained as the sum of the multiplication of the contact pressure at each face by the area of that face and the unit vector defining the face normal vector.

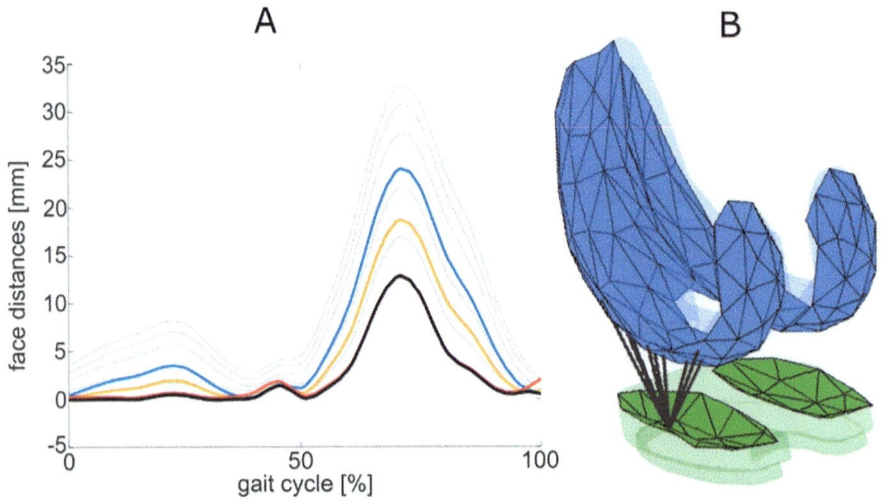

Fig. 1. A. Example of the smooth calculation of minimum distance between faces of two bodies. **B.** Map of potential contact pairs between a face of the tibial tray and all the other potential contact pairs of the femoral component.

All experimental data were from the Knee Grand Challenge to predict in vivo knee contact forces [6], corresponding to a normal gait trial of a subject implanted with a force-measuring tibial prosthesis at the right knee (gender: male, age: 88 yrs, mass: 65 kg, and height: 166 cm). The model had 34 degrees of freedom (including the 6 degrees of freedom at the right knee) and was driven by 94 muscles. Joint passive moments were included to represent the passive joint structures. Residual knee contact forces and residual joint moments were included to facilitate the convergence (though ideally these values should be zero at the optimal solution).

The contact model was tested within a tracking problem to ensure dynamic consistency among measured kinematics, ground reaction forces and knee contact forces. This was formulated as an optimal control problem using direct collocation and automatic differentiation [7], solved in CasADi [8]. The weighted cost function had tracking and minimization terms. Specifically, joint coordinates, ground reaction forces and knee contact forces (medial and lateral separately) were tracked. Muscle activations, residual knee contact forces and residual joint moments were minimized. Joint accelerations and the derivatives of muscle activations and normalized tendon forces were also minimized with a lower weight. Dynamic constraints were included in the optimization formulation. Implicit formulations of activation and contraction dynamics were also included as constraints. Musculoskeletal dynamic equations (the sum of muscle moments equal to inverse dynamics loads) were included as path constraints.

3 Results

The optimal solution was obtained in 3 h and 20 min (in a PC-desktop Intel Xeon W2123 CPU 3.6 GHz, 32 GB RAM). The obtained knee contact forces were accurate (RMSE medial = 50.8 N, RMSE lateral = 81.6 N) (Fig. 2). Kinematics (mean RMSE joint angles = $2.1 \pm 4.1°$) and ground reaction forces (mean RMSE GRF = 66.0 ± 71.2 N) were also accurate. Musculoskeletal equations (set as constraints) were satisfied with an error lower than 10^{-4}.

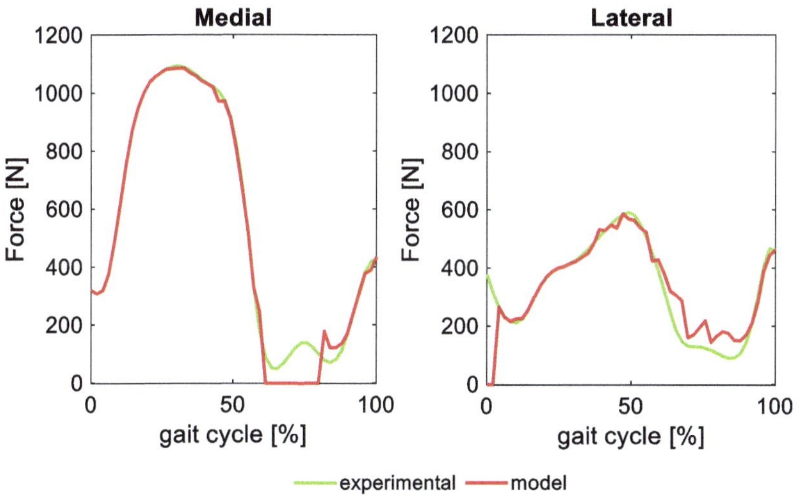

Fig. 2: Estimation of knee contact forces (experimental in green, model in red) of medial (left) and lateral (right) compartments).

4 Conclusions

The results show that the presented mesh-based contact model can be used within optimal control problems using gradient-based optimizations and automatic differentiation with a regular computer. Kinematics, muscle, ground reaction and knee contact forces and joint moments were estimated simultaneously. The tracking of knee contact forces was comparable to other studies [9–11].

This contact model tackles the typical issues of compliant contact models. The use of smooth functions allows working with gradient-based optimizations avoiding discontinuity issues, and also avoids formulating multiple branches of the expression graph when using automatic differentiation.

Future work is oriented to formulate predictive simulations of muscle-driven full-body models to simultaneously predict kinematics and dynamics data of subjects post-surgery. These methods open the door to simulate orthopedic surgeries in silico integrating full-body biomechanical data.

Acknowledgements. G. Serrancolí acknowledges the support from Serra Húnter Program and the grant EIN2020-112411, MICINN.

References

1. Falisse, A., Serrancolí, G., Dembia, C.L., Gillis, J., Jonkers, I., De Groote, F.: Rapid predictive simulations with complex musculoskeletal models suggest that diverse healthy and pathological human gaits can emerge from similar control strategies. J. R. Soc. Interface **16**(157), 20190402 (2019). https://doi.org/10.1098/rsif.2019.0402
2. Serrancolí, G., et al.: Subject-exoskeleton contact model calibration leads to accurate interaction force predictions. IEEE Trans. Neural Syst. Rehabil. Eng. **27**(8), 1 (2019). https://doi.org/10.1109/tnsre.2019.2924536
3. Haralabidis, N., Serrancolí, G., Colyer, S., Bezodis, I., Salo, A., Cazzola, D.: Three-dimensional data-tracking simulations of sprinting using a direct collocation optimal control approach. PeerJ **9**, 1–33 (2021). https://doi.org/10.7717/peerj.10975
4. Bishop, P.J., et al.: Computational modelling of muscle fibre operating ranges in the hindlimb of a small ground bird (Eudromia elegans), with implications for modelling locomotion in extinct species, vol. 17, no. 4 (2021)
5. Bei, Y., Fregly, B.J.: Multibody dynamic simulation of knee contact mechanics. Med. Eng. Phys. **26**(9) SPEC.ISS, 777–789 (2004). https://doi.org/10.1016/j.medengphy.2004.07.004
6. Fregly, B.J., et al.: Grand challenge competition to predict in vivo knee loads. J. Orthop. Res. **30**(4), 503–513 (2012). https://doi.org/10.1002/jor.22023
7. Falisse, A., Serrancolí, G., Dembia, C.L., Gillis, J., DeGroote, F.: Algorithmic differentiation improves the computational efficiency of OpenSim-based trajectory optimization of human movement. PLoS ONE **14**(10), 1–19 (2019). https://doi.org/10.1371/journal.pone.0217730. (Falisse, A., Serrancolí, G., Dembia, C.L.C.L., Gillis, J., De Groote, F.: Algorithmic differentiation improves the computational efficiency of OpenSim-based optimal control simulations of movement. bioRxiv, 644245 (2019). https://doi.org/10.1101/644245)
8. Andersson, J.A.E., Gillis, J., Horn, G., Rawlings, J.B., Diehl, M.: CasADi: a software framework for nonlinear optimization and optimal control. Math. Program. Comput. **11**(1), 1–36 (2018). https://doi.org/10.1007/s12532-018-0139-4
9. Kinney, A.L., Besier, T.F.: Update on grand challenge competition to predict in vivo knee loads **135**, 1–4 (2014). https://doi.org/10.1115/1.4023255

10. Serrancolí, G., Kinney, A.L., Fregly, B.J.: Influence of musculoskeletal model parameter values on prediction of accurate knee contact forces during walking. Med. Eng. Phys. **85**, 35–47 (2020). https://doi.org/10.1016/j.medengphy.2020.09.004
11. Serrancolí, G., Kinney, A.L., Fregly, B.J., Font-Llagunes, J.M.: Neuromusculoskeletal model calibration significantly affects predicted knee contact forces for walking. J. Biomech. Eng. **138**(8), 081001 (2016). https://doi.org/10.1115/1.4033673

Lower Limb Joint Load Comparison from Subject Specific Musculoskeletal Model Simulation and Direct Measurements on Different Subject with Instrumented Implant During Normal and Abnormal Gait

Carlos Rodrigues[1,2](✉), Miguel Correia[1,2], João Abrantes[3], Marco Rodrigues[4], and Jurandir Nadal[5]

[1] Faculty of Engineering of the University of Porto, Porto, Portugal
c.rodrigues@fe.up.com
[2] Institute for Systems and Computer Engineering, Technology and Science, Porto, Portugal
[3] CICANT, MovLab, Lusófona University, Lisbon, Portugal
[4] Federal University of Pernambuco, Recife, Brazil
[5] Federal University of Rio de Janeiro, Rio de Janeiro, Brazil

Abstract. This study presents lower limb joint load comparison from subject specific musculoskeletal model simulation (MSK-MS) and direct measurements from instrumented implants on post-operative (PO) patients. A case study was considered for MSK-MS gait analysis of a 40-year-old healthy male with 70 kg and 1.86 m height. Reflective adhesive markers were applied on skin surface of selected anatomical points at right and left lower limbs. Orthostatic and dynamic acquisition on normal gait (NG), stiff-knee gait (SKG) and slow running (SR) was performed from ground reaction forces with two force plates at 2 kHz and trajectories of skin markers with eight-camera system at 100 Hz. Subject specific MSK-MS was performed using AnyGait and morphed Twente Lower Extremity Model (TLEM), matching the size and joint morphology of the stick-figure model. Over-determinate kinematic analysis was performed, and motion equations solved with hard and soft constraints. Representative MSK-MS gait cycles were selected at NG, SKG and SR lower limb joint vertical force components at the hip, the knee, and the ankle normalized to body weight (*JFz/BW*). Internal joint direct measurements of four PO patients', 61–83 years, average weight 808 N and 1.71 m height, with telemetric Hip I (4-channel), Hip II (8-channel) and knee (9-channel) instrumented implants were selected from Orthoload database with comparable gait to NG, SKG and SR. Statistical measurements presented similar mean *JFz/BW* at right/left hip, knee, ankle MSK-MS and asymmetric peak values with dominant NG, SKG and SR different variances ($p < 0.05$). Direct *JFz/BW* measures contrasted NG with similar hip and knee mean and variance from SKG and SR with different mean and variance. Peak *JFz/BW* direct measurements presented higher hip and knee values on SR and NG than SKG, with higher values at the knee than the hip on NG and SKG, and the opposite on SR. Direct *JFz/BW* measurements presented at the hip and the knee lower values than their corresponding MSK-MS on NG, SKG and SR.

J. M. R. S. Tavares et al. (Eds.): CMBBE 2021, LNCVB 38, pp. 250–270, 2023.
https://doi.org/10.1007/978-3-031-10015-4_22

Keywords: Musculoskeletal model · Instrumented implant · Joint load

1 Introduction

Knowledge of joint internal loads is essential for gait impairment diagnosis, intervention planning and post-operative rehabilitation [1]. Nevertheless, direct measurement of joint internal loads can only be performed invasively, pointing for the need of subject specific musculoskeletal models and input data for estimation of specific subject joint internal loads at different gait conditions [2]. Despite main reason for gait study is related with gait disorders, dominant information is available on normal gait with the need of subject specific model and simulation for abnormal gait due to specific condition of each subject disability. For this reason, musculoskeletal modeling, and simulation (MSK-MS) presents as a key tool for estimation of subject specific internal joint loads based on external measurement of dynamic contact forces and kinematics of selected anatomical points.

MSK-MS has evolved as a computer-aided engineering (CAE) tool as in other classic engineering, with additional challenge on multibody dynamics representation of human living being and the need for subject, body system, and task specific modelling a challenging problem [3]. Despite the advantages of computational model at the situations when direct measurement data is not feasible or is difficult to obtain, along with the possibility of multiple scenario approach, application of subject specific musculoskeletal modeling for diagnosis, treatment planning and intervention point for the need of model verification and validation. While verification process essentially concerns with the model implementation and numerical accuracy of the solution, validation relates to the ability of the computer simulation to predict the variables of interest [3]. Thus, model evaluation for validation can be performed by comparing computational and experimental results, extrapolation of the model prediction to the intended application and determination of the accuracy for the study purpose.

For this reason, there is an increasing need to validate model estimated joint internal loads, comparing computed values with direct measured loads of different subjects and gait conditions, contributing to improve musculoskeletal model predictions. Present work corresponds to a musculoskeletal modeling and simulation study comparison of subject specific dynamic analysis on human abnormal gait. Specific event refers to stiff knee with the reduction of the knee range of motion and the adoption of stiff knee gait (SKG) for lower velocities and slow running (SR) for higher velocities in relation to normal gait (NG).

2 Materials and Methods

2.1 Subject Specific Musculoskeletal Modeling

Case study of a 40-year-old healthy male voluntary with 70 kg mass and 1.86 m height was considered for subject specific musculoskeletal modeling and gait analysis at human movement lab. Ethical committee approval (N-20130014) and written informed consent were obtained in agreement with the ethical guidelines of the North Denmark Region Committee on Health Research Ethics.

Reflective adhesive markers for the musculoskeletal model were attached to skin surface of palped anatomical points at the right and left anterior superior iliac spines (RAsis/LAsis), right and left posterior superior iliac spines (RPsis/LPsis), right and left thigh superior (RThighSuperior/LThighSuperior), right and left knee lateral (RKneeLateral/LKneeLateral), right and left knee medial (RKneeMedial/LKneeMedial), right and left shank superior (RShankSuperior/LShankSuperior), right and left ankle lateral (RAnkleLateral/LAnkleLateral), right and left ankle medial (RAnkleMedial/LAnkleMedial), right and left heel (RHeel/LHeel), and right and left third metatarsal (RToe/LToe) represented at Fig. 1 of marker protocol with filed circles. Additional reflective markers were attached to the subject for the motion capture session, namely right and left thigh lateral (RThighLateral/LThighLateral), right and left thigh inferior (RThighInferior/LThighInferior), right and left shank lateral (RShankLateral/LShankLateral), right and left shank inferior (RShankInferior/LShankInferior), right and left first metatarsal (RToeMedial/LToeMedial), right and left fifth metatarsal (RToeLateral/LToeLateral), and right and left navicular bone (RNavicular/LNavicular), represented with unfilled circles at Fig. 1 marker protocol.

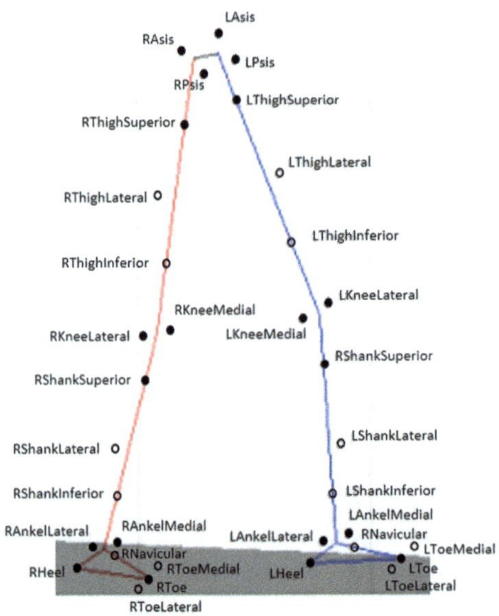

Fig. 1. Marker protocol with filed circles corresponding to main reflective markers and unfilled circles corresponding to additional reflective markers.

Four experimental trials were considered on (*i*) standing reference static trial with subject instructed to assume orthostatic position; (*ii*) normal gait (NG) at self-selected velocity; (*iii*) stiff-knee gait (SKG) with subject instructed to maintain right knee strait during gait at self-selected velocity; and (*iv*) slow running (SR) at minimum velocity to keep a phase with both feet off the ground. During trial tests ground reaction forces were acquired with two AMTI force plates at 2000 Hz and trajectories of attached skin markers to the subject recorded at 100 Hz with an eight-camera Qualisys system. Figure 2 presents 3D view of lower limb stick-figure strobe at NG, SKG and SR gait cycle, with assessed average horizontal progression velocity 1.8 m/s, 1.4 m/s and 2.5 m/s respectively.

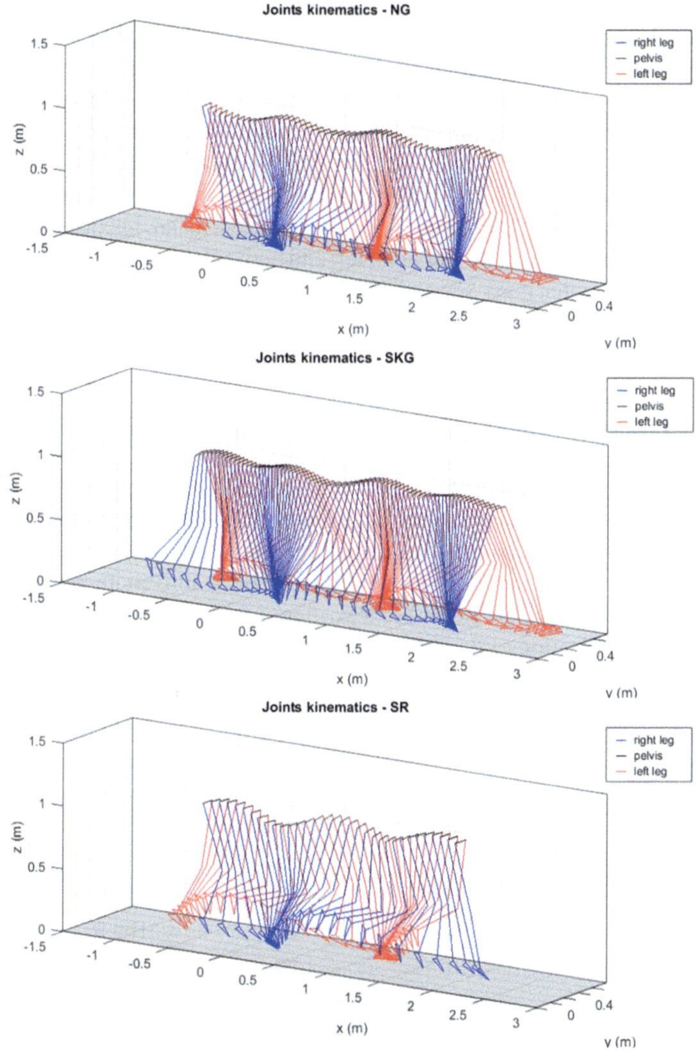

Fig. 2. Three-dimensional perspective of lower limb stick-figure strobe during representative gait cycle at normal gait (NG), stiff-knee gait (SKG) and slow running (SR).

Subject specific musculoskeletal analysis was performed using AnyGait v.0.92 configured for the experimental setup. Stick-figure was generated based on static trial and over-determinate kinematic analysis performed over NG, SKG and SR dynamic trials obtaining lower limb joint angles at each trial. Stick-figure and Twente Lower Extremity Model (TLEM) [4] were morphed, matching the size and joint morphology of the stick-figure model. Inverse dynamics was then performed using computed joint angles and kinetic boundary conditions [5] obtaining internal joint forces and force moments.

Over-determinate kinematic analysis was performed using complete set of captured marker coordinates and model scaling. Local marker coordinates were used, and model scaling performed using joint positions and orientations, with parameter identification for model calibration from motion data, minimizing the differences between the model and measured motion [6]. Motion equations were solved as an optimization problem after equation split in two sets, with first set solved exactly and second set used as a minimum objective function using general computational-efficient solution algorithm [7].

Representative gait cycles were selected at NG, SKG and SR musculoskeletal model of lower limb joint vertical force (*JFz*) components at the hip, the knee, and the ankle normalized to body weight (BW), Fig. 3–5, with gait cycle mean, maximum and variance compared among joints, gait type and right, left and both sides jointly considered, Figs. 3–5 and Tables 3–8. For this purpose, statistical principles were applied to entire gait cycle at each joint signal and condition, assuming periodic gait nature and thus ergodic random process [8].

2.2 Direct Measurements on Instrumented Implants

Internal joint direct measurements of post-operative (PO) patients with instrumented telemetric implants were obtained from Orthoload [9] the free public database of in vivo load measurements, with safety of instrumented implants approved by the legal responsible authorities, clinical studies approved by the local university ethics committee (Charité Universitaetsmedizin Berlin) and written informed consent from patients for pseudonymised publication of their data and images.

Selected subjects for direct measurements correspond to four male PO patients with hip and knee instrumented implants due to osteoarthritis indication, ages at implantation from 61 to 83 years, average weight 808 N and 1.71 m height, Table 1.

Table 1. Patient data with instrumented implants.

Patient	Age (Years)	Weight (N)	Height (m)	Implant	Side
KWR	61	700	1.68	Hip II	Right
K1L	63	1010	1.77	Knee	Left
K2L	71	870	1.71	Knee	Left
EBL	83	650	1.68	Hip I	Left

Considered activities of PO patients for comparison of lower limb joint internal loads with similar gait conditions at subject specific musculoskeletal modeling simulations were at NG, trial 1 (KWRWS1) of KWR patient with 10-month PO at right hip implant walking free with slow velocity and trial 2 (K1L_040707_1_24p) of K1L patient with

4-month PO at left knee implant on level walking, whereas at SKG trial 3 (KWR0153A) corresponds to KWR patient with 8-month PO at right hip implant walking supported by crutches in both hands and trial 4 (K2L_140607_1_63P) to K2L patient with 0.3-month PO of left knee implant on level walking with three-point support. Finally, SR was associated to trial 5 (EBL284A) of EBL patient with 30-month PO left hip implant jogging on treadmill at 8 km/h constant speed and trial-6 (K1L_261110_5_41p) of K1L patient with 44-month PO left knee implant walking on treadmill at 4 km/h with 5 mm lateral wedged insoles and ankle orthosis, Table 2.

Table 2. Gait trial data of patient with instrumented implants.

Trial	Activity	Patient	Implant	Side	PO (month)
T1	Walking free with slow velocity	KWR	Hip	Right	10
T2	Level walking	K1L	Knee	Left	4
T3	Walking supported by crutches in both hands	KWR	Hip	Right	8
T4	Level walking with three-point support	K2L	Knee	Left	0.3
T5	Walking on treadmill at 8 km/h	EBL	Hip	Right	30
T6	Walking on treadmill at 4 km/h	K1L	Knee	Left	40

Data from Orthoload database [10] was selected as the most adequate available for comparison with the subject specific musculoskeletal model and simulation (MSK-MS) estimated hip and knee joint vertical forces during NG, SKG and SR. Since there was no SKG trial on Orthoload database, measurements of the hip joint forces for walking supported by crutches in both hands were selected for comparison with the hip joint reaction forces estimated by MSK-MS and level walking with three-point support for comparing with MSK-MS estimated knee joint reaction forces on SKG. Also, because no running trial was available on Orthoload database, measurements of the hip forces for walking on a treadmill at constant velocity of 8 km/h were selected for comparison with the hip joint reaction forces estimated by MSK-MS and measurements of the knee forces for walking on a treadmill at constant velocity of 4 km/h with lateral wedged 5 mm insoles and ankle orthosis were selected from Orthoload, for comparing with the knee joint reaction forces estimated by MSK-MS on SR. Due to the absence of the ankle instrumented implant on the Orthoload database, comparison of MSK-MS estimated internal forces and direct measurement with instrumented implant were restricted to the hip and knee joints. Entire gait cycle was selected at the hip and the knee joint internal forces from direct measurements on instrumented implants with time delimited on trials T1–T6 to the interval between consecutive contacts of the same feet with the ground, Figs. 7–9. Absolute vertical joint forces (JFz) were normalized to subject body weight (BW) for comparison with MSK-MS.

3 Results

3.1 Subject Specific Musculoskeletal Modeling

The hip, the knee, and the ankle joint vertical forces normalized to body weight (*JFz/BW*) presented similar mean gait cycle values for the right and left sides at NG without statistical significative differences ($p \geq 0.05$), and differences at the right and left sides in relation to the resultant support on both sides at each lower limb joint ($p < 0.05$), Fig. 3 and Tables 3–4. Comparing at the right, left and both sides mean normalized vertical support forces at each joint on NG, the hip-knee presented at the left and both sides together statistical significative differences ($p < 0.05$), without ssd ($p \geq 0.05$) at the remaining joints and sides, Table 5.

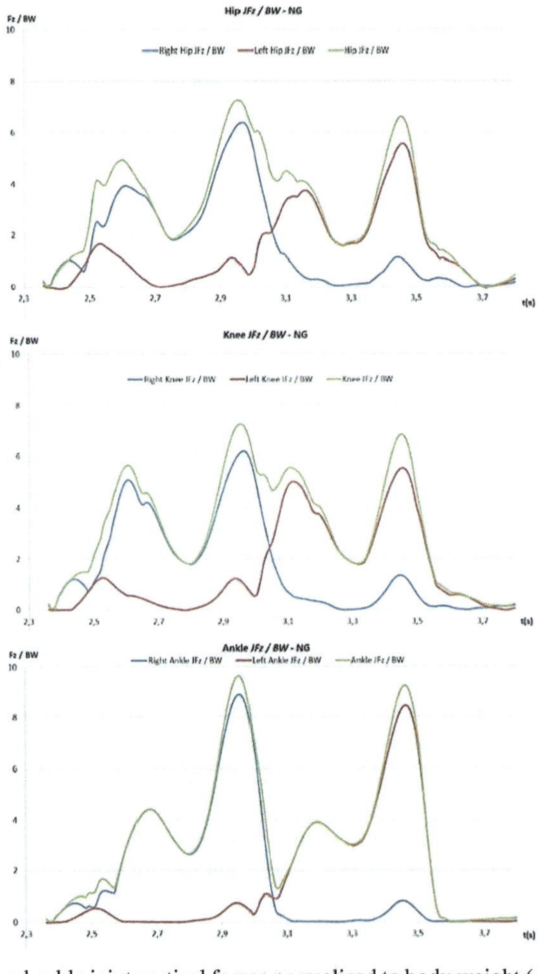

Fig. 3. Hip, knee, and ankle joint vertical forces normalized to body weight (*JFz/BW*) at the right (blue), left (red) and both (green) sides during normal gait (NG) cycle.

Also, the hip, the knee, and the ankle *JFz*/*BW* presented at SKG, Fig. 4 and Tables 3–4, similar mean gait cycle at the right and left sides without ssd ($p \geq 0.05$), whereas the comparison of the left and right mean *JFz*/*BW* with mean resultant vertical force on both sides presented ssd ($p < 0.05$) at the hip, the knee, and the ankle. Regarding the comparison of the right, left and both sides mean vertical support forces at each joint on SKG, the hip, the knee and the ankle presented no ssd ($p \geq 0.05$), with exception of the right knee-ankle, Table 5.

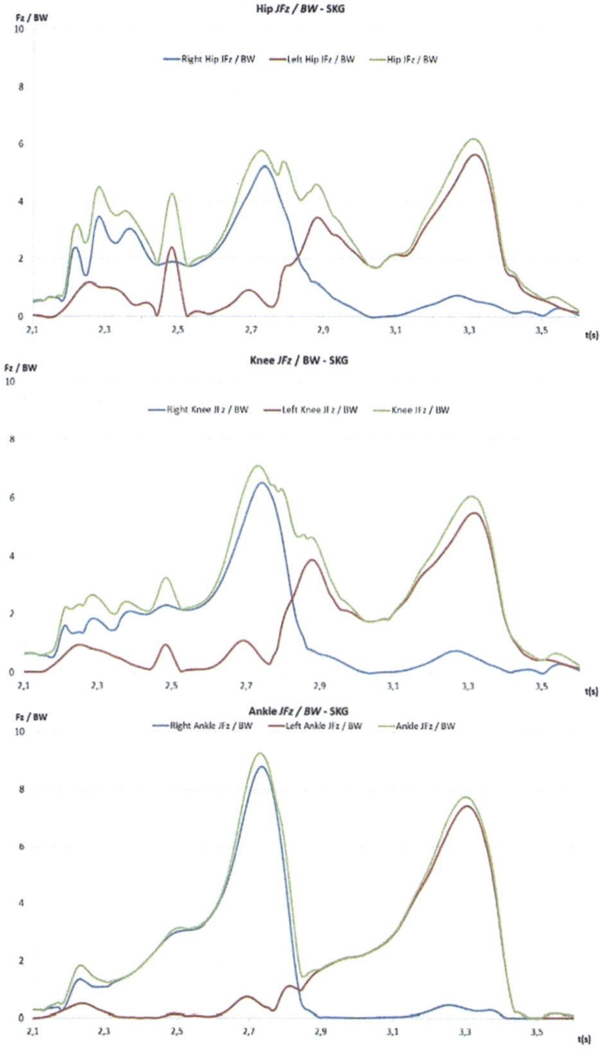

Fig. 4. Hip, knee, and ankle joint vertical forces normalized to body weight (*JFz*/*BW*) at the right (blue), left (red) and both (green) sides during stiff-knee gait (SKG) cycle.

Concerning the hip, the knee and the ankle *JFz/BW*, the right and left sides presented at SR, Fig. 5 and Tables 3–4, similar mean values without ssd ($p \geq 0.05$), and differences between the right and left with resultant support *JFz/BW* of both sides with ssd ($p < 0.05$). Unlike NG and SKG, SR presented on comparison at the right, left and both sides between the hip, the knee, and the ankle dominant ssd ($p < 0.05$) of mean *JFz/BW* with exception of the knee-ankle comparison at the right and left sides with no ssd ($p \geq 0.05$), Table 5.

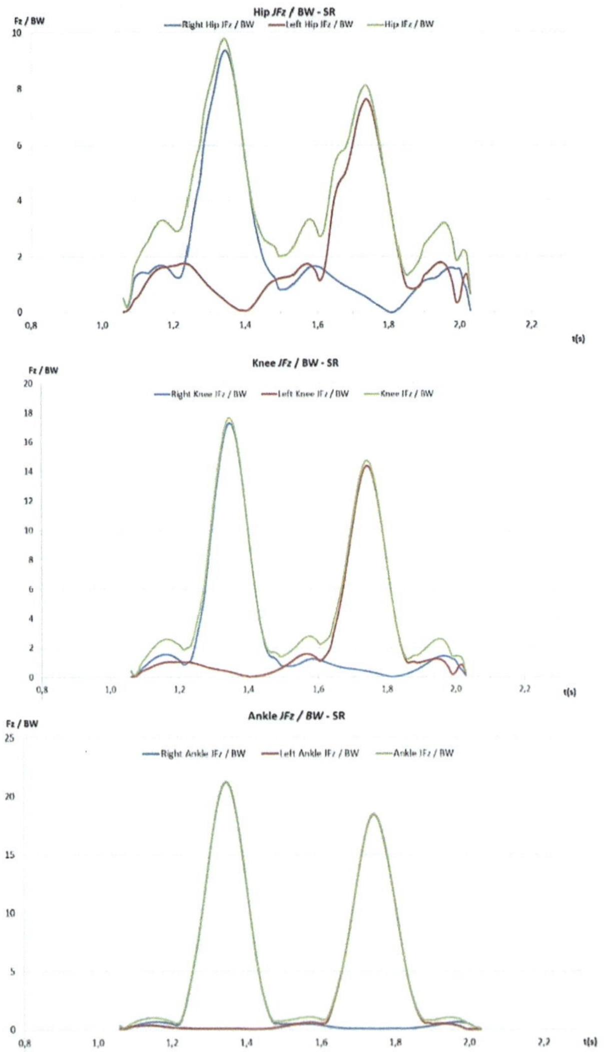

Fig. 5. Hip, knee, and ankle joint vertical forces normalized to body weight (JFz/BW) at the right (blue), left (red) and both (green) sides during slow running (SR) gait cycle.

Table 3. Mean values of *JFz/BW* at the hip, knee, ankle right (R), left (L) and both (B) sides.

JFz/BW	Hip			Knee			Ankle		
Mean	Right	Left	Both	Right	Left	Both	Right	Left	Both
NG	1.63	1.44	3.07	1.65	1.56	3.22	1.70	1.58	3.27
SKG	1.44	1.62	3.06	1.44	1.58	3.02	1.58	1.61	3.19
SR	2.12	1.92	4.04	2.79	2.43	5.22	3.00	2.67	5.67

Table 4. Standard deviation *JFz/BW* hip, knee, ankle right (R), left (L) and both (B) sides.

JFz/BW	Hip			Knee			Ankle		
Std. dev	Right	Left	Both	Right	Left	Both	Right	Left	Both
NG	1.77	1.45	2.02	1.82	1.67	2.11	2.38	2.28	2.73
SKG	1.39	1.48	1.63	1.64	1.54	1.84	2.30	2.16	2.57
SR	2.40	1.93	2.40	4.51	3.75	4.98	5.86	5.12	6.72

Table 5. p-value significance of the mean hip, knee, ankle *JFz/BW* comparison at right (R), left (L) and both (B) sides.

JFz/BW	Hip - Knee			Knee - Ankle			Hip - Ankle		
p	Right	Left	Both	Right	Left	Both	Right	Left	Both
NG	0.319	$<10^{-3}$	$<10^{-3}$	0.631	0.874	0.670	0.369	0.148	0.104
SKG	0.989	0.097	0.432	0.020	0.729	0.082	0.148	0.865	0.308
SR	0.003	0.012	$<10^{-3}$	0.134	0.104	0.020	0.015	0.015	$<10^{-3}$

Comparing at each side on each joint of different gait modes, the right, left and jointly both sides of the hip, the knee, and the ankle presented similar *JFz/BW* average gait cycles at NG and SKG without ssd ($p \geq 0.05$), and dominant ssd ($p < 0.05$) between NG-SR and SKG-SR, with exception at the right hip and left ankle on NG-SR, as well as the left hip and left ankle on SKG-SR, Table 6, and Fig. 6.

The hip and the knee presented at the right-side and at both sides jointly considered statistically different variances ($p < 0.05$) of *JFz/BW* at NG in relation to SKG, whereas the left side at the hip and the knee presented no statistically significant differences ($p \geq 0.05$) at NG in relation to SKG, with the same at the ankle for the right and left sides, separately and jointly considered ($p \geq 0.05$), Tables 4 and 7.

As regards to the comparison of *JFz/BW* variance at NG and SR, the hip presented no statistically significant differences ($p \geq 0.05$), with the knee and the ankle presenting at the right, left, and both sides jointly considered statistically significative different variances ($p < 0.05$), Tables 4 and 7. Concerning the comparison of the *JFz/BW* variance at

Table 6. p-value significance of the mean hip, knee, ankle *JFz/BW* comparison at NG, SKG and SR.

JFz/BW	NG - SKG			NG - SR			SKG - SR		
p	Hip	Knee	Ankle	Hip	Knee	Ankle	Hip	Knee	Ankle
Right	0.313	0.283	0.686	0.068	0.021	0.039	0.012	0.006	0.025
Left	0.290	0.926	0.911	0.028	0.035	0.051	0.169	0.036	0.055
Both	0.972	0.387	0.791	$<10^{-3}$	$<10^{-3}$	0.001	$<10^{-3}$	$<10^{-3}$	$<10^{-3}$

SKG and SR, only the left side of the hip presented no statistically significant differences ($p \geq 0.05$), with the remaining ssd ($p < 0.05$), Tables 4 and 7.

Table 7. p-value significance of the hip, knee, ankle *JFz/BW* variance comparison at NG, SKG and SR.

JFz/BW	NG - SKG			NG - SR			SKG - SR		
p	Hip	Knee	Ankle	Hip	Knee	Ankle	Hip	Knee	Ankle
Right	0.002	0.023	0.321	0.185	$<10^{-3}$	$<10^{-3}$	$<10^{-3}$	$<10^{-3}$	$<10^{-3}$
Left	0.569	0.263	0.437	0.103	$<10^{-3}$	$<10^{-3}$	0.208	$<10^{-3}$	$<10^{-3}$
Both	$<10^{-3}$	0.006	0.834	0.076	$<10^{-3}$	$<10^{-3}$	$<10^{-3}$	$<10^{-3}$	$<10^{-3}$

NG presented at the hip, the knee, and the ankle higher *JFz/BW* peaks than SKG, with exception of the hip left side and the right knee presenting higher *JFz/BW* peaks on SKG in relation to NG, Table 8, pointing for lateral asymmetry of *JFz/BW* peaks at the hip and the knee, with higher peaks at the ankle in relation to the hip and the knee on NG and SKG. Also, SR presented at the hip, the knee, and the ankle higher *JFz/BW* peaks than NG and SKG.

Table 8. Maximum *JFz/BW* at the hip, knee, ankle right (R), left (L) and both (B) sides.

JFz/BW	Hip			Knee			Ankle		
Max	Right	Left	Both	Right	Left	Both	Right	Left	Both
NG	6.39	5.54	7.26	6.20	5.52	7.28	8.91	8.45	9.65
SKG	5.24	5.64	6.21	6.53	5.51	7.11	8.80	7.42	9.27
SR	9.38	7.61	9.79	17.26	14.32	17.64	21.10	18.37	21.20

The hip, the knee and the ankle presented higher maximum *JFz/BW* at the right than the left side at NG and SR, as well as the knee and the ankle at SKG, with the hip presenting at SKG higher maximum *JFz/BW* at left than the right side, Table 8. The

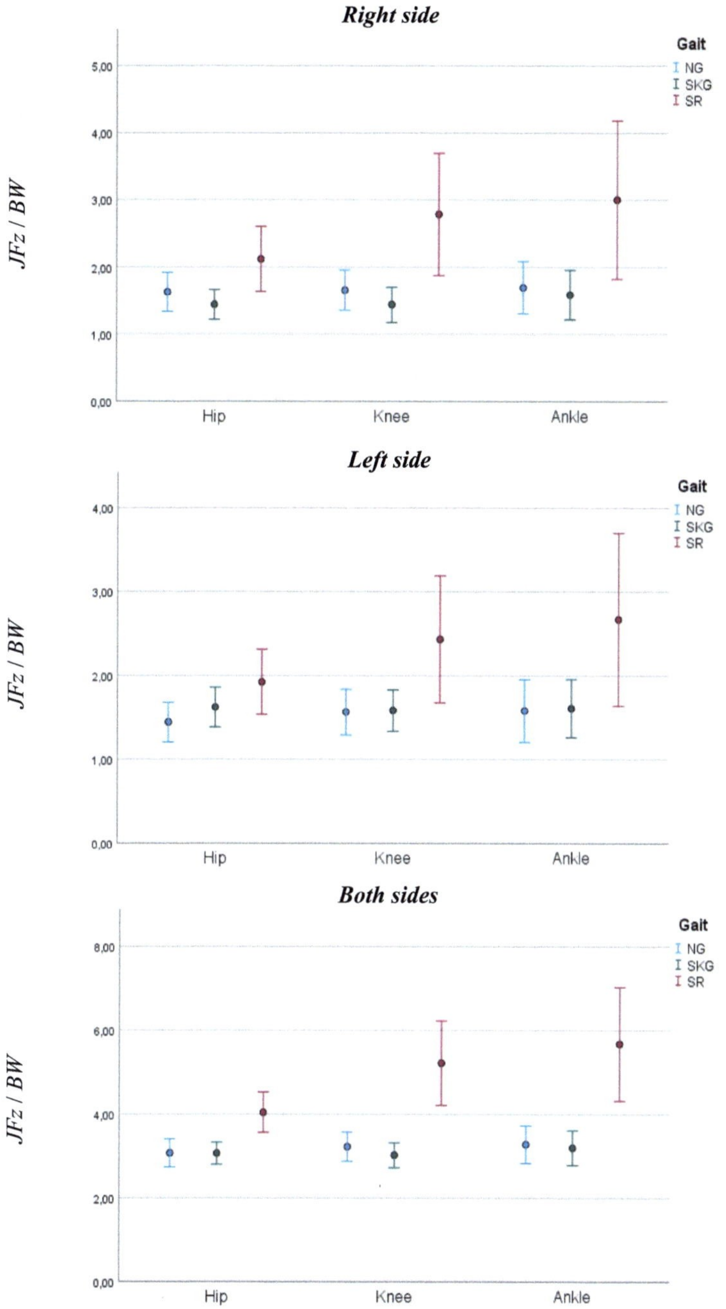

Fig. 6. Right, left and both sides error bars 95% confidence interval of the hip, knee and ankle joint vertical forces normalized to body weight (JFz/BW) during NG, SKG and SR gait cycle.

right and left hip, knee and ankle joints presented on NG, SKG and SR lower maximum *JFz/BW* than the resultant corresponding values of both sides jointly considered.

The hip presented at the right and both sides jointly considered higher maximum *JFz/BW* at NG than SKG, with the opposite at the left side and higher values at SKG than NG, Table 8. The knee presented at the right side higher maximum *JFz/BW* at SKG than NG, with the opposite at the left and both sides jointly considered higher maximum of *JFz/BW* at NG than SKG. The ankle presented at the right, left and the resultant on both sides higher maximum *JFz/BW* at NG than SKG.

3.2 Direct Measurements on Instrumented Implants

Direct measurements on instrumented implant of vertical forces normalized to body weight (*JFz/BW*) presented at the KWR patient right hip during T1 walking free with slow velocity, Fig. 7, similar gait cycle time profile to the corresponding computed right hip MSK-MS on NG, Fig. 3, with increasing value to a first local maximum followed by an absolute maximum and decrease with minimum value during contralateral limb support. The same was registered on direct measurements of vertical forces normalized to body weight (*JFz/BW*) for the left knee of K1L patient instrumented implant during T2 level walking, Fig. 7, with similar gait cycle time profile on NG corresponding left knee MSK-MS, Fig. 3, and increasing value from absolute minimum to a local maximum, followed by an absolute maximum and decrease intensity with gait periodicity.

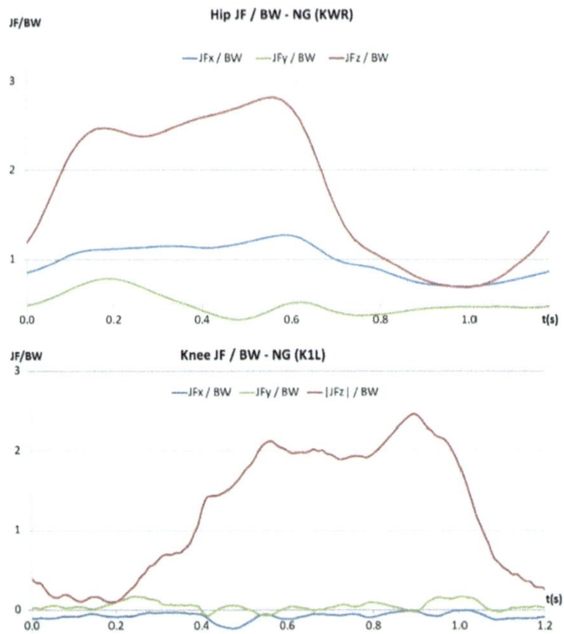

Fig. 7. Gait cycle hip and knee joint force components (*JFx* - anteroposterior; *JFy* - mediolateral; *JFz* - vertical) normalized to body weight (BW) on trials T1–T2 associated to NG.

The right hip instrumented implant presented for KWR patient direct measurements of the vertical forces normalized to body weight (*JFz*/*BW*) during T3 walking supported by crutches in both hands, Fig. 8, similar *JFz*/*BW* gait cycle time profile to the associated right hip MSK-MS on SKG, Fig. 4, with ripple corresponding to stiff-knee on SKG and crutches support at KWR, preceding maximum value. Also, the left knee direct measurements on instrumented implant, presented for K2L patient during T4 level walking with three-point support, Fig. 8, similar *JFz*/*BW* gait cycle time profile to the SKG left knee MSK-MS, Fig. 4, with an absolute and a local maximum at K2L by the opposite order at MSK-MS.

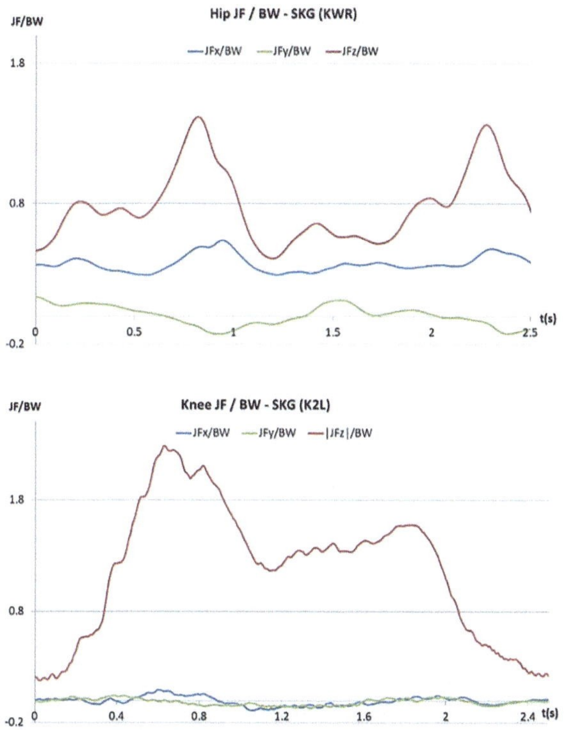

Fig. 8. Gait cycle hip and knee joint force components (*JFx* - anteroposterior; *JFy* - mediolateral; *JFz* - vertical) normalized to body weight (BW) on trials T3-T4 associated to SKG.

Direct measurements of the vertical forces normalized to body weight *JFz*/*BW* at the right hip instrumented implant presented for EBL patient on T5 walking on a treadmill at 8 km/h, Fig. 9, similar gait cycle time profile to the associated right hip MSK-MS on SR, Fig. 5, with corresponding sharp peak and minor subsequent plateau on EBL T5 and preceding ripple on MSK-MS SR. As regards to direct measurements of the *JFz*/*BW* at K1L left knee during T6 walking on treadmill at 4 km/h, Fig. 9, gait cycle time profile presents different two successive maximum, whereas the left knee *JFz*/*BW* presents on left knee MSK-MS a single sharp peak, Fig. 5, and this differences likely related with different subject and task assessment, along with MSK-MS difficult prediction on SR lower limb joint internal loads.

Distribution of the hip and the knee joint internal loads direct measurements from instrumented implants vertical forces normalized to body weight during NG, SKG and SR gait cycle are represented by the corresponding box plot at Fig. 10, with location and dispersion measurements assessed by the gait cycle mean, maximum and standard deviation presented at Table 9.

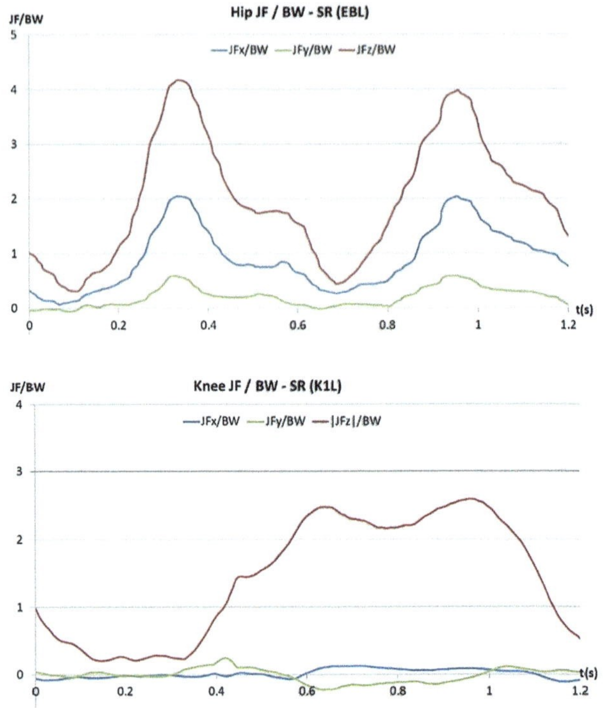

Fig. 9. Gait cycle hip and knee joint force components (*JFx* - anteroposterior; *JFy* - mediolateral; *JFz* - vertical) normalized to body weight (BW) on trials T5–T6 associated to SR.

Direct measurements from instrumented joint implants normalized to body weight *JFz/BW* presented at the hip and the knee higher peak values at the trials associated to SR than at NG, both higher than SKG, with the knee presenting higher *JFz/BW* peak than the hip at NG, SKG and opposite higher *JFz/BW* peak at the hip than the knee on SR, Table 9 and Fig. 10.

Table 9. Location and dispersion of *JFz/BW* at the hip and knee instrumented implants.

JFz/BW	Hip NG (KWR)	Knee NG (K1L)	Hip SKG (KWR)	Knee SKG (K2L)	Hip SR (EBL)	Knee SR (K1L)
Mean	1.28	1.19	0.70	1.20	1.95	1.45
St-dev	0.78	0.79	0.29	0.59	1.09	0.87
Max	2.31	2.45	1.46	2.28	4.46	2.67

The hip and the knee presented at NG selected trials similar *JFz/BW* mean and variances without statistical significative differences ($p \geq 0.05$) and statistical significative differences ($p < 0.05$) at SKG and SR, Table 10.

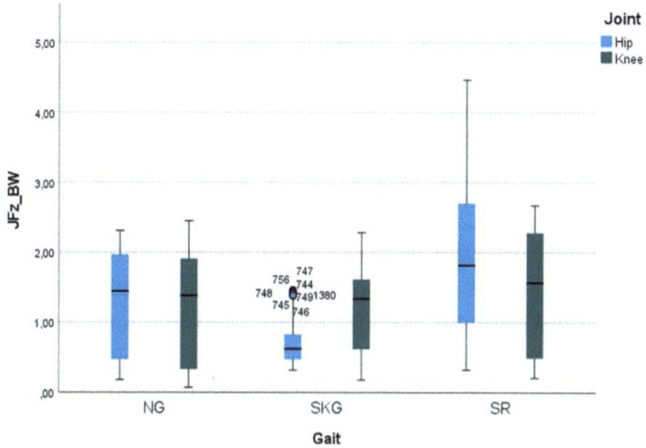

Fig. 10. Hip and knee box plot of direct measurements from vertical joint forces normalized to body weight (JFz/BW) during NG, SKG and SR gait cycle.

Table 10. p-value significance of the hip and knee *JFz/BW* mean and variance comparison on instrumented implants at similar gait mode.

JFz/BW *p*	Hip NG - Knee NG (KWR) (K1L)	Hip SKG - Knee SKG (KWR) (K2L)	Hip SR - Knee SR (EBL) (K1L)
Mean	0.185	$<10^{-3}$	$<10^{-3}$
Variance	0.914	$<10^{-3}$	$<10^{-3}$

The hip presented at different gait trials statistical significative differences ($p < 0.05$) of mean *JFz/BW* and variances at NG, SGK and SR, Table 11, without statistical significative differences of the knee mean *JFz/BW* between NG-SKG ($p \geq 0.05$) and statistical significative differences on NG-SR and SKG-SR ($p < 0.05$), as well as variance statistical significative differences ($p < 0.05$) at NG, SK and SR, Table 12.

Table 11. p-value significance of the hip *JFz/BW* mean and variance comparison on instrumented implants at different gait mode.

JFz/BW / *p*	Hip NG - Hip SKG (KWR) (KWR)	Hip NG - Hip SR (KWR) (K1L)	Hip SKG - Hip SR (KWR) (K1L)
Mean	$<10^{-3}$	$<10^{-3}$	$<10^{-3}$
Variance	$<10^{-3}$	$<10^{-3}$	$<10^{-3}$

Table 12. p-value significance of the knee *JFz/BW* mean and variance comparison on instrumented implants at different gait mode.

JFz/BW / *p*	Knee NG - Knee SKG (K1L) (K2L)	Knee NG - Knee SR (K1L) (K1L)	Knee SKG - Knee SR (K2L) (K1L)
Mean	0.802	$<10^{-3}$	$<10^{-3}$
Variance	$<10^{-3}$	0.010	$<10^{-3}$

3.3 Musculoskeletal Modeling and Direct Measurements

Musculoskeletal model vertical joint forces normalized to body weight *JFz/BW* presented at the right and left sides of the hip and knees higher peak values in relation to the corresponding direct measurements with instrumented implants during NG, SKG and SR, Tables 8–9 and Fig. 11.

Direct measurements of the vertical joint forces normalized to body weight *JFz/BW* presented at the right and left sides of the hip and knees lower variance than the corresponding musculoskeletal model estimation on NG, SKG and SR, with statistical significative differences ($p < 0.05$), Fig. 11 and Tables 13–15.

Also mean vertical joint forces normalized to body weight *JFz/BW* presented at direct measurements lower values than their corresponding estimated by musculoskeletal model, with statistical significative differences ($p < 0.05$), Fig. 11 and Tables 13–15. Nevertheless, some exceptions were detected without statistical significative differences ($p \geq 0.05$) between direct measurements and corresponding MSK-MS estimation of meant *JFz/BW*, at the left hip on NG, Table13, the right knee on SKG, Table 14, and the right and left hip on SR, Table 15.

Direct measurements presented also lower peak values of *JFz/BW* at the hip and the knee comparatively to the corresponding musculoskeletal model simulation values, Tables 8–9 and Fig. 11.

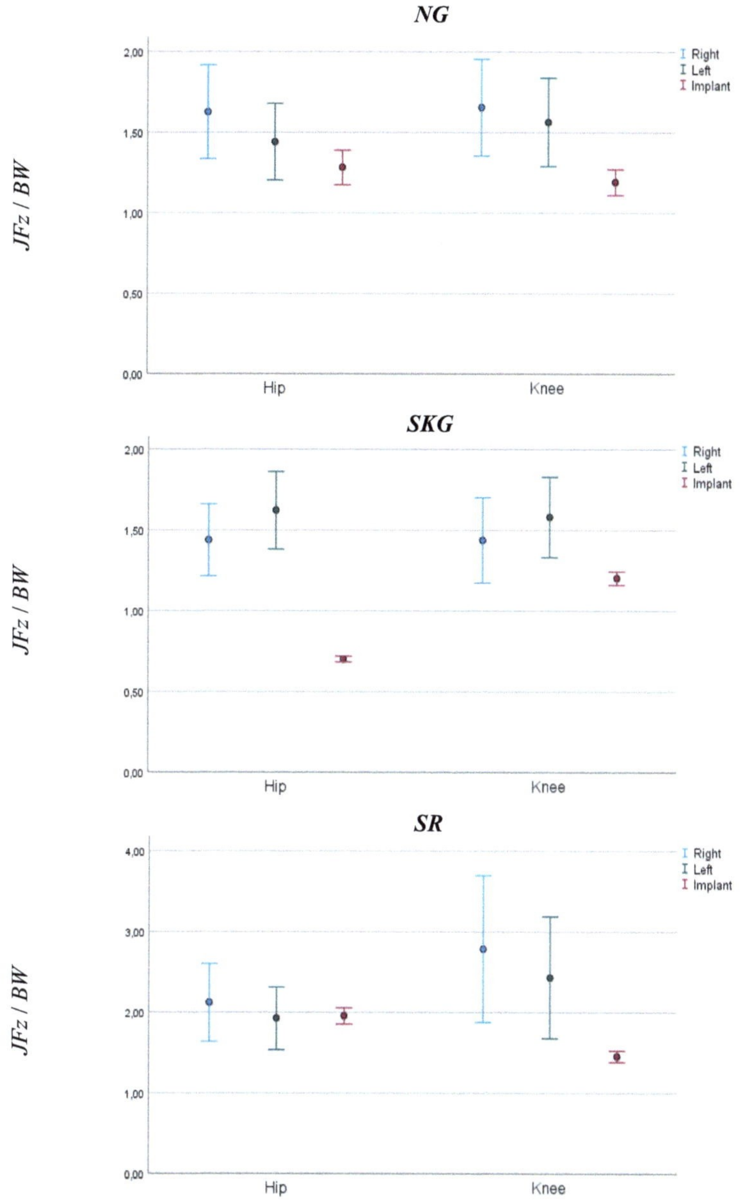

Fig. 11. Vertical joint forces error bars 95% confidence interval normalized to body weight (*JFz/BW*) at musculoskeletal model simulation right/left hip, knee, and instrumented implant.

Table 13. p-value significance of the hip *JFz/BW* mean and variance comparison of musculoskeletal modeling instrumented implants at the hip and the knee on NG.

JFz/BW *p*	Hip NG (Right) - (Implant)	Hip NG (Left) - (Implant)	Knee NG (Right) - (Implant)	Knee NG (Left) - (Implant)
Mean	0.030	0.229	0.003	$<10^{-3}$
Variance	$<10^{-3}$	$<10^{-3}$	$<10^{-3}$	$<10^{-3}$

Table 14. p-value significance of the hip *JFz/BW* mean and variance comparison of musculoskeletal modeling instrumented implants at the hip and the knee on SKG.

JFz/BW *p*	Hip SKG (Right) - (Implant)	Hip SKG (Left) - (Implant)	Knee SKG (Right) - (Implant)	Knee SKG (Left) - (Implant)
Mean	$<10^{-3}$	$<10^{-3}$	0.084	0.003
Variance	$<10^{-3}$	$<10^{-3}$	$<10^{-3}$	$<10^{-3}$

Table 15. p-value significance of the hip *JFz/BW* mean and variance comparison of musculoskeletal modeling instrumented implants at the hip and the knee on SR.

JFz/BW *p*	Hip SR (Right) - (Implant)	Hip SR (Left) - (Implant)	Knee SR (Right) - (Implant)	Knee SR (Left) - (Implant)
Mean	0.505	0.882	0.005	0.012
Variance	$<10^{-3}$	$<10^{-3}$	$<10^{-3}$	$<10^{-3}$

4 Discussion

Subject specific MSK-MS allowed estimation of lower limb joint internal loads at NG, SKG and SR, namely the hip, the knee and the ankle *JFz/BW* gait cycle time profile and corresponding location and dispersion measurements. The hip, the knee and the ankle presented similar mean *JFz/BW* at the right and left sides on NG, SKG and SR, with differences of individual sides to both sides resultant load and dominant differences between the hip, the knee and the ankle joints at both sides jointly considered with higher mean differences at each joint between NG-SR and SKG-SR in relation to NG-SKG. As regards to the variance of MSK-MS *JFz/BW*, dominant statistical significative differences ($p < 0.05$) were detected at the right, left and both joints between NG, SKG, SR apart from the ankle on NG-SKG and the hip on NG-SR as well as the left hip, knee on NG-SKG and the left hip on SKG-SR pointing at these cases for similar variances. Despite mean and variance featured location and dispersion of *JFz/BW* underlying distribution, maximum values play a determinant role on joint loads with implication on joint physical integrity. In this regard the hip, the knee and the ankle joints presented dominant higher maximum *JFz/BW* values at NG than SKG with exception of the left hip and right knee, pointing thus SKG as protection mechanism reducing *JFz/BW* maximum in relation to NG. Also, SR imposed at both sides of the hip, the knee and the ankle, higher maximum *JFz/BW* values in relation to NG and SKG, suggesting that SR must be adopted with caution if joint physical integrity needs to be preserved. In particular, the ankle joint must be carefully preserved since it presents for both sides higher maximum values of *JFz/BW* than the hip and the knee on NG, SKG and SR. MSK-MS joint loads presented also clear asymmetry expressed by higher maximum *JFz/BW* at the right side in relation to the left side for the hip, the knee and the ankle on NG, SKG and SR, apart from the hip at SKG with lower values at the right in relation to the left side, associated to reduced right knee flexion.

As regards to direct measurements from joint instrumented implants, whereas the trials associated to NG presented at the hip and the knee similar *JFz/BW* gait cycle mean and variance, the trials associated to SKG and SR presented at the hip and the knee statistical significative differences ($p < 0.05$) for *JFz/BW* gait cycle mean and variance. Additionally, the hip presented at NG, SKG and SR associated trials different *JFz/BW* gait cycle mean and variances, whereas the knee presented at NG and SKG similar gait cycle *JFz/BW* mean, with different mean at NG-SR, SKG-SR and different variances at NG, SKG and SR. The hip and the knee joints presented also clear distinction with higher *JFz/BW* gait cycle maximum on SR than NG both higher than SKG, and higher maximum *JFz/BW* at the knee in relation to the hip on NG, SKG with the opposite higher *JFz/BW* maximum at the hip than the knee on SR. Direct measurements from joint instrumented implants presented lower variances of *JFz/BW* at the hip and the knee in relation to the corresponding MSK-MS computed values, as well as dominant lower *JFz/BW* mean values apart from the left hip at NG, the right knee on SKG and the left and right hip on SR. Maximum direct measurements of *JFz/BW* gait cycle presented also lower values at the hip and the knee than the corresponding MSK-MS computed values at NG, SKG and SR, pointing for careful loading on instrumented implanted patient as a protection mechanism.

5 Conclusion

Subject specific MSK-MS proved adequate for estimation of comparable lower limb joint internal loads with direct measurements on post-operative instrumented implants. Nevertheless, more realistic models require specific subject data of internal structures, their static and dynamic mechanical properties. Also, MSK-MS require validation of computed values with direct measurement of subject specific internal loads. Whereas model information is generally available on healthy subjects, direct measurements were performed after intervention on post-operative patients at different subjects. Thus, MSK-MS results should be compared with the same subject direct measurement. Since patients are generally intervened with only one implant, direct load measurement from instrument implant comparison with the same subject MSK-MS computed loads can be performed only at the instrumented joint implant, with the need for different patients with different implants and corresponding MSK-MS for comparison of different joint loads. Also, if MSK-MS is intended for surgery planning, personalized models and load prediction must be performed on pre-operative specific subject, allowing prospective what-if scenarios and optimization analysis.

References

1. Bergmann, G., et al.: Hip contact forces and gait patterns from routine activities. J. Biomech. **34**(2001), 859–871 (2001)
2. Erdemir, A., McLean, S., Herzog, W., van den Bogert, A.J.: Model-based estimation of muscle forces exerted during movements. Clin. Biomech. **22**(2007), 131–154 (2007)
3. Lund, M.E., de Zee, M., Andersen, M.S., Rasmussen, J.: On validation of multibody musculoskeletal models. Proc. Inst. Mech. Eng. H J. Eng. Med. **226**(2), 82–94 (2012)
4. Horsman, M.D.K.: The Twente lower extremity model: consistent dynamic simulation of the human locomotor apparatus. PhD thesis University of Twente, Netherlands (2012)
5. Andersen, M.S., de Zee, M., Rasmussen, J.: Assignment for the Course in Musculoskeletal Modeling by Multibody Biomechanics, 2nd edn. Department of Mechanical and Manufacturing Engineering Aalborg University, Aalborg (2012)
6. Andersen, M.S., Damsgaard, M., Rasmussen, J.: Kinematic analysis of over-determinate biomechanical systems. Comput. Methods Biomech. Biomed. Engin. **12**(4), 371–384 (2009)
7. Andersen, M.S., Damsgaard, M., MacWilliams, B., Rasmussen, J.: A computationally efficient optimization-based method for parameter identification of kinematically determinate and over-determinate biomechanical systems. Comput. Methods Biomech. Biomed. Engin. **13**(2), 171–183 (2010)
8. Bendat, J.S., Piersol, A.G.: Random Data Analysis and Measurement Procedures, 3rd edn. Wiley, New York (2000)
9. Bergmann, G.: OrthoLoad. Charité Universitaetsmedizin Berlin, Berlin (2008)
10. Bergmann, G. (ed.): Charité Universitaetsmedizin Berlin. http://www.OrthoLoad.com. Accessed 11 May 2012

An Anisotropic Hyperelastic Model for Human Skin: Finite Element Modeling, Identification of Parameters, Mechanical Tests

Wael Alliliche(✉), Christine Renaud, Jean-Michel Cros, and Zhi-Qiang Feng

Université Paris-Saclay, Univ Evry, LMEE, 91020 Evry, France
wael.alliliche@univ-evry.fr

Abstract. The skin is a living tissue that behaves in a hyperelastic anisotropic way. To model the skin, we propose an accurate, constitutive law called *HGO-Yeoh*. This model is implemented in a finite element code FER "Finite Element Research" [4] in order to benefit from its tools, including the bi-potential contact method: a method coupling contact and friction which is more efficient than traditional ones. Identification of parameters of the skin-related material is made with an optimization procedure. A tensile test is simulated using the numerical code FER and ANSYS. The results are compared with the experimental data. A simulation of an indentation test using a bipotential contact law is also performed.

Keywords: Biomechanics · Contact · Anisotropic hyperelasticity

1 Introduction

The skin is the organ that covers the human body to maintain its equilibrium; it protects it from external nuisances and attacks. So it is essential to study its mechanical behavior. To model the behavior of the skin, it is necessary to know its anatomy and structure, which is composed of three main layers: Epidermis, Dermis, Hypodermis. We consider only the dermis because it is the most crucial layer of the skin. It has a thickness between 0.6 and 4 mm, and it is mainly composed of collagen and elastin fibers. It contains a basic material called the matrix, which is less critical than fibers. Collagen fibers give the skin the ability to resist traction [7,12].

The skin behaves as a hyperelastic material like almost animal biomaterials. It is almost incompressible because it is mainly made of water [18]. Moreover, it is anisotropic due to the collagen fibers. Many constitutive laws are used to model skin and soft tissue in the literature. Some of them represent isotropic behavior, such as neo-Hookean [9], Mooney-Rivlin [14], 8-chain [1], Yeoh [19], and Gent [5] models. The main difference between them is the trend of the stress-strain curve. Other laws describe both the isotropic and anisotropic behaviors, for example, the HGO model [6].

J. M. R. S. Tavares et al. (Eds.): CMBBE 2021, LNCVB 38, pp. 271–280, 2023.
https://doi.org/10.1007/978-3-031-10015-4_23

In this paper, we present the new energy density energy of strain by coupling the isotropic *Yeoh* model with the anisotropic *HGO* model. Next, we introduce the bipotential method to model the contact between the indenter and the skin. After, we explain how to identify our model parameters. Finally, we compare tensile test simulation results on FER and ANSYS. We perform several 2D indentation simulations on the skin. We change the angle direction of the dominant fiber family each time, and we analyze the influence of the anisotropy on the skin behavior.

2 Skin Model Analysis

2.1 Used Invariants

For soft tissue in biomechanics, the materials are nonlinear, hyperelastic, and almost incompressible. To take into account the anisotropy of the material, we choose a single direction (**a**) of fibers whose components are:

$$\mathbf{a} = \{\cos(\alpha), 0, \sin(\alpha)\} \tag{1}$$

α is the angle between the collagen fibers and the longitudinal axis of the skin $\vec{x}$ (Fig. 1).

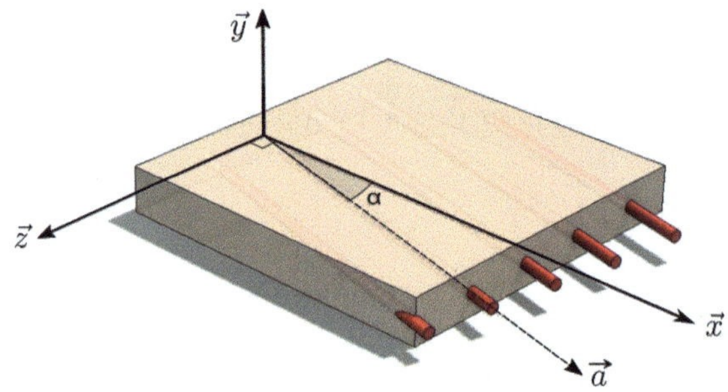

Fig. 1. Direction of the fibers according to the angle α.

We note **M** as the orientation matrix of the fibers.

$$\mathbf{M}_{ij} = (\mathbf{a} \otimes \mathbf{a})_{ij} \tag{2}$$

The transformation tensor **F** and the Cauchy-Green strain **C** are defined as:

$$\mathbf{F} = \mathbf{I} + \frac{\partial \mathbf{u}}{\partial \mathbf{X}}, \quad \mathbf{C} = \mathbf{F}^T \mathbf{F} \tag{3}$$

where **u** is the displacement vector and **I** is the identity matrix. We use the tensor **C** (3) and the matrix **M** (2) to calculate the invariants:

$$I_1 = \mathrm{Tr}(\mathbf{C}), \quad I_3 = \det(\mathbf{C}), \quad I_4 = \mathrm{Tr}(\mathbf{C}\,\mathbf{M}), \quad J = \det(\mathbf{F}) = I_3^{\frac{1}{2}} \tag{4}$$

We use augmented invariants of I_1 and I_4 in our model, to take into account the variation of compression by introducing I_3. J represents the constraint of compression, and for isochoric deformation $J \approx 1$ [10].

$$\bar{I_1} = I_1\, I_3^{-\frac{1}{3}}, \quad \bar{I_4} = I_4\, I_3^{-\frac{1}{3}} \tag{5}$$

2.2 HGO-Yeoh Model

The anisotropy of the dermis is due to the collagen fibers, and the isotropy is due to the matrix. So, the behavior law W is a sum of three deformation energies: isotropic W_{iso}, anisotropic W_{aniso} and volumetric W_{v}.

$$W = W_{\text{iso}} + W_{\text{aniso}} + W_{\text{v}} \tag{6}$$

To build the model, we start from HGO model whose isotropic part is the neo-Hookean model [$W_{iso} = a(I_1 - 3)$]. We replace it by the Yeoh model (7). This modification allows us to obtain relevant results during the identification in Sect. 3. To expand on each term of the Eq. (6) :

$$\begin{aligned} W_{\text{iso}}(\bar{I_1}) &= \textstyle\sum_{i=1}^{3} a_i(\bar{I_1} - 3)^i \\ W_{\text{aniso}}(\bar{I_4}) &= \tfrac{c_1}{2\,c_2}(e^{c_2\,(\bar{I_4}-1)^2} - 1) \\ W_{\text{v}}(J) &= \tfrac{1}{d}(J-1)^2 \end{aligned} \tag{7}$$

The constants a_1, a_2, a_3, c_1, c_2 are the material parameters to be defined, d is an expansion parameter that is chosen very small ($d = 10^{-6}$) in incompressible cases, it models numerically the hydrostatic pressure.

In our constitutive model, the fibers only work in tensile. Therefore, there is no effect on the compression, so the part W_{aniso} is under these conditions :

$$\begin{cases} \text{if} \quad I_4 < 1 \quad W_{\text{aniso}}(\bar{I_4}) = 0 & \textit{Compression} \\ \text{if} \quad I_4 \geq 1 \quad W_{\text{aniso}}(\bar{I_4}) = \tfrac{c_1}{2\,c_2}(e^{c_2\,(\bar{I_4}-1)^2} - 1) & \textit{Tensile} \end{cases} \tag{8}$$

2.3 Implementation

For the implementation, we calculate the second Piola-Kirchhoff stress tensor $\mathbf{S}$ [11] by deriving the elastic potential W (6) with respect to the strain tensor $\mathbf{C}$ (3):

$$\mathbf{S} = 2\,\frac{\partial W}{\partial \mathbf{C}} \tag{9}$$

Then, we develop the calculus by using the partial derivative of $\mathbf{S}$ (9):

$$\mathbf{S} = 2\left(\frac{\partial W}{\partial I_1}\frac{\partial I_1}{\partial \mathbf{C}} + \frac{\partial W}{\partial I_3}\frac{\partial I_3}{\partial \mathbf{C}} + \frac{\partial W}{\partial I_4}\frac{\partial I_4}{\partial \mathbf{C}} + \frac{\partial W}{\partial J}\frac{\partial J}{\partial \mathbf{C}}\right) \tag{10}$$

Finally, we deduce the Cauchy stress tensor $\boldsymbol{\sigma}$ from $\mathbf{S}$ (9) and transformation tensor $\mathbf{F}$ (3):

$$\boldsymbol{\sigma} = \tfrac{1}{J}\,\mathbf{F}\,\mathbf{S}\,\mathbf{F}^T \tag{11}$$

$\mathbb{D}$ is the elastic constitutive tensor that is used to compute the tangent matrix $\mathbf{K}_T$. This tensor derives from $\mathbf{S}$ [15].

$$\mathbb{D} = 4\,\frac{\partial^2 W}{\partial \mathbf{C}^2} \tag{12}$$

2.4 The Bipotential (Contact, Friction) Method

We use the bipotential method to simulate a skin indentation in FER. This law considers two potentials; the first is associated with Signorini conditions and Coulomb's law. These are coupled in a single non-differentiated potential called bipotential [3]. This method is more efficient because it takes into account the coupling of the normal force (contact) and the tangential force (friction) in the same formulation, and it does not use the penalty term that induces interpenetration [13].

According to the contact reaction $\mathbf{R}$ given by this method, the internal forces $\mathbf{F}_{\text{int}}$ and the external forces $\mathbf{F}_{\text{ext}}$ stade by the problem, we can write the static equation as follows:

$$\mathbf{F}_{\text{int}} - \mathbf{F}_{\text{ext}} - \mathbf{R} = 0 \tag{13}$$

All these methods are implemented within the finite element code *FER*, developed by the LMEE *"Laboratoire de Mécanique et d'Énergétique d'Évry"*. The nonlinear geometric analysis can be described using the total Lagrangian formulation, where we build the tangent stiffness matrix using the original configuration, which stays constant [2]. A Newton-Raphson iterative procedure is applied for nonlinear analysis to resolve this kind of problem [16].

2.5 Parameters Identification

In this paragraph, the procedure used to determine the material parameters of the *HGO-Yeoh* model (7) from experimental data is explained. To achieve this goal we rely on experimental data ($f^{\text{exp}} = \sigma_{xx}^{\text{exp}}(\varepsilon_{xx})$) from a uniaxial tensile test performed on a skin sample [17]. We also use an analytic development of the Cauchy stress ($f^{\text{th}} = \sigma_{xx}^{\text{th}}(\varepsilon_{xx})$) using (11), we can write it as:

$$\sigma_{xx}^{\text{th}} = \frac{2F_{11}^2}{J}\left[\left(\frac{-I_1\, I_3^{-\frac{1}{3}}\, C_{11}^{-1}}{3} + I_3^{-\frac{1}{3}}\right)\left(a_1 + 2\,a_2\,(I_1\, I_3^{\frac{1}{3}} - 3) + 3\,a_3(I_1\, I_3^{\frac{1}{3}} - 3)^2\right)\right.$$
$$\left. + c_1\left(\frac{1}{3}\, I_4\, I_3^{-\frac{4}{3}} + I_3^{-1/3}\cos^2(\alpha)\right)\left((I_4\, I_3^{-\frac{1}{3}} - 1)\, e^{c_2 (I_4\, I_3^{-\frac{1}{3}} - 1)^2}\right)\right] \tag{14}$$

We consider the material constants a_1, a_2, a_3, c_1, c_2 ordered in a vector as the variables of the function (15).

$$f = \sigma_{xx}^{\text{th}}(a_1, a_2, a_3, c_1, c_2) \tag{15}$$

To obtain these parameters, we proceed by an optimization in order to find a minimal value of the least square error Φ [8]. The Matlab function "fmincon" is used.

$$\Phi = \sum_{i=1}^{n} (\sigma_{xx}^{\exp}(\varepsilon_i) - \sigma_{xx}^{\text{th}}(\varepsilon_i))^2 \qquad (16)$$

3 Results

The optimization algorithm converges after 187 iterations. The minimization of (16) gives us the following material parameters related to the skin:
$a_1 = 269.125$ kPa, $a_2 = 376.065$ kPa, $a_3 = 415.963$ kPa, $c_1 = -317.012$ kPa, $c_2 = 0.319$. Figure 2 represents the experimental and the analytic (15) curves using these results.

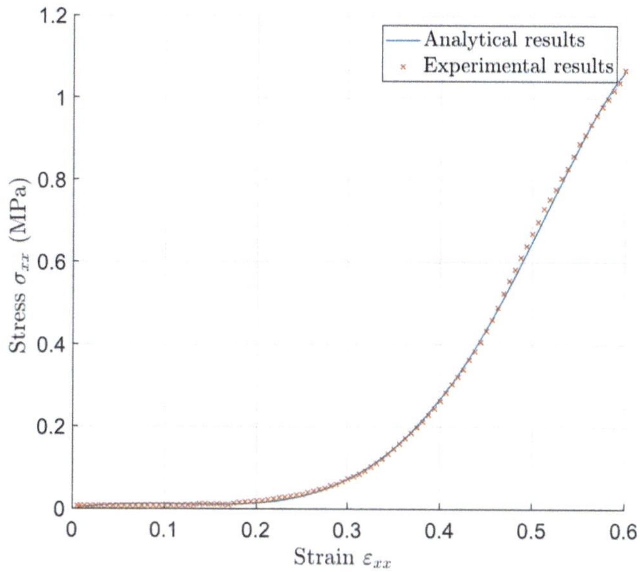

Fig. 2. Coincidence of the experimental curve with the analytical curve.

3.1 Uniaxial Tensile Test on the Skin

In this part, we perform a simulation of the experimental trial extracted from [17]. Only 1/8 of the skin sample is discretized. Symmetry conditions are applied, and a displacement $u_{xx} = 18$ mm is imposed along with the $\vec{x}$ axis. This test is simulated with FER and ANSYS with an 8-node hexahedral element. We can draw the curves, σ_{xx}^{FER} and $\sigma_{xx}^{\text{ANSYS}}$ (Fig. 3), according to the deformation ε_{xx}, we can then compare the obtained curves with the experimental one. Figure 4 shows us the stress at different strains.

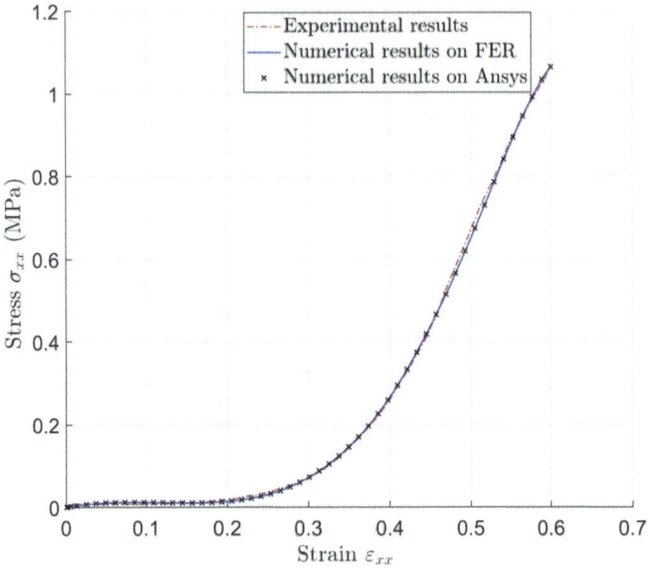

Fig. 3. Comparison of experimental results with the FER and ANSYS results.

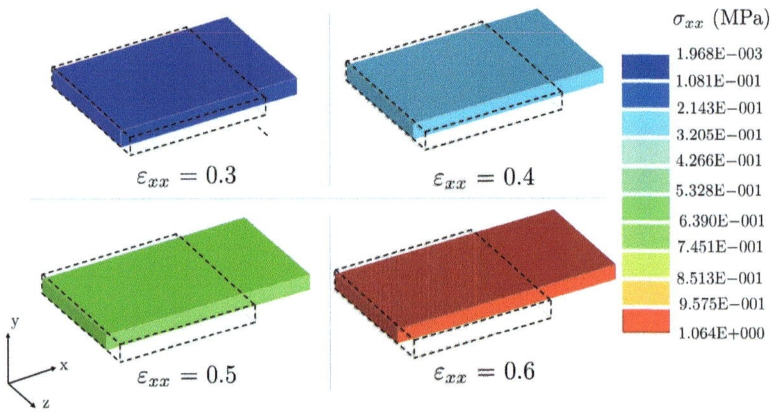

Fig. 4. Stress and deformation of the skin simulated with FER.

3.2 Indentation Test on the Skin

For this test we apply an indentation of a rigid circular cylinder into the human skin. The radius of the cylinder is 2.5 mm while the dimension of the deformed body is 14 mm × 4 mm. For discritization we use 1798 four-node isoparametric plane strain elements and 1903 nodes. Twenty load steps are applied to the cylinder by imposing a vertical displacement of 50 µm in each increment. Zhang [20] could identify the coefficients of Coulomb friction μ of the skin, considering the indenter made of aluminum. We also assume that the abdomen has the same friction properties as the anterior legs. Therefore, $\mu \approx 0.34$. We refined the contact zone with thin elements to have more details about the contact conditions. Figure 5 shows the distribution of the von Mises stress for different fiber orientations. Finally, we can give the evolution of stress with respect to the orientation angle for different expansion parameter values "d" (Fig. 6).

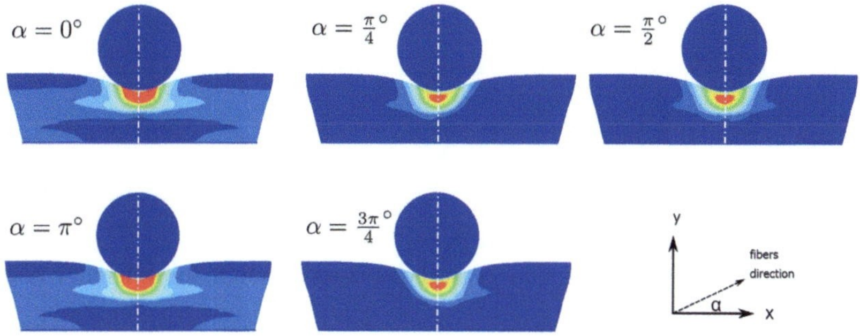

Fig. 5. von Mises stress distribution at different angles.

4 Discussion

From the optimization results (Fig. 2), we can obtain the error between analytic data and the experimental ones, and we see that the maximum value is less than 2.5% (Fig. 7). This error occurs in areas where the experimental curve has critical points that do not perfectly follow the trend. After the simulation of our sample with ANSYS and FER, we notice that for both simulations, the numerical result curves are in agreement with the experimental ones and are in perfect coincidence with the analytic curves (Fig. 3). We noticed in ANSYS 2019 R2 that the compression was poorly managed. The program took into account the compression of the fibers contrary to their natural functioning.

Then for indentation simulation, since our contact method is interpenetration free, the results are more realistic than those obtained with commercial codes such as ANSYS and ABAQUS. From this simulation, we notice that the stress concentration is generally slightly concentrated towards the indenter (Fig. 5). It is important to note that the stress concentration is located inside the body rather than at the contact point. We performed several simulations by changing the orientation angle of the fibers in each test. For the angles 0 and π, we obtain precisely the same results because the orientation vector has only a non-zero component on $\vec{x}$. Moreover, we can easily see the influence of anisotropy on stress distribution. The fiber distribution is symmetrical to the axis normal to the contact surface. For the angle $\frac{\pi}{4}$, the stress distribution is asymmetrical with a higher concentration on the left side. The results obtained with the $\frac{3\pi}{4}$ angle are

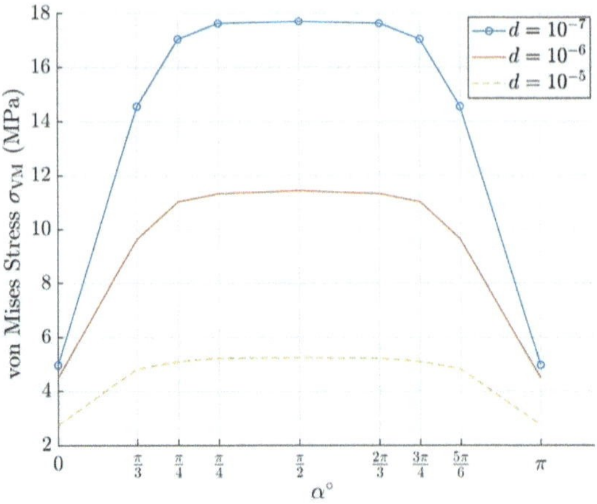

Fig. 6. Evolution of stress with respect to the orientation angle for different expansion parameter values

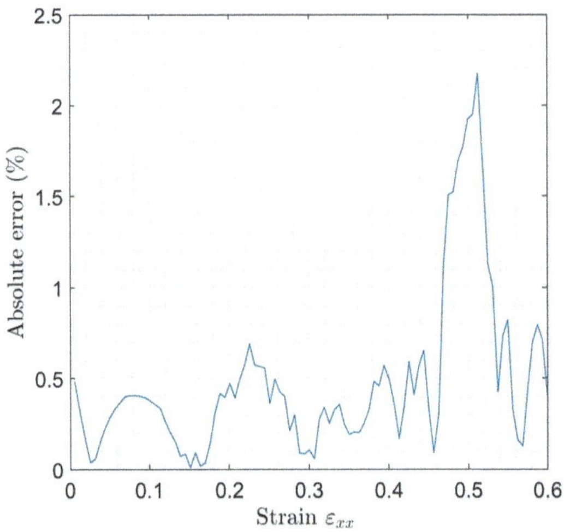

Fig. 7. The absolute error between the experimental and the analytical curves.

symmetric to the results for the $\frac{\pi}{4}$ angle, this time with a greater concentration of stress on the right side. The stress concentration for the $\frac{\pi}{2}$ angle is symmetric. Indeed, the fibers almost do not work in compression. We plot the maximum stress as a function of the angles varying between 0 and π (Fig. 6). The curve looks like a symmetrical Gaussian function with a maximum in $\frac{\pi}{2}$. We also compared these results as a function of the expansion parameter d. The compression evolves conversely to this parameter. This parameter represents the influence of hydrostatic pressure numerically. Therefore, we notice that the more this coefficient decreases, the stronger the stress are because the hydrostatic pressure is the biggest. We want to point out that the coefficient of the anisotropic term is negative, contrary to the constants related to the isotropic terms. It means that the more the fibers are solicited, the more the stress decreases. Biomechanically, this describes the following working mechanism: when the skin matrix is in tension, the fibers work in compression to reduce the stress on the skin.

5 Conclusion

In this work, we have succeeded in identifying the parameters used in the *HGO-Yeoh* model to simulate mechanical actions on the skin. In nature, the skin differs between individuals according to age, ethnicity, gender, and health status. Our study aims to show that our model is usable for modeling human skin. If we wanted to go further, we just had to take other experimental data from other specimens. The identification process of the *HGO-Yeoh* behavior law from experimental data in a uniaxial tensile test is validated. Nevertheless, it remains underlined that our modeling remains limited to small-scale examples. We have taken into account some simplifications; we have considered that the skin is made up of only one layer, whereas in reality, it must be modeled in a material stratified in three layers. This simplification is due to the difficulty of carrying out mechanical tests that consider these different layers. The variation of the direction of the fibers, giving the skin the property of anisotropy, presents an obstacle in large-scale skin modeling; we must use a very fine mesh with different directions of anisotropy with a local reference frame for each element.

References

1. Arruda, E.M., Boyce, M.C.: A three-dimensional constitutive model for the large stretch behavior of rubber elastic materials. J. Mech. Phys. Solids **41**(2), 389–412 (1993)
2. Crisfield, M.A.: Nonlinear Finite Element Analysis of Solids and Structures. Volume 1: Essentials. Wiley Professional Software (1991)
3. De Saxcé, G., Feng, Z.Q.: The bipotential method: a constructive approach to design the complete contact law with friction and improved numerical algorithms. Math. Comput. Model. **28**(4), 225–245 (1998)
4. Feng, Z.Q.: FER/View - an interactive finite element post-processor. In: Computational Mechanics WCCM VI in Conjunction with APCOM 2004, Beijing, 5–10 September (2004)
5. Gent, A.N.: A new constitutive relation for rubber. Rubber Chem. Technol. **69**(1), 59–61 (1996)
6. Holzapfel, G.A., Gasser, T.C., Ogden, R.W.: A new constitutive framework for arterial wall mechanics and a comparative study of material models **61**, 1–48 (2001)
7. Lin, J., Shi, Y., Men, Y., Wang, X., Ye, J., Zhang, C.: Mechanical roles in formation of oriented collagen fibers. Tissue Eng. Part B Rev. **26**(2), 116–128 (2020)
8. Marquardt, D.W.: An algorithm for least-squares estimation of nonlinear parameters. J. Soc. Ind. Appl. Math. **11**(2), 431–441 (1963)
9. Mooney, M.: A theory of large elastic deformation. J. Appl. Phys. **11**(9), 582–592 (1940)

10. Ogden, R.W.: Non-linear elastic deformations. ZAMM - J. Appl. Math. Mech. **65**(9), 404–404 (1984)
11. Peyraut, F., Renaud, C., Labed, N., Feng, Z.Q.: Modélisation de tissus biologiques en hyperélasticité anisotrope - Étude théorique et approche éléments finis. Comptes Rendus Mécanique **337**(2), 101–106 (2009)
12. Remache, D.: Contribution à l'étude expérimentale et numérique du comportement hyperélastique et anisotrope de la peau humaine. Institut FEMTO-ST - Université Franche-Comté, Thèse (2013)
13. Renaud, C., Cros, J.M., Feng, Z.Q., Yang, B.: The Yeoh model applied to the modeling of large deformation contact/impact problems. Int. J. Impact Eng **36**(5), 659–666 (2009)
14. Rivlin, R.S.: Large elastic deformations of isotropic materials iv. further developments of the general theory. Philos. Trans. Roy. Soc. London Ser. A Math. Phys. Sci. **241**(835), 379–397 (1948)
15. Schröder, J., Neff, P., Balzani, D.: A variational approach for materially stable anisotropic hyperelasticity. Int. J. Solids Struct. **42**(15), 4352–4371 (2005)
16. Simo, J.C., Hughes, T.: Computational Inelasticity. Springer, New York (1998). https://doi.org/10.1007/b98904
17. Socci, L., Pennati, G., Gervaso, F., Vena, P.: An axisymmetric computational model of skin expansion and growth. Biomech. Modeling Mechanobiol. **6**, 177–88 (2007)
18. Xu, F., Lu, T.: Introduction to Skin Biothermomechanics and Thermal Pain. Springer, Heidelberg (2011). https://doi.org/10.1007/978-3-642-13202-5
19. Yeoh, O.: Some forms of the strain energy function for rubber. Rubber Chem. Technol. **66**, 754–771 (1993)
20. Zhang, M., Mak, A.F.T.: In vivo friction properties of human skin. Prosthet. Orthot. Int. **23**(2), 135–141 (1999)

Elastic Properties of Normal Breast Tissues Using an Indentation Protocol - Preliminary Study

Ana Margarida Teixeira[1], António André[1], Rossana Correia[2], Maria da Luz Barroso[3], Horácio Costa[3], and Pedro Martins[1,4(✉)]

[1] INEGI, Porto, Portugal
{aandre,rcorreia@ipatimup.pt}
[2] Histology and Electron Microscopy Service, i3s, Porto, Portugal
rcorreia@ipatimup.pt
[3] Centro Hospitalar de Gaia/Espinho, Serviço de Cirurgia Plástica, Reconstrutiva, Craneomaxilofacial, Mão e Unidade de Microcirurgia, Vila Nova deGaia, Portugal
[4] LAETA, Portugal; ARAID, i3A, Universidad de Zaragoza, Zaragoza, Spain

Abstract. The mechanical properties of breast tissues are important for medical diagnose since the stiffness changes with a pathology. The mechanical behavior of breast tissues has been investigated, however the results in literature are not coherent.s The experimental protocols are not standardized and the need of preconditioning is still debatable, which might be one of the reasons for the variety of results. Thus, this study aims to evaluate the elastic properties of normal breast tissues, considering the effect of preconditioning. Indentation tests were performed and hysteresis was observed. The Young's modulus ranged from 3.2 ± 0.4 kPa to 8.8 ± 1.1 kPa and from 43.0 ± 5.2 kPa to 117.6 ± 14.2 kPa in the first and second linear regions, respectively. Moreover, the preconditioning increased the stiffness on the first linear region. Despite being a preliminary study, important considerations about the experimental protocol were obtained.

Keywords: Breast tissues · Indentation tests · Mechanical properties

1 Introduction

The breast is a vital organ for women, which is composed by different types of tissues, such as adipose, fibrous and glandular tissues, and by the suspensory ligaments called Cooper's ligaments. This heterogeneous structure changes along the life of the woman by the influence of many inherited factors, including age, hormonal state, menopause, menstrual cycle, pregnancy, lactation and in a presence of a disorder [1].

The most severe pathology associated to the breast is cancer. In 2020, worldwide, the female breast cancer had the highest incidence, and for women, it was the main cause of deaths as well [2].

J. M. R. S. Tavares et al. (Eds.): CMBBE 2021, LNCVB 38, pp. 281–289, 2023.
https://doi.org/10.1007/978-3-031-10015-4_24

Therefore, since we know that mechanical properties of the breast change in the presence of a disease, such as breast cancer, it is crucial to improve the knowledge about the breast tissues. When compared to their healthy state, breast tissues get stiffer due to pathology.

This property of the tissues is used by physicians to detect a dysfunction. Therefore, quantifying the stiffness is very important to aid the development or improvement of medical devices, either for diagnose or treatment.

Accordingly, several authors [3–11] aim to understand normal and pathological breast tissues.

Sarvazyan et al. [3] compared pathological with normal breast tissues and concluded that benign and malignant tissues were four times and seven times stiffer than normal tissues, respectively. Krouskop et al. [4] studied adipose, glandular and fibrous tissues as well as intraductal and infiltrating ductal carcinomas. The results showed that adipose tissue was the softest and infilltrating ductal carcinoma was the stiffest tissue, which also had the highest nonlinear increase in stiffness. Wellman et al. [5] had the same conclusion, showing that the difference in stiffness from normal tissues to tumors was higher at high strains. The group of Samani et al. developed a lot of studies in this field [6–9] using indentation tests and finite element analysis. They measured the Young's modulus, showing that it is sensitive to preconditioning, and also confirmed the non linearity of breast tissues. They observed that malignant tumors were the stiffest, followed by benign tumors, being normal breast tissues the softest. Within normal tissues, the elastic modulus between adipose and fibroglandular tissues was similar. Using compression tests, Matsumura et al. [10] measured the elasticity of breast tissues under the stress range usually applied in elastography. They noticed a significant non-linearity in tissue elasticity and a difference in the Young's modulus depending on the compression status. Comparing normal tissues with carcinomas, the first ones were softer. Adipose and glandular tissues showed a similar stress-strain curve, with a larger non linearity than lesions. The same conclusion was obtained by Umemoto et al. [11], showing that the difference in stiffness between normal tissues and lesions tended to gradually decrease with increasing applied stress, due to the nonlinear properties.

There's some evidence in literature that normal breast tissues are the softest, while malignant cancers are the stiffest, showing that more invasive tumors have greater elastic moduli.

Despite existing research, experimental protocols are not standardized and the need for preconditioning is still debatable.

According to some studies, precompression has influence on the measured stiffness [4,6,12,13], however this is not yet a generally accepted conclusion. This incoherence in the mechanical testing protocols is one of the reasons for the variety of results found in literature, not only between the different types of tissues but also within each type. Thus, this study, besides the evaluation of the elastic properties of breast tissues, aims to assess the effect of preconditioning.

2 Materials and Methods

To investigate the mechanical properties of normal breast tissues, samples were provided by the physicians who collaborate with our research group. The samples were collected from one patient, who underwent mammary reduction surgery. From the original sample, two adjacent specimens could be obtained for testing. The specimens were cut by embedding them in an agar solution, which allowed to encapsulate the specimens and avoid undesired movement while cutting in the correct geometry with standard dimensions (Fig. 1). The agar solution had a concentration of $\approx$3.5% and it was poured into the mold with the specimen at $\approx 40^o$C to avoid damage the tissue.

When it was solid, the tissue was cut with a $12 \times 12\,\text{mm}^2$ punch and the height was $\approx$5 mm. This methodology was based on the procedure developed by Samani et al. [6].

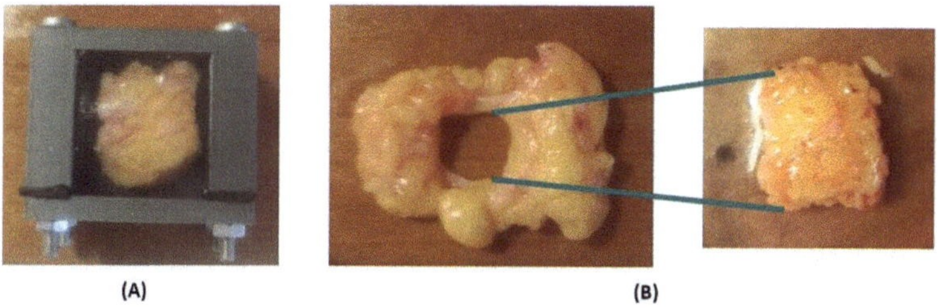

Fig. 1. Specimen preparation. From left to right: (A) tissue embedded in agar solution; (B) tissue after being cut, including the leftover and the final sample.

After specimens' preparation, they were mechanically tested using a load cell of 10 N and a flat-ended cylindrical indenter with 5 mm of diameter. The mechanical apparatus is composed also by an actuador with a load capacity of 12 kg and a resolution of 3.05×10^{-4} mm. The sample was immersed in a saline bath at 37ºC to mimic the physiological environment (Fig. 2). The indentation was set at 10% strain amplitude with a constant speed rate of 1.2 mm/min. In order to study the preconditioning, it was applied 14 cycles in one of the samples and 20 cycles in the other. The force-displacement data was recorded for each cycle and the stiffness was analyzed in the loading curve of the last cycle.

Most of the indentation models determine the reduced or effective modulus, E^*, which is defined as:

$$\frac{1}{E^*} = \frac{1-\nu_S^2}{E_S} + \frac{1-\nu_i^2}{E_i} \tag{1}$$

where ν is the Poisson coefficient, E is the Young's modulus and the subscripts S and i correspond to the sample and indenter, respectively. Since the indenter is too rigid (e.g. steel) compared to the sample (e.g. soft tissue), the contributions from the indenter are negligible [14].

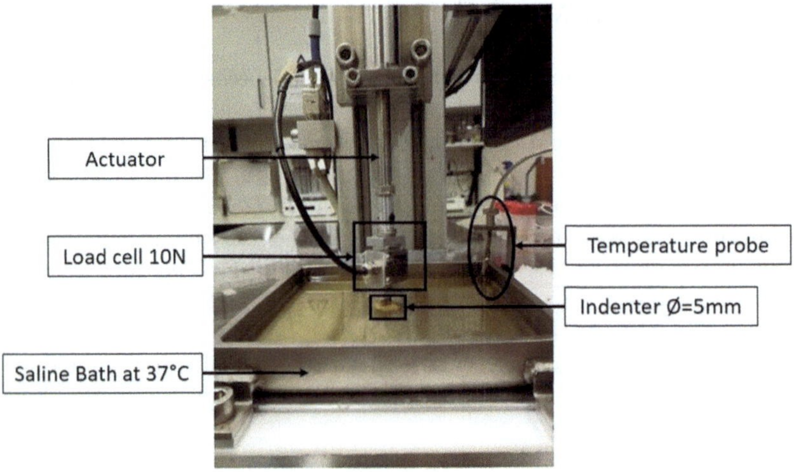

Fig. 2. Mechanical apparatus for the indentation tests.

Delaine-Smith et al. [14] used four indentation models, based on literature, to calculate the reduced/effective modulus (Eqs. 2, 3 and 4) or the indentation modulus (Eq. 5) of each trial. In this work, these four models were also used and they are described by the following equations, being represented in ascending order of the model (Model 1 → Model 4):

$$E^* = \frac{S}{2ak} \tag{2}$$

$$E^* = \frac{S}{2a} \tag{3}$$

$$E^* = \frac{S}{\pi a/2} \tag{4}$$

$$E = S \times (3/\pi 2a) \tag{5}$$

where E^* is the reduced/effective modulus, E is the indentation modulus, S is the indentation stiffness, which was obtained from the slope of the load-displacement curve, k is a geometric correction factor [15,16] and a is the indenter radius.

All the models assume isotropy in the samples and negligible friction. Moreover, the samples were assumed to be nearly incompressible due to their high fluid content (e.g. Poisson ratio equals to 0.5).

3 Results and Discussion

Following the methodology outlined in the previous section, the force-displacement data was recorded for all the cycles and hysteresis was observed

(Fig. 3). This effect is characterized by a difference in the loading and unloading curves as well as from one cycle to another. These deviations are related to the viscoelastic behavior of the tissues and to the energy dissipation from the tissues when they recover from contraction or elongation due to shear stress [1].

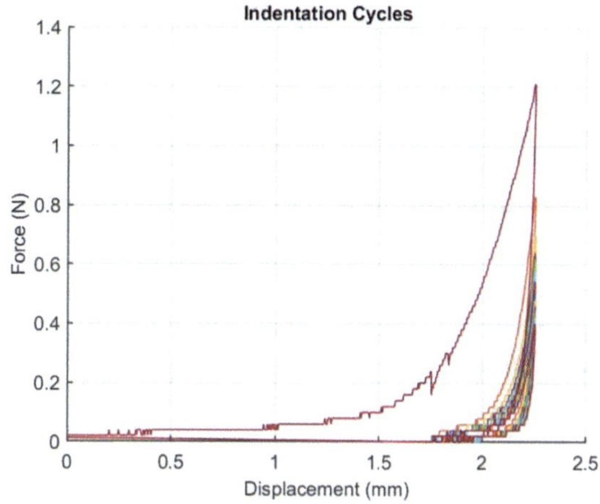

Fig. 3. Force-displacement data for 14 cycles.

The experimental data was further analysed applying the models proposed by Delaine-Smith et al. [14] at the two linear regions of the loading curve at cycles 14 and 20. For each sample, the results obtained at cycle 14 are presented in Table 1. The Young's modulus (mean±standard deviation) ranged from 3.21 ± 0.4kPa to 8.8 ± 1.1 kPa and from 43.0 ± 5.2 kPa to 117.6 ± 14.2 kPa in the first and second linear regions, respectively.

Table 1. Young's modulus at cycle 14 for samples 1 and 2, for each model, in the 1st and 2nd linear regions of the loading curve

Sample	E (kPa)							
	Model 1		Model 2		Model 3		Model 4	
	1st	2nd	1st	2nd	1st	2nd	1st	2nd
1	3.51	46.67	7.53	100.22	9.59	127.60	6.10	81.23
2	2.90	39.32	6.23	84.44	7.93	107.51	5.05	68.44

These results are consistent with literature [3–6]. The values for normal breast tissues in literature ranged from 0.69 ± 0.19 kPa to 24 ± 6 kPa for adipose tissue and from 0.73 ± 0.18 kPa to 271.8 ± 167.7 kPa for glandular tissue [4–6,8,10,11]. There is a discrepancy in the values of stiffness found in literature since the experimental protocols differ. Differences in the test conditions and/or in the test parameters, such as strain amplitude or loading frequency, influence the measurement of the stiffness, which emphasizes the importance of getting standardize protocols. However, the stiffness range found in literature is wide and accommodates the results obtained in this study.

In order to evaluate the parameters of the experimental protocol, 6 more cycles were added to one of the samples to assess if the preconditioning influences the stiffness of the sample.

The goal of preconditioning is to reduce the effect of hysteresis and soften the tissue and it consists in repeated loading and unloading of the tissue until steady state is reached [1]. Therefore, the influence of additional preconditioning cycles to the stiffness, in both linear regions, was investigated for each model considered. The Young's modulus for 14 and 20 cycles are presented in Table 2.

Table 2. Young's modulus per model for sample 2 at 14 and 20 cycles in the 1st and 2nd linear regions of the loading curve

Cycles	E (kPa)							
	Model 1		Model 2		Model 3		Model 4	
	1st	2nd	1st	2nd	1st	2nd	1st	2nd
14	2.90	39.32	6.23	84.44	7.93	107.51	5.05	68.44
20	6.29	38.89	13.50	83.52	17.19	106.34	10.94	67.70

Regarding the effect of preconditioning, adding 6 cycles to one of the samples only had an impact on the first linear region, increasing the stiffness, for any models considered. It was not observed any influence in the results of the second linear region. This indicates that the preconditioning has more importance at low strains, since the sample at high strains reach a steady state with a fewer number of cycles.

Moreover, an histological analysis, using H&E stain, was performed to evaluate the main components of each sample. In image 4, it is possible to see adipose tissue, represented by a star, glandular tissue, indicated by a triangle and connective tissue (called stroma) colored with light pink and represented by a circle.

Through the histological images obtained, it is not accurate to mention the main tissue present in each sample. However, Dempsey et al. [17], recently, found out that the differences between the types of normal breast tissues were statistically insignificant. They studied the viscoelastic properties of adipose, fibroglandular and mixed tissues and concluded that it is correct to assume that the breast is an homogeneous structure under large strains, an observation in good agreement with the results of the current study.

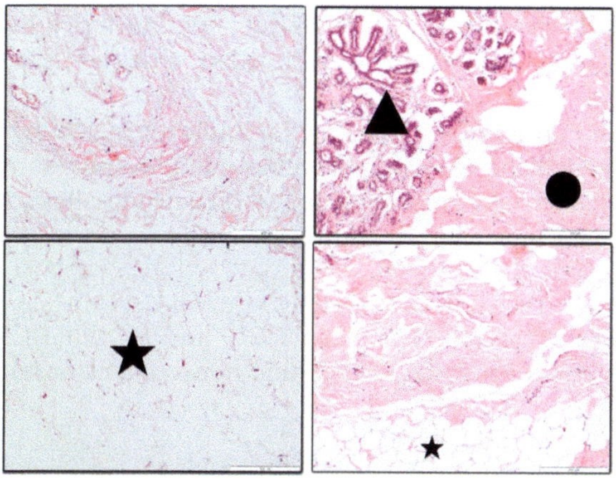

Fig. 4. Histological images for each sample. Both images at the left side correspond to sample 1; and both images at the right side correspond to sample 2. Symbols: star - adipose tissue; triangle - glandular tissue; circle - fibrous tissue (stromal). Magnification of 200 μ m

4 Conclusion

In this study, the mechanical properties of normal breast tissues were investigated. It was observed the hysteresis effect induced by the cyclic loading of the tissues, which is related to the viscoelastic behavior of soft tissues. Moreover, the Young's modulus was calculated from the force-displacement data and the results were in accordance with the literature. Concerning the preconditioning, it was concluded that increasing the number of cycles, has an impact in the stiffness of the tissues, mainly in the first linear region of the loading curve.

However, there were some limitations on this preliminary study. Since it was the first approach, there were some difficulties associated to the preparation and handling of the samples, which increased the time of preparation. Also, at the time of this work, samples were provided from only one surgery and thus only two specimens from the same patient were tested. Therefore, the number of samples was very restricted. Nevertheless, some insights for future work were achieved, particularly regarding the experimental protocol.

In future iterations, this study will improve by increasing the number of samples. It is known that the preconditioning is important, therefore it will be taking into account for future experimental protocols that aim to understand the viscoelastic behavior, which occur at large strains. Regarding the Young's modulus estimation, it can also be improved by using finite element analysis.

The inconsistencies found in literature, regarding the breast tissues' behavior, points to the need for further research.

There are already several intrinsic factors that affect the stiffness of the tissues, such as the age of the woman, hormonal state, pregnancy, lactation, etc., which make it more important to minimize external factors, such as those related with mechanical characterization. Therefore, experimental protocols should be standardized to reduce the variability associated with the parameters and conditions of the tests.

The better the information about the mechanical properties of breast tissues, the most accurate will be the computational models and the developments in tissue engineering and regenerative medicine. This could allow the physicians to have better solutions for diagnose and treatment.

Acknowledgments. The authors gratefully acknowledge funding from FCT, Portugal, under grant 2020.08718.BD, and from project MImBI - PTDC/EME-APL/29875/2017 financed through FEDER and from research unit LAETA.

References

1. Ramião, N., et al.: Biomechanical properties of breast tissue, a state-of-the-art review. Biomech. Model. Mechanobiology **15**(5), 1307–1323 (2016)
2. Sung, H., et al.: Global cancer statistics 2020: globocan estimates of incidence and mortality worldwide for 36 cancers in 185 countries. CA: A Cancer J. Clinicians **71**(3), 209–249 (2021)
3. Sarvazyan, A.P., et al.: Biophysical bases of elasticity imaging. Springer, Boston, pp. 223-240 (1995). https://doi.org/10.1007/978-1-4615-1943-0_23
4. Krouskop, T.A., et al.: Elastic moduli of breast and prostate tissues under compression. Ultrason. Imaging **20**(4), 260–274 (1998)
5. Wellman, P.S., et al.: Breast Tissue Stiffness in Compression is Correlated to Histological Diagnosis, Technical Report. Harvard BioRobotics Laboratory, pp. 1-15 (1999)
6. Samani, A., et al.: Measuring the elastic modulus of ex vivo small tissue samples. Phys. Med. Biol. **48**, 2183–2198 (2003)
7. Samani, A., Plewes, D.: A method to measure the hyperelastic parameters of ex vivo breast tissue samples. Phys. Med. Biol. **49**(18), 4395–4405 (2004)
8. Samani, A., Zubovits, J., Plewes, D.: Elastic moduli of normal and pathological human breast tissues: an inversion-technique-based investigation of 169 samples. Phys. Med. Biol. **52**(6), 1565–1576 (2007)
9. Samani, A., Plewes, D.: An inverse problem solution for measuring the elastic modulus of intact ex vivo breast tissue tumours. Phys. Med. Biol. **52**(5), 1247–1260 (2007)
10. Matsumura, T. et al.: Measurement of elastic property of breast tissue for elasticity imaging. In: Proceedings - IEEE Ultrasonics Symposium, pp. 1451-1454 (2009)
11. Umemoto, T., et al.: Ex vivo and in vivo assessment of the non-linearity of elasticity properties of breast tissues for quantitative strain elastography. Ultrasound Med. Biol. **40**(8), 1755–1768 (2014)
12. Barr, R.G., Zhang, Z.: Effects of precompression on elasticity imaging of the breast: development of a clinically useful semiquantitative method of precompression assessment. J. Ultrasound Med. **31**(6), 895–902 (2012). https://doi.org/10.7863/jum.2012.31.6.895
13. Han, L., Noble, J.A., Burcher, M.: A novel ultrasound indentation system for measuring biomechanical properties of in vivo soft tissue. Ultrasound Med. Biol. **29**(6), 813–823 (2003). https://doi.org/10.1016/S0301-5629(02)00776-7

14. Delaine-Smith, R.M., et al.: Experimental validation of a flat punch indentation methodology calibrated against unconfined compression tests for determination of soft tissue biomechanics. J. Mech. Behav. Biomed. Mater. **60**, 401–415 (2016)
15. Zhang, M., Zheng, Y.P., Mak, A.F.T.: Estimating the effective Young ' s modulus of soft tissues from indentation tests - nonlinear finite element analysis of effects of friction and large deformation. Med. Eng. Phys **19**(6), 512–517 (1997)
16. Hayes, W.C., et al.: A mathematical analysis for indentation tests of articular cartilage. J. Biomech. **5**, 541–551 (1972)
17. Dempsey, S.C.H., O'Hagan, J.J., Samani, A.: Measurement of the hyperelastic properties of 72 normal homogeneous and heterogeneous ex vivo breast tissue samples. J. Mech. Behav. Biomed. Mater. **124**, 104794 (2021)

Author Index

A
Abrantes, João, 250
Acot, Bianca, 90
Alliliche, Wael, 271
Andersen, Michael Skipper, 65
André, António, 281
André, António Diogo, 50
Arockia Xavier Annie, R., 215

B
Barbosa, M. I. A., 123
Barroso, Maria da Luz, 281
Bäumler, Kathrin, 143
Belinha, J., 123
Bini, Fabiano, 74
Blase, Christopher, 153
Böhme, Max, 65

C
Carvalho, A., 123
Carvalho, Violeta, 101
Cavatorta, Maria Pia, 65
Chander, Divyaksh Subhash, 65
Correia, Miguel, 250
Correia, Rossana, 281
Costa, Horácio, 281
Cros, Jean-Michel, 271

D
Dettwiler, Annegret, 90

E
Elangovan, Taranya, 215
Ennis, Daniel B., 143

F
Feng, Zhi-Qiang, 271
Ferreira, Louis, 183
Fichera, Fulvia, 36
Flanagan, Riley, 208
Fleischmann, Christopher, 194
Fleischmann, Dominik, 143
Fornito, Michael, 208
Fritzen, Claus-Peter, 153

G
Gagliardo, Cesare, 74
Gilchrist, Michael D., 90
Glisic, Branko, 90

H
Hing, Caroline, 170
Huß, Armin, 153

K
Knowles, Nikolas, 183
Kusins, Jonathan, 183

L
Lamb, Christopher, 59
Lim, Yeram, 24
Lima, Rui, 101
Lu, Jenny, 1

M
Marinozzi, Franco, 74
Marrale, Maurizio, 74
Marsden, Alison L., 143
Martensson, Nicole, 183

J. M. R. S. Tavares et al. (Eds.): CMBBE 2021, LNCVB 38, pp. 291–292, 2023.
https://doi.org/10.1007/978-3-031-10015-4

Martins, Pedro, 50, 281
Miehling, Jörg, 194
Monllau, Joan Carles, 244

N
Nadal, Jurandir, 250
Nasr, Firas, 170
Natal Jorge, R. M., 123
Nelson, Andrew, 183

O
Ovesy, Marzieh, 128

P
Peng, Q., 112
Perelli, Simone, 244
Pica, Andrada, 74
Pradhakshya, J. D., 215

R
Rasmussen, John, 65
Renaud, Christine, 271
Rivas, Tania, 208
Rodrigues, Carlos, 250
Rodrigues, Marco, 250
Rodrigues, Nelson, 101
Röhrle, Oliver, 234
Romanato, Marco, 36
Rosenstein, Seth, 24

S
Saini, Harnoor, 234
Sawacha, Zimi, 36
Scherb, David, 194
Serrancolí, Gil, 244
Sesselmann, Stefan, 194
Spolaor, Fabiola, 36
Sundaresan, Keerthana, 215

T
Targosinski, Jakub, 183
Teixeira, Ana Margarida, 50, 281
Teixeira, José C., 101
Teixeira, Senhorinha, 101
Torner, Jordi, 244

V
Vermolen, F. J., 112
Vermolen, Fred, 1
Vilches, Pablo, 59
Volpe, Daniele, 36

W
Walck, Christine, 24, 59, 208
Wartzack, Sandro, 194
Weihs, D., 112
Wittek, Andreas, 153

Z
Zimmermann, Judith, 143
Zysset, Philippe K., 128

MIX
Papier aus verantwortungsvollen Quellen
Paper from responsible sources
FSC® C105338

If you have any concerns about our products,
you can contact us on
ProductSafety@springernature.com

In case Publisher is established outside the EU,
the EU authorized representative is:
Springer Nature Customer Service Center GmbH
Europaplatz 3, 69115 Heidelberg, Germany

Printed by Libri Plureos GmbH
in Hamburg, Germany